Immunosenescence

Jos A. Bosch • Anna C. Phillips • Janet M. Lord
Editors

Immunosenescence

Psychosocial and Behavioral Determinants

Foreword by Keith W. Kelley, Ph.D.

Springer

Editors
Jos A. Bosch
University of Amsterdam
Amsterdam
The Netherlands

Janet M. Lord
University of Birmingham
Birmingham
United Kingdom

Anna C. Phillips
University of Birmingham
Birmingham
United Kingdom

ISBN 978-1-4614-4775-7 ISBN 978-1-4614-4776-4 (eBook)
DOI 10.1007/978-1-4614-4776-4
Springer New York Dordrecht Heidelberg London

Library of Congress Control Number: 2012952265

Printed on acid-free paper

Springer is part of Springer Science+Business Media (www.springer.com)

Foreword

The post-World War II generation of babies born in the USA between 1946 and 1964 began to reach their 65th birthday earlier this year. This explosion in births, amounting to 76 million American children, temporarily stemmed the tide in the USA's declining birth rate. Similarly, most western European countries experienced an increase in births immediately following the end of World War II. As these baby boomers join organizations like the American Association of Retired Persons (AARP), they want clearer answers and more effective treatments for their medical ailments. This book, *Immunosenescence: Psychosocial and behavioural determinants*, specifically addresses those needs.

Arguably, baby boomer kids across the United States and Western Europe grew up to become the largest, wealthiest, most rebellious, optimistic and free-thinking generation that the modern world has ever known. Baby boomers redefined traditional values of the Western world. In 1946, the year following the end of World War II, Winston Churchill described an "Iron Curtain" that had fallen over the European continent. The baby boomer generation grew up under this dark shadow and the constant threat of a nuclear Armageddon, only to witness dismantling of the infamous Berlin Wall in 1989 and an end to the Cold War in 1992. During this time, major events occurred. Baby boomers throughout the world remember names and movements like Fidel Castro, the Suez Canal crisis, Charles de Gaulle, Neil Armstrong, John F. Kennedy, Martin Luther King, freedom riders of the civil rights movement, the Chicago Seven, Elvis Presley, anti-Vietnam protests that pervaded Europe and the USA, Kent State killings, Watergate scandal, Beatles, hippies, free love, Woodstock, Three Mile Island and Chernobyl, just to name a few. These names stir the collective memory of baby boomers because this was their generation.

The post-World War II era witnessed incredible discoveries in medicine. The medical community saw birth of the first test tube baby, hip transplants, the birth control pill, heart-lung transplants, polio vaccine, genetic engineering, pacemakers, lasers, MRI scanners, stem cells, Prozac and the discovery of and treatment for human immunodeficiency virus. These advances encouraged development of specialists, such as surgeons who were experts in orthopedic, vascular, neurological, maxillofacial, cardiovascular, colon/rectal, hand, thoracic and plastic surgery. Other medical specialties developed, such as interventional cardiologists, nuclear medicine specialists, addiction psychiatrists, sleep disorders specialists, sports medicine specialists, reproductive endocrinologists, preventative medicine specialists, pain management specialists, medical geneticists, hyperbaric physicians and nuclear medicine specialists. These kinds of doctors remind of the Pete Seeger song of the early '60s, "Where have all the flowers gone?" Baby boomers did not grow up with these kinds of physicians. Today, as they begin to turn 65 years of age, they are asking, "Where have all the doctors gone?"

Baby boomers were born into a world that considered doctors to be what is now known as "infectious disease specialists." But, perhaps because of the logarithmic explosion in medical specialists during the past 30 years, a new "specialist" has appeared on the scene. Just two years ago, it became possible for M.D.s to be certified and boarded as Diplomates of the American Board of Hospital Medicine (ABHM; http://abpsus.org/hospital-medicine). Hospitalists are interdisciplinary since they care for all cases of acutely ill patients in hospitals and serve to coordinate and manage medical care between patients and medical specialists. It could be that despite all the wonderful advances that have been made in medicine since the birth of the first baby boomer, medical "providers" are finally being asked to consider patients as more than simply a human body with some sort of dysfunction. This reminds of the oft-repeated phrase, "The whole is greater than the sum of its parts."

It is the baby boomer recognition of the need for interdisciplinary and integrative medicine that created a niche for this book, *Immunosenescence: Psychosocial and behavioural determinants*. It is ironic that this idea in nothing more than a return to their roots. When boomers were babies, most physicians were well-trained in all aspects of physiology. This integrative science is derived from the Greek prefix "*physis*" which means "nature or origin" and the suffix "*logia*," or "the study of." For me, systemic physiology means the study of function, and it focuses on how the major organs of the body, such as the heart, lungs, kidneys and brain, not only function independently but also communicate with each other. Lungs depend on the heart to pump blood through oxygen-rich alveoli, and the heart depends on kidneys to regulate plasma volume. The study of regulatory physiology conveys a sound understanding of the numerous routes of communication among different organ systems. One strength of this book is that it reminds us of the important theme of immunophysiology because it emphasizes the concept that the immune system is just another organ system, like the heart, lungs and kidneys. It is a diffuse system that wanders throughout the entire body searching for foreigners. Once a stranger is recognized by cells of the immune system, it sends an alert message to the brain and other organs. The five classical sensory systems of touch, sight, smell, taste and sound did not evolve to recognize a foreign substance that invades the body. But the innate immune system can detect these foreigners. Indeed, cells of the immune system are now known to send both neuronal and humoral signals to inform the brain that an infection has occurred in the periphery. As Ed Blalock correctly hypothesized in his seminal paper published in the *Journal of Immunology* in 1984 (132:1067), the innate immune system is our "sixth sense."

The 13 chapters in this timely book highlight the multiple physiological systems that are impacted by the immune system during aging. As the immune system grows older, its dialogue with the brain changes. These changes affect how the aging baby boomer population responds to stress, how well they sleep, how they adjust to loss of independence that often accompanies aging, the rate at which they will heal after surgery and how they will deal with losing a spouse or living alone. The authors are internationally-respected scholars in the interdisciplinary field of brain, behavior and immunity. They have all published cutting edge, peer-reviewed papers in leading journals on the multitude of issues that confront gerontologists on topics that range from health psychology to host resistance to infections.

Immunosenescence: Psychosocial and behavioural determinants helps to push back the frontiers of immunophysiology. It espouses the philosophy that a better understanding of the aging immune system will be achieved only if we learn how its afferent and efferent communication signals affect other organs, particularly the brain. As a child who was born at the beginning of the baby boom, I applaud the direction and learning trajectory that this book is taking us.

August 9, 2011
University of Illinois at
Urbana-Champaign,
Champaign, USA

Keith W. Kelley, Ph.D.
Professor of Immunophysiology
Editor-in-Chief,
Brain, Behavior, and Immunity

Contents

Chapter 1
Introduction to Ageing of the Innate Immune System

Niharika A. Duggal and Janet M. Lord

1.1 Introduction

Over the past century, human life expectancy has doubled in developed countries and is continuing to increase at a rate of 2 years per decade. However, although life expectancy has increased, advanced age is accompanied by an increase in susceptibility towards infection and development of chronic illnesses that have a negative impact on an individual's quality of life. Even adults considered to have undergone healthy ageing show a significant decline in immune competence, termed immunosenescence, which is responsible for the increased rate of infections with advancing age (DiCarlo et al. 2009). Recent studies have also reported a reduced ability to mount a robust immune response to vaccination, combat new pathogens or maintain immunity to persistent infections such as *Herpes zoster* in older adults (Gavazzi and Krause 2002b; Trzonkowski et al. 2009; Weinberger et al. 2008). Delaying or reversing the effects of ageing on the immune system may therefore be extremely beneficial to the health and quality of life of the elderly population (Dorshkind et al. 2009).

1.2 Inflammaging

A universal feature of physiological ageing is a higher basal production of pro-inflammatory cytokines (IL-1β, IL-6, IL8, TNFα and IL-15), accompanied by a reduced production of anti-inflammatory cytokines (IL-10) known as 'Inflammaging' (Franceschi et al. 2007). Importantly, inflammaging is a predictor of frailty and mortality in aged humans. Studies in centenarians (Di Bona et al. 2009) and extremely long-lived mice (Arranz et al. 2010) show that long-lived individuals maintain the cytokine profile of younger adults. In addition to increased cytokines, other inflammatory markers including C-reactive protein (CRP) and clotting factors (fibrinogen)

J. M. Lord (✉) · N. A. Duggal
MRC Centre for Immune Regulation, School of Immunity and Infection,
University of Birmingham, Edgbaston, Birmingham B15 2TT, UK
e-mail: J.M.Lord@bham.ac.uk

J. A. Bosch et al. (eds.), *Immunosenescence*,

DOI 10.1007/978-1-4614-4776-4_1, © Springer Science+Business Media New York 2013

are also increased with age. The underlying mechanisms driving inflammaging are thought to be varied (Krabbe et al. 2004) and include: the increase in adipose tissue which is a significant source of inflammatory cytokines and pro-inflammatory hormones termed adipokines (Fantuzzi 2005); decreased production of sex steroids many of which are anti-inflammatory (Nilsson 2007); sub-clinical infections (Effros 2001); a sedentary lifestyle (Lavoie et al. 2010) and constitutive low-level production of cytokines by monocytes (discussed below).

The rest of this chapter will focus upon the age-related changes to the functioning of the cells of the innate immune system as it is these changes that dramatically affect the ability to combat bacterial and viral infections in old age.

1.3 Age-Related Changes to Innate Immune Cell Production

The innate immune system acts as the first barrier encountered by pathogens and is responsible for immediate and robust responses to micro-organisms. In aged subjects, a breakdown in the integrity of innate physical barriers such as the skin, gastrointestinal tract and lung occurs, resulting in increased susceptibility to invasion by pathogens and an increased burden on the cells of the innate immune system.

All immune cells are formed from differentiation of the multipotent haematopoietic stem cell (HSC) which is responsible for continuously replenishing the immune system. HSCs differentiate into multipotent progenitor cells, which can produce cells of either myeloid (neutrophils, monocytes, macrophages, dendritic cells, eosinophils and megakaryocytes/platelets) or lymphoid (T cells, B cells and NK cells) lineage. HSCs from older donors have a reduced ability for self-renewal and a myeloid skewing of their differential potential; suggesting profound changes in multiple levels of HSC differentiation during ageing (Chambers et al. 2007; Dykstra and de Haan 2008). Molecular factors contributing to HSC ageing include accumulation of DNA damage, altered gene expression patterns and epigenetic deregulation (Warren and Rossi 2009). Skewing of the HSCs with age is thought to be responsible for declining immuno-competence, increased autoimmunity, anemia, diminished stress response and increased predisposition to a spectrum of diseases including myeloid leukemia (Warren and Rossi 2009).

1.4 Neutrophils

Neutrophils are the most abundant leukocyte and a key element of the innate immune system and are one of the first cells to migrate into the site of infection. They are responsible for recognising, ingesting and destroying pathogens, most importantly rapidly dividing bacteria, yeast and fungi. The increased incidence of bacterial infections in older adults suggests (Bonomo 2002; Gavazzi and Krause 2002a) that neutrophil numbers and/or function are reduced with age.

Neutrophils have a very short life span (5.4 days) in peripheral blood and are lost due to spontaneous apoptosis, and as a consequence they are produced in large numbers (1–2 $\times$ 10^{11} per day) (Wessels et al. 2010). The bone-marrow content of precursors for the neutrophil lineage as well as their number in blood remains unchanged in the healthy elderly (Chatta et al. 1993; Gomez et al. 2008). During an infection, a large increase in neutrophil production is observed (neutrophilia) in order to combat infection. Studies have shown that older adults are able to mount a normal neutrophilia in response to an infection (Lord et al. 2001). Neutrophil's life span is controlled by growth factors (IL-3), chemokines and lineage specific cytokines, specifically granulocyte-colony stimulating factor (G-CSF) and granulocyte-macrophage colony stimulating factor (GM-CSF) (Chatta et al. 1993). Even though neutrophils from older donors have a diminished ability to respond to anti-apoptotic mediators such as G-CSF, their responses to GM-CSF and IL-3 are unaltered with age; resulting in an ability to maintain neutrophil numbers with age (Fulop et al. 2002).

During an infection, neutrophils leave the circulation (extravasation), roll along and then adhere to vascular endothelial cells, followed by migration to the site of infection along a gradient of chemotactic factors. Once the neutrophils have reached the site of infection via chemotaxis, they engulf the invading microbes by phagocytosis. Pathogen phagocytosis by neutrophils is followed by microbial killing by several mechanisms, including generation of reactive oxygen and nitrogen species, release of proteolytic enzymes from intracellular granules and the extrusion of DNA coated with antimicrobial compounds, termed neutrophil extracellular traps (NETs).

Adherence of neutrophils to endothelium, which is essential for neutrophil extravasation, is slightly increased or unaltered with age. The expression of markers CD11a and CD11b, necessary for neutrophil extravasation, remains unaltered with age. However, an increase in the expression of CD15, responsible for binding of E-selectin on endothelium, has been reported in older adults (Butcher et al. 2001; Esparza et al. 1996), suggesting that reduced extravasation of neutrophils is not responsible for the increase rate of bacterial infections in older adults.

Chemotactic molecules (IL-8, fMLP, complement C5a) are responsible for attracting neutrophils to the site of infection. There is a significant degree of discordance in the literature regarding chemotaxis of neutrophils from the elderly; some studies have reported that it remains unchanged with age (MacGregor and Shalit 1990), whilst others have reported reduced chemotaxis (directional movement) of neutrophils in response to fMLP and GM-CSF in the elderly, whilst chemokinesis (non-directional movement) remains unaltered (Antonaci et al. 1984; Niwa et al. 1989; Wenisch et al. 2000). The decrease in chemotaxis may be particularly important, as it would compromise efficiency of recruitment of neutrophils to the infection focus. Even though neutrophils are capable of migrating towards the infection site, the data suggest that they have reduced directional control and will take a longer route to the infection site. This will result in an increased exposure of tissue to proteases that neutrophils release during the migratory process, consequently damaging healthy tissue. Further research is required to determine the mechanisms underlying reduced directional migration in neutrophils from older donors.

An age-related decline has also been reported in the percentage of phagocytosing neutrophils and the ability of neutrophils to ingest opsonised bacteria and yeast (Butcher et al. 2001; Wenisch et al. 2000). However, neutrophil ability to phagocytose unopsonised bacteria remains unaltered with age (Emanuelli et al. 1986), suggesting that it is the response via opsonin receptors (Fcγ and C3b complement receptors) that is affected by ageing. A number of intracellular defects that contribute to reduced phagocytic ability with age may include: impaired calcium mobilisation, reduced hexose transport (phagocytosis is an energy-dependent process), reduced CD16 expression, impaired downstream signalling events (PI-3 kinase, MAP kinase activation) and reduced ligand-induced actin polymerisation (Butcher et al. 2001; Rao 1986; Wessels et al. 2010). In addition to these intrinsic factors, extrinsic factors such as altered circulating cytokine levels also contribute towards reduced neutrophil functioning efficiency with age.

Ingested microbes are destroyed by two different microbicidal mechanisms; production of reactive oxygen species and degranulation, both of which lead to the release of cytotoxic proteases. Regarding the first mechanism, studies of neutrophil respiratory burst have reported unaltered superoxide production in response to phorbol myristate acetate (PMA) (Braga et al. 1998; Butcher et al. 2000; Esparza et al. 1996). Antibody Fc-receptor-mediated neutrophil superoxide generation in response to gram-positive bacteria such as *Staphylococcus aureus* has been shown to be reduced (Wenisch et al. 2000), whereas superoxide generation in response to *Escherichia coli* is unaltered with age (Panda et al. 2009; Wessels et al. 2010) or reduced (Tortorella et al. 2000). Studies have also reported a decline in degranulation in response to microbial products (Fulop et al. 1986).

In addition to phagocytosing pathogens and destroying them by generation of ROS, neutrophils are also capable of engulfing and destroying pathogens via extrusion of NETs (Brinkmann et al. 2004). So far, no studies have examined the effect of age on NET generation.

Impaired neutrophil functioning detected in older adults can be a result of modifications in signalling pathways and receptor distribution with age. Alterations in physiochemical properties of neutrophil membranes such as reduced membrane fluidity due to altered cholesterol/phospholipid content and changes in actin cytoskeleton have been reported with age in rodents (Alvarez et al. 2001). Altered membrane fluidity results in impaired neutrophil signal transduction as the functioning of lipid rafts is affected and this could explain, to a large extent, the reduced signalling through PI-3 kinase, MAP kinase, protein kinase B, Jak-STAT and SHP-1 signalling seen in neutrophils with increasing age (Shaw et al. 2010). Ca^{2+} ion mobilisation plays a central role in intracellular signal transduction resulting in activation of cytosolic proteins. Resting neutrophils of older adults have an increased level of Ca^{2+} (Wenisch et al. 2000). As a result, older adults are unable to mount a proper Ca^{2+} mobilisation response (Fulop et al. 1989). TREM-1 is a recently identified receptor, responsible for initiating neutrophil inflammatory responses including phagocytosis, superoxide production and degranulation. TREM-1 engagement is impaired in the neutrophils of aged donors resulting in altered neutrophil early activation events; potentially due to alterations in lipid raft engagement of TREM-1 with age (Fortin et al. 2007).

1.5 Macrophages/Monocytes

Macrophages are widely distributed innate immune cells that act as 'pathogen sensors' and are essential for the initiation of immunological responses (against bacteria, viruses, parasites and tumour cells) as well as regulation of adaptive immune responses. Monocytes leave the bone marrow and enter the bloodstream to migrate into body tissues including skin, liver, lungs, spleen, where they differentiate into macrophages. Numerous studies have reported alterations in macrophage functioning with age; altered gene expression, accumulation of DNA mutations and impaired DNA repair appear to be the basis of this impairment in functioning (Lloberas and Celada 2002). Although the absolute number of blood monocytes remains unaltered with age (Gomez et al. 2008; Takahashi et al. 1985), the bone marrow of the elderly has lower numbers of $CD68^+$ macrophage precursors (Lloberas and Celada 2002; Ogawa et al. 2000). A decline in haematopoietic cell proliferation due to replicative senescence and increased apoptosis could explain the reduction in macrophage precursors. However, the number, size, DNA content and cell-surface markers expressed during macrophage maturation (e.g., Mac 1) are similar in macrophages from aged to young mice (Sebastian et al. 2005).

Interestingly, monocytes share some functions with neutrophils they can directly resolve infection by migrating towards invading pathogens followed by phagocytosis and killing. A decline in the ability of macrophages to migrate towards the site of infection has been reported with age. Impaired chemotactic responses of macrophages in older adults towards complement factors may contribute towards delayed pathogen clearance (Fietta et al. 1993). However, the data examining the effect of ageing on macrophage phagocytic ability are contradictory. Studies of murine and human cells have reported a decline in macrophage phagocytic ability with age (Fietta et al. 1994; Khare et al. 1996; Plowden et al. 2004; Swift et al. 2001). In contrast, studies in rats have reported an increase in the phagocytic ability of specialised liver macrophages (kuppfer cells) and alveolar macrophages with age (Hilmer et al. 2007; Mancuso et al. 2001). The impact of ageing on phagocytosis-promoting receptors (CD14, CD36, mannose receptor, scavenger receptor, MARCO and MER) and signal transduction pathways still remain unreported. Further research in this direction might be helpful in developing a better understanding of the effect of ageing on macrophage phagocytosis.

Macrophages from old donors also demonstrate impaired intracellular killing of pathogens due to lower production of reactive oxygen intermediates in response to IFNγ, PMA or opsonised zymogen stimulation with age (Davila et al. 1990; Ding et al. 1994). Studies examining an alternative macrophage anti-microbial mechanism, reactive nitrogen intermediate production, have reported conflicting results. Reduced levels of inducible nitric oxide synthase (iNOS) mRNA have been shown in response to lipopolysaccharide (LPS), resulting in decreased production of reactive nitrogen intermediates (Kissin et al. 1997). Conversely, an increase in iNOS levels and NO_2- and other intermediates in response to LPS stimulation has also been reported (Chen et al. 1996). Macrophages display different production of reactive nitrogen intermediates in response to different stimuli (LPS, IFNγ), and the different experimental protocols and conditions used during these studies might be responsible for the conflicting data.

Importantly, macrophages activate other inflammatory cells via secretion of a range of cytokines (TNFα, IL-6), chemokines (IL-8), growth factors (GM-CSF, M-CSF), coagulation factors and prostaglandin E2. As stated above, ageing is associated with an increase in constitutive levels of serum pro-inflammatory cytokines. In contrast, macrophage production of pro-inflammatory cytokines (IL-6, TNFα, IL12, IL-1β) upon LPS stimulation is less in aged mice and humans in comparison with the young (Boehmer et al. 2004; Chelvarajan et al. 2005; Delpedro et al. 1998; Renshaw et al. 2002), which may contribute to reduced immune responses. Reduced expression of several genes including MHC class II, TLR1 and TLR4 in aged macrophages, due to epigenetic modifications and reduced levels of transcription factors with age, contribute to the observed reduction in cytokine production upon stimulation (Chelvarajan et al. 2005; Gomez et al. 2005; Renshaw et al. 2002; van Duin and Shaw 2007). Studies have also reported impairments in TLR-mediated signalling pathways including lower expression of MAP kinases p38 and Jun Kinases in macrophages with age resulting in lowered activation upon LPS stimulation, which will also contribute to the decline in macrophage pro-inflammatory cytokine production (Boehmer et al. 2004; Chelvarajan et al. 2005). In addition to impaired TLR expression on monocytes with age, a decline in their ability to upregulate co-stimulatory molecules such as CD80 upon TLR engagement has been reported with age (van Duin and Shaw 2007).

Interestingly, an up-regulation of the production of the anti-inflammatory cytokine (IL-10) has been observed in macrophages by some groups (Chelvarajan et al. 2005; Kelly et al. 2007), which may also contribute to poor macrophage-based immunity. Activated macrophages produce higher levels of prostaglandin E2 with age, which is responsible for increased IL-10 production and decreased expression of class II major histocompatibility (MHC) on macrophages resulting in lower antigen presentation and poorer $CD4^+$ T-cell responses (Herrero et al. 2001; Solana et al. 2006).

Macrophages also play an essential role during wound repair by keeping the site free from infection and secreting growth factors to promote angiogenesis (VEGF) and infiltration of fibroblasts (Barbul and Regan 1995). Studies in human and rodent models have reported a significant decline in wound healing with age (Gosain and DiPietro 2004). A delay in macrophage infiltration in the wound site due to lower expression of adhesion molecules (VCAM-1, ICAM-1), occurs with age (Ashcroft et al. 1998), and a 37 % decline in VEGF production has been reported in macrophages from aged mice resulting in diminished angiogenesis, delayed wound closure and a higher risk of developing an infection at site of injury (Swift et al. 1999; Thomas 2001).

1.6 Dendritic Cells

Dendritic cells (DC) represent a rare population of circulating cells that form about 1 % of total peripheral blood cells and play a key role in bridging the adaptive and innate immune systems (Schuurhuis et al. 2006). They are major antigen-presenting cells interacting with T and B cells and modulating the composition of cytokines produced by T cells (Th1, Th2, Th17), which in turn influences the nature of the immune response. DCs in humans have been divided into two categories—those that are present in peripheral blood (myeloid and plasmacytoid DC) and those that are present in tissues such as mucosa, skin and internal organs. Developmental and functional alterations in DCs with age are additional factors contributing towards immune malfunctioning.

Using multiple cell-surface marker analysis to define DC's one study reported no significant differences in numbers of circulating myeloid DCs and plasmacytoid DCs between aged and young humans (Agrawal et al. 2007). Also, a decreased frequency of tissue DCs has been reported in Peyer's patches with age, which might contribute towards age-associated mucosal dysregulation (Fujihashi and McGhee 2004). However, there have been contradictory reports that have failed to find alterations in numbers of DCs with age (Grolleau-Julius et al. 2006). Furthermore, another study reported that a dense network of DCs pervaded brain areas in aged mice (Stichel and Luebbert 2007). In addition to alterations in DC numbers, their phenotype appears to alter with age, with a decline in expression of MHC class II, CD86, CD40 and CD54 expression reported in DCs from older individuals (Varas et al. 2003).

Immature DCs are remarkably efficient in capturing antigens, which is essential for generating specific immune responses. DCs capture antigens by several mechanisms including: macropinocytosis (non-selective endocytosis), receptor-mediated endocytosis and phagocytosis (Agrawal et al. 2007). DCs are capable of recognising pathogens through the presence of pathogen recognition receptors (PRRs) such as TLRs, NODs and C-type lectin receptors. According to a recent study, normal TLR function and expression has been reported in circulating mDCs and bone marrow derived DCs from aged mice (Tesar et al. 2006). Even though expression of PRRs is unaltered in DCs with age, a decline in DC's ability to uptake antigens has been reported (Tesar et al. 2006), which may have an effect on antigen presentation by MHC class II molecules to T cell resulting in decreased T-cell priming and reduced vaccination responses.

DCs are also responsible for phagocytosing self-antigens such as apoptotic cells and for maintaining peripheral tolerance to the body's own tissues. With age, inefficient removal of apoptotic cells by DCs has been reported, which may be responsible for higher risk of autoimmunity and reduced capacity to clear infections associated with ageing (Agrawal et al. 2007, 2008). In addition to antigen presentation, DCs are also capable of secreting cytokines that stimulate T-cell proliferation. LPS activated DCs in the elderly produce increased levels of IL6 and TNF-α but IL-10 levels remain unaltered (Agrawal et al. 2008). Increased secretion of pro-inflammatory cytokines by DCs might be one of the factors responsible for chronic inflammation (inflammaging) in the elderly.

Phosphatidylinositol 3 Kinase (PI3 K) acts as a negative feedback regulator of inflammation (Guha and Mackman 2002). PI3 K is responsible for blocking p38 MAPK activation in DCs, whilst AKT acts as a negative regulator of MAPK activation. A decline in AKT activation has been reported in DCs from older adults, which can be associated with upregulation of p38MAPK activity in DCs (Fukao et al. 2002).

PI3 K-AKT pathway positively regulates DC migration and phagocytosis, lowered PI3 K activation results in impaired phagocytic ability and migration of DCs with age (Agrawal et al. 2007).

DCs regulate T and B cells via co-stimulatory and inhibitory molecules expressed on their surface. With age, no significant alteration in the expression of co-stimulatory molecules has been observed. While the effect of ageing on expression of inhibitory molecules has not been examined (Agrawal et al. 2007). DCs are capable of driving T-cell polarisation via cytokine production, secretion of pro-inflammatory cytokines (IL-12) drives TH1 responses while IL-10 drives Th2 responses. A decline in TNFα and IL-6 output and increased IL-10 secretion upon LPS stimulation has been observed in DCs of aged subjects (Agrawal et al. 2009). The stimulatory capacity of DCs to induce proliferation of T cells is preserved with age (Steger et al. 1997). With respect to B-cell stimulation, aged DCs have lower cell surface expression of complement receptors (CD21) and lower antigen: antibody complexes, resulting in lower B-cell stimulation (Plackett et al. 2004). Follicular DC of aged mice often remain trapped within the sub-capsular sinus of lymph nodes and are unable to reach B-cell follicles resulting in decreased germinal centre formation (Kapsenberg 2003).

1.7 NK/NKT Cells

1.7.1 NK Cells

Natural killer (NK) cells are non-T-lymphocyte cytotoxic cells capable of recognising and killing cells infected with intracellular pathogens (predominantly viruses) and tumour cells. NK cells account for about 10–20 % of peripheral blood lymphocytes and have been divided into two separate subsets on the basis of relative expression of the adhesion molecule CD56 and the Fcγ receptor CD16, namely CD16+, $CD56^{dim}$ and the CD16+, $CD56^{bright}$ (Cooper et al. 2001). Approximately 90 % of NK cells are $CD56^{dim}$; these cells have high cytotoxicity towards virus-infected cells and tumour cells. Even though only 10 % of NK cells in the blood are $CD56^{bright}$; these cells make up to 90 % of NK cells in lymph nodes. $CD56^{bright}$ cells have immunomodulatory properties and are capable of initiating adaptive immune responses, activation of DCs and production of immunoregulatory cytokines (IL-10, IL-13, TNFα, IFNγ; Cooper et al. 2001; Vitale et al. 2004).

NK cell functioning and dynamics are affected by ageing. There is a 40–60 % decline in NK-cell proliferation with age in response to IL-2 due to lower Ca^{2+} mobilisation (Borrego et al. 1999), but this is balanced by an increased number of circulating NK cells (Almeida-Oliveira et al. 2011; Franceschi et al. 1995; Sansoni et al. 1993) possibly due to a higher proportion of long-lived NK cells in aged donors. The increase in the number of NK cells results from an expansion of the $CD56^{dim}$ subset but there is also a decline in the number of $CD56^{bright}$ NK cells (Chidrawar et al. 2006; Vitale et al. 1992) suggesting a phenotypic and functional shift in NK cells with age.

A decline in NK cell cytotoxicity on a per cell basis has been consistently reported with age (Di Lorenzo et al. 1999; Mariani et al. 1998), though the mechanisms are only partially understood. NK cells' ability to recognise and bind to target cells and perforin (a key component of cytotoxic granules) expression in NK cells is preserved with age (Mariani et al. 1992; Vitale et al. 1992), but defective transmembrane signalling; especially involving protein kinase C (PKC)-dependent pathways has been reported. Antibody-dependant cellular cytotoxicity (ADCC) in NK cells is mediated via signalling molecule CD16, and CD16 expression on NK cells is unaltered with age resulting in unaltered ADCC in NK cells with age (Mariani et al. 1998; Sansoni et al. 1993).

In addition to cytotoxic functions during early responses against infection, NK cells are also responsible for promoting adaptive responses against infections via secretion of immunoregulatory cytokines. Cytokine (IFNγ, IFNα) and chemokine (MIP-1α, Rantes, IL-8) production by NK cells in response to IL-2 is lower in elderly subjects compared with the young (Mariani et al. 2002). However, an increase in IL-4 and IL-10 production by NK cells and unaltered TNFα production has been reported with age (Solana et al. 1999). NK cells in the elderly show lower proliferation responses to IL-2 and a parallel impaired expression of CD69 activation antigen (Solana et al. 1999). The decreased proliferation response has been ascribed to reduced Ca^{2+} mobilisation with age (Borrego et al. 1999).

Natural cytotoxicity receptors (NCRs) such as NKp46 play an important role in NK cell cytotoxicity and cytokine production. NKp30 plays a central role in the crosstalk between NK cells and DCs. Recently, a study has reported a decline in NKp46 and NKp30 receptor expression with age (Almeida-Oliveira et al. 2011). CD94/NKG2 and KIR receptors have been known to regulate NK cell activation. An increase in KIR expression and decreased CD94/NKG2 A expression has been reported with age (Lutz et al. 2005). Altered NK cell receptor expression pattern with age may have an influence on susceptibility to infections and inflammatory diseases.

1.7.2 NKT Cells

NKT cells are "innate immune lymphocytes" that in addition to NK cell markers express T-cell receptor (TCR). The origin and function of NKT cells is poorly understood, although they have been suggested to play an important role in immune regulation via cytokine production. In addition to immune regulation, NKT cells have cytotoxic potential and are directly capable of killing target cells. Studies done in rodents have reported an increase in the number of NKT cells with age (Dubey et al. 2000; Faunce et al. 2005), and a decline in NKT cell cytotoxicity and impaired IFNγ production has been reported in aged mice and humans (Mocchegiani and Malavolta 2004). Thus, the limited data available suggests that the age-related changes in NKT cells are similar to those of NK cells.

Table 1.1 Significant changes in innate immune cells with advancing age

Cell population	Changes with age	Direction of change
Neutrophils	Numbers at site of infection	↔
	Chemotaxis	↓
	Phagocytic ability	↓
	Superoxide production	↓
	Ability to be rescued from apoptosis	↓
Monocytes	Absolute numbers	↔
	Migratory ability	↓
	Phagocytic ability	↓
	Intracellular killing of pathogens	↓
	Production of pro-inflammatory cytokines upon activation	↓
	Wound-healing function	↓
Dendritic cells	Number of circulating DCs	↓
	Migratory ability	↓
	Phagocytic ability	↓
	Ability to induce T-cell proliferation	↔
	Production of pro-inflammatory cytokines	↑
NK cells	NK cell numbers in circulation	↑
	NK cell cytotoxicity	↓
	Cytokine and chemokine production	↓
NKT cells	Absolute numbers	↑
	NKT cell cytotoxicity	↓
	Cytokine production (IFNγ)	↓

1.8 Concluding Remarks

Over the past decades, extensive evidence concerning a decline in immune system functioning has increased in parallel with other biological systems with age. Age-related alterations in several immunological parameters in both adaptive and innate immune systems have been reported. Given this decline in immune functioning in the aged, their ability to mount an adequate immune response to infection is questionable, resulting in increased susceptibility towards infections. Defects have been reported in functioning of all cells of the innate immune system as a result of intrinsic defects and altered extrinsic environment, summarised in Table 1.1. Impaired immune responses in the elderly thus appear to be a major factor responsible for the increased risk of bacterial and viral infection in the aged, resulting in increased morbidity and mortality. Although the past decade has seen a rapid explosion in research in the field of immunosenescence, the effect of ageing on innate immune responses is still incompletely understood, notably our understanding of the underlying molecular causes is incomplete. Moreover, our understanding of how the changes in the immune system with age are influenced by psychosocial factors mediated via neuroendocrine-immune interactions is a field that is still in its relative infancy. A detailed understanding of the impact of ageing on the innate immune system and its interaction with endocrine factors will enable us to develop therapeutic strategies; both lifestyle and pharmaceutical, to provide protection against infections and develop better strategies to control diseases in older adults.

Acknowledgments Niharika Duggal is supported by a grant from the United Kingdom cross-research council initiative New Dynamics of Ageing.

References

Agrawal A, Agrawal S, Gupta S (2007) Dendritic cells in human aging. Exp Gerontol 42:421–426.

Agrawal A, Agrawal S, Tay J, Gupta S (2008) Biology of dendritic cells in aging. J Clin Immunol 28:14–20

Agrawal A, Tay J, Ton S, Agrawal S, Gupta S (2009) Increased reactivity of dendritic cells from aged subjects to self-antigen, the human DNA. J Immunol 182:1138–1145

Almeida-Oliveira A, Smith-Carvalho M, Porto LC, Cardoso-Oliveira J, Ribeiro Ados S, Falcao RR et al (2011) Age-related changes in natural killer cell receptors from childhood through old age. Human Immunol 72:319–329

Alvarez E, Ruiz-Gutierrez V, Sobrino F, Santa-Maria C (2001) Age-related changes in membrane lipid composition, fluidity and respiratory burst in rat peritoneal neutrophils. Clin Exp Immunol 124:95–102

Antonaci S, Jirillo E, Ventura MT, Garofalo AR, Bonomo L (1984) Non-specific immunity in aging: deficiency of monocyte and polymorphonuclear cell-mediated functions. Mech Ageing Dev 24:367–375

Arranz L, Lord JM, De la Fuente M (2010) Preserved ex vivo inflammatory status and cytokine responses in naturally long-lived mice. Age 32:451–466

Ashcroft GS, Horan MA, Ferguson MW (1998) Aging alters the inflammatory and endothelial cell adhesion molecule profiles during human cutaneous wound healing. Lab Invest 78:47–58

Barbul A, Regan MC (1995) Immune involvement in wound healing. Otolaryngol Clin North Am 28:955–968

Boehmer ED, Goral J, Faunce DE, Kovacs EJ (2004) Age-dependent decrease in Toll-like receptor 4-mediated proinflammatory cytokine production and mitogen-activated protein kinase expression. J Leukoc Biol 75:342–349

Bonomo RA (2002) Resistant pathogens in respiratory tract infections in older people [Review]. J Amer Geriatrics Soc 50(Suppl):S236–S241

Borrego F, Alonso MC, Galiani MD, Carracedo J, Ramirez R, Ostos B et al (1999) NK phenotypic markers and IL2 response in NK cells from elderly people. Exp Gerontol 34:253–265

Braga PC, Sala MT, Dal Sasso M, Mancini L, Sandrini MC, Annoni G (1998) Influence of age on oxidative bursts (chemiluminescence) of polymorphonuclear neutrophil leukocytes. Gerontology 44:192–197

Brinkmann V, Reichard U, Goosmann C, Fauler B, Uhlemann Y, Weiss DS et al (2004) Neutrophil extracellular traps kill bacteria. Science 303:1532–1535

Butcher S, Chahal H, Lord JM (2000) Ageing and the neutrophil: no appetite for killing? Immunology 100:411–416

Butcher SK, Chahal H, Nayak L, Sinclair A, Henriquez NV, Sapey E et al (2001) Senescence in innate immune responses: reduced neutrophil phagocytic capacity and CD16 expression in elderly humans. J Leukoc Biol 70:881–886

Chambers SM, Shaw CA, Gatza C, Fisk CJ, Donehower LA, Goodell MA (2007) Aging hematopoietic stem cells decline in function and exhibit epigenetic dysregulation. PLoS Biol 5:e201

Chatta GS, Andrews RG, Rodger E, Schrag M, Hammond WP, Dale DC (1993) Hematopoietic progenitors and aging: alterations in granulocytic precursors and responsiveness to recombinant human G-CSF, GM-CSF, and IL-3. J Gerontol 48:M207–M212

Chelvarajan RL, Collins SM, Van Willigen JM, Bondada S (2005) The unresponsiveness of aged mice to polysaccharide antigens is a result of a defect in macrophage function. J Leukoc Biol 77:503–512

Chen LC, Pace JL, Russell SW, Morrison DC (1996) Altered regulation of inducible nitric oxide synthase expression in macrophages from senescent mice. Infect Immun 64:4288–4298

Chidrawar SM, Khan N, Chan YL, Nayak L, Moss PA (2006) Ageing is associated with a decline in peripheral blood CD56bright NK cells. Immun Ageing 3:10

Cooper MA, Fehniger TA, Caligiuri MA (2001) The biology of human natural killer-cell subsets. Trends Immunol 22:633–640

Davila DR, Edwards CK, 3rd, Arkins S, Simon J, Kelley KW (1990) Interferon-gamma-induced priming for secretion of superoxide anion and tumor necrosis factor-alpha declines in macrophages from aged rats. Faseb J 4:2906–2911

Delpedro AD, Barjavel MJ, Mamdouh Z, Faure S, Bakouche O (1998) Signal transduction in LPS-activated aged and young monocytes. J Interferon Cytokine Res 18:429–437

Di Bona D, Vasto S, Capurso C, Christiansen L, Deiana L, Franceschi C et al (2009) Effect of interleukin-6 polymorphisms on human longevity: a systematic review and meta-analysis. Ageing Res Rev 8:36–42

Di Lorenzo G, Balistreri CR, Candore G, Cigna D, Colombo A, Romano GC et al (1999) Granulocyte and natural killer activity in the elderly. Mech Ageing Dev 108:25–38

DiCarlo AL, Fuldner R, Kaminski J, Hodes R (2009) Aging in the context of immunological architecture, function and disease outcomes. Trends Immunol 30:293–294

Ding A, Hwang S, Schwab R (1994) Effect of aging on murine macrophages. Diminished response to IFN-gamma for enhanced oxidative metabolism. J Immunol 153:2146–2152

Dorshkind K, Montecino-Rodriguez E, Signer RA (2009) The ageing immune system: is it ever too old to become young again? Nat Rev Immunol 9:57–62

Dubey DP, Husain Z, Levitan E, Zurakowski D, Mirza N, Younes S et al (2000) The MHC influences NK and NKT cell functions associated with immune abnormalities and lifespan. Mech Ageing Dev 113:117–134

Dykstra B, de Haan G (2008) Hematopoietic stem cell aging and self-renewal. Cell Tissue Res 331:91–101

Effros RB (2001) Ageing and the immune system. Novartis Found Symp 235:130–139

Emanuelli G, Lanzio M, Anfossi T, Romano S, Anfossi G, Calcamuggi G (1986) Influence of age on polymorphonuclear leukocytes in vitro: phagocytic activity in healthy human subjects. Gerontology 32:308–316

Esparza B, Sanchez H, Ruiz M, Barranquero M, Sabino E, Merino F (1996) Neutrophil function in elderly persons assessed by flow cytometry. Immunol Invest 25:185–190

Fantuzzi G (2005) Adipose tissue, adipokines, and inflammation. J Allergy Clin Immunol 115:911–919

Faunce DE, Palmer JL, Paskowicz KK, Witte PL, Kovacs EJ (2005) CD1d-restricted NKT cells contribute to the age-associated decline of T cell immunity. J Immunol 175:3102–3109

Fietta A, Merlini C, De Bernardi PM, Gandola L, Piccioni PD, Grassi C (1993) Non specific immunity in aged healthy subjects and in patients with chronic bronchitis. Aging 5:357–361

Fietta A, Merlini C, Dos Santos C, Rovida S, Grassi C (1994) Influence of aging on some specific and nonspecific mechanisms of the host defense system in 146 healthy subjects. Gerontol 40:237–245

Fortin CF, Lesur O, Fulop T, Jr (2007) Effects of TREM-1 activation in human neutrophils: activation of signaling pathways, recruitment into lipid rafts and association with TLR4. Int Immunol 19:41–50

Franceschi C, Capri M, Monti D, Giunta S, Olivieri F, Sevini F et al (2007) Inflammaging and anti-inflammaging: a systemic perspective on aging and longevity emerged from studies in humans. Mech Ageing Dev 128:92–105

Franceschi C, Monti D, Sansoni P, Cossarizza A (1995) The immunology of exceptional individuals: the lesson of centenarians. Immunol Today 16:12–16
Fujihashi K, McGhee JR (2004) Mucosal immunity and tolerance in the elderly. Mech Ageing Dev 125:889–898
Fukao T, Tanabe M, Terauchi Y, Ota T, Matsuda S, Asano T et al (2002) PI3 K-mediated negative feedback regulation of IL-12 production in DCs. Nat Immunol 3:875–881
Fulop T, Jr., Komaromi I, Foris G, Worum I, Leovey A (1986) Age-dependent variations of intralysosomal enzyme release from human PMN leukocytes under various stimuli. Immunobiol 171(3):302–310
Fulop T, Jr, Larbi A, Linteau A, Desgeorges S, Douziech N (2002) The role of Mcl-1 and Bax expression alteration in the decreased rescue of human neutrophils from apoptosis by GM-CSF with aging. Ann N Y Acad Sci 973:305–308
Fulop T, Jr, Varga Z, Csongor J, Foris G, Leovey A (1989) Age related impairment in phosphatidylinositol breakdown of polymorphonuclear granulocytes. FEBS Lett 245:249–252
Gavazzi G, Krause KH (2002a) Ageing and infection. Lancet Infect Dis 2:659–666
Gavazzi G, Krause KH (2002b) Ageing and infection. Lancet Infect Dis 2:659–666
Gomez CR, Boehmer ED, Kovacs EJ (2005) The aging innate immune system. Curr Opin Immunol 17:457–462
Gomez CR, Nomellini V, Faunce DE, Kovacs EJ (2008) Innate immunity and aging. Exp Gerontol 43:718–728
Gosain A, DiPietro LA (2004) Aging and wound healing. World J Surg 28:321–326
Grolleau-Julius A, Garg MR, Mo R, Stoolman LL, Yung RL (2006) Effect of aging on bone marrow-derived murine CD11c+ CD4-CD8alpha-dendritic cell function. J Gerontol A: Biol Sci Med Sci 61:1039–1047
Guha M, Mackman N (2002) The phosphatidylinositol 3-kinase-Akt pathway limits lipopolysaccharide activation of signaling pathways and expression of inflammatory mediators in human monocytic cells. J Biol Chem 277(35):32124–32132
Herrero C, Marques L, Lloberas J, Celada A (2001) IFN-gamma-dependent transcription of MHC class II IA is impaired in macrophages from aged mice. J Clin Invest 107:485–493
Hilmer SN, Cogger VC, Le Couteur DG (2007) Basal activity of Kupffer cells increases with old age. J Gerontol A: Biol Sci Med Sci 62:973–978
Kapsenberg ML (2003) Dendritic-cell control of pathogen-driven T-cell polarization. Nat Rev Immunol 3:984–993
Kelly J, Ali Khan A, Yin J, Ferguson TA, Apte RS (2007) Senescence regulates macrophage activation and angiogenic fate at sites of tissue injury in mice. J Clin Invest 117:3421–3426
Khare V, Sodhi A, Singh SM (1996) Effect of aging on the tumoricidal functions of murine peritoneal macrophages. Nat Immun 15:285–294
Kissin E, Tomasi M, McCartney Francis N, Gibbs CL, Smith PD (1997) Age-related decline in murine macrophage production of nitric oxide. J Infect Dis 175:1004–1007
Krabbe KS, Pedersen M, Bruunsgaard H (2004) Inflammatory mediators in the elderly. Exp Gerontol 39:687–699
Lavoie ME, Rabasa-Lhoret R, Doucet E, Mignault D, Messier L, Bastard JP et al (2010) Association between physical activity energy expenditure and inflammatory markers in sedentary overweight and obese women. Int J Obesity 34:1387–1395
Lloberas J, Celada A (2002) Effect of aging on macrophage function. Exp Gerontol 37:1325–1331
Lord JM, Butcher S, Killampali V, Lascelles D, Salmon M (2001) Neutrophil ageing and immunosenescence. Mech Ageing Dev 122:1521–1535
Lutz CT, Moore MB, Bradley S, Shelton BJ, Lutgendorf SK (2005) Reciprocal age related change in natural killer cell receptors for MHC class I. Mech Ageing Dev 126:722–731
MacGregor RR, Shalit M (1990) Neutrophil function in healthy elderly subjects. J Gerontol 45:M55–M60

Mancuso P, McNish RW, Peters-Golden M, Brock TG (2001) Evaluation of phagocytosis and arachidonate metabolism by alveolar macrophages and recruited neutrophils from F344xBN rats of different ages. Mech Ageing Dev 122:1899–1913
Mariani E, Gemma Monaco MC, Cattini L, De Zwart JF, Tripodi G, Orengo A et al (1992) Effect of aging on distribution and functional activity of GL183+ and EB6+ NK cells. Arch Gerontol Geriatr 15(Suppl 1):249–256
Mariani E, Mariani AR, Meneghetti A, Tarozzi A, Cocco L, Facchini A (1998) Age-dependent decreases of NK cell phosphoinositide turnover during spontaneous but not Fc-mediated cytolytic activity. Int Immunol 10:981–989
Mariani E, Meneghetti A, Neri S, Ravaglia G, Forti P, Cattini L et al (2002) Chemokine production by natural killer cells from nonagenarians. Eur J Immunol 32:1524–1529
Mocchegiani E, Malavolta M (2004) NK and NKT cell functions in immunosenescence. Aging Cell 3:177–184
Nilsson BO (2007) Modulation of the inflammatory response by estrogens with focus on the endothelium and its interactions with leukocytes. Inflammation Res 56:269–273
Niwa Y, Kasama T, Miyachi Y, Kanoh T (1989) Neutrophil chemotaxis, phagocytosis and parameters of reactive oxygen species in human aging: cross-sectional and longitudinal studies. Life Sci 44:1655–1664
Ogawa T, Kitagawa M, Hirokawa K (2000) Age-related changes of human bone marrow: a histometric estimation of proliferative cells, apoptotic cells, T cells, B cells and macrophages. Mech Ageing Dev 117:57–68
Panda A, Arjona A, Sapey E, Bai F, Fikrig E, Montgomery RR et al (2009) Human innate immunosenescence: causes and consequences for immunity in old age. Trends Immunol 30:325–333
Plackett TP, Boehmer ED, Faunce DE., Kovacs EJ (2004) Aging and innate immune cells. J Leukoc Biol 76:291–299
Plowden, J, Renshaw-Hoelscher M, Engleman C, Katz J, Sambhara S (2004) Innate immunity in aging: impact on macrophage function. Aging Cell 3(4):161–167
Rao KM (1986) Age-related decline in ligand-induced actin polymerization in human leukocytes and platelets. J Gerontol 41:561–566
Renshaw M, Rockwell J, Engleman C, Gewirtz A, Katz J, Sambhara S (2002) Cutting edge: impaired Toll-like receptor expression and function in aging. J Immunol 169:4697–4701
Sansoni P, Cossarizza A, Brianti V, Fagnoni F, Snelli G, Monti D et al (1993) Lymphocyte subsets and natural killer cell activity in healthy old people and centenarians. Blood 82:2767–2773
Schuurhuis DH, Fu N, Ossendorp F, Melief CJ (2006) Ins and outs of dendritic cells. Int Arch Allergy Immunol 140:53–72
Sebastian C, Espia M, Serra M, Celada A, Lloberas J (2005) MacrophAging: a cellular and molecular review. Immunobiology 210:121–126
Shaw AC, Joshi S, Greenwood H, Panda A, Lord JM (2010) Aging of the innate immune system. Curr Opin Immunol 22:507–513
Solana R, Alonso MC, Pena J (1999) Natural killer cells in healthy aging. Exp Gerontol 34:435–443
Solana R, Pawelec G, Tarazona R (2006) Aging and innate immunity. Immunity 24:491–494
Steger, MM, Maczek C, Grubeck-Loebenstein B (1997) Peripheral blood dendritic cells reinduce proliferation in vitro aged T cell populations. Mech Ageing Dev 93:125–130
Stichel CC, Luebbert H (2007) Inflammatory processes in the aging mouse brain: participation of dendritic cells and T-cells. Neurobiol Aging 28:1507–1521
Swift ME, Burns AL, Gray KL, DiPietro LA (2001) Age-related alterations in the inflammatory response to dermal injury. J Invest Dermatol 117:1027–1035
Swift ME, Kleinman HK, DiPietro LA (1999) Impaired wound repair and delayed angiogenesis in aged mice. Lab Invest 79:1479–1487
Takahashi I, Ohmoto E, Aoyama S, Takizawa M, Oda Y, Nonaka K et al (1985) Monocyte chemiluminescence and macrophage precursors in the aged. Acta Med Okayama 39:447–451

Tesar BM, Walker WE, Unternaehrer J, Joshi NS, Chandele A, Haynes L et al (2006) Murine [corrected] myeloid dendritic cell-dependent toll-like receptor immunity is preserved with aging. Aging Cell 5:473–486

Thomas DR (2001) Age-related changes in wound healing. Drugs Aging 18(8):607–620

Tortorella C, Piazzolla G, Spaccavento F, Vella F, Pace L, Antonaci S (2000) Regulatory role of extracellular matrix proteins in neutrophil respiratory burst during aging. Mech Ageing Dev 119:69–82

Trzonkowski P, Mysliwska J, Pawelec G, Mysliwski A (2009) From bench to bedside and back: the SENIEUR Protocol and the efficacy of influenza vaccination in the elderly. Biogerontolgy 10:83–94

van Duin D, Shaw AC (2007) Toll-like receptors in older adults. J Am Geriatr Soc 55:1438–1444

Varas A, Sacedon R, Hernandez-Lopez C, Jimenez E, Garcia-Ceca J, Arias-Diaz J et al (2003) Age-dependent changes in thymic macrophages and dendritic cells. Microsc Res Tech 62:501–507

Vitale M, Della Chiesa M, Carlomagno S, Romagnani C, Thiel A, Moretta L et al (2004) The small subset of CD56brightCD16- natural killer cells is selectively responsible for both cell proliferation and interferon-gamma production upon interaction with dendritic cells. Eur J Immunol 34:1715–1722

Vitale M, Zamai L, Neri LM, Galanzi A, Facchini A, Rana R et al (1992) The impairment of natural killer function in the healthy aged is due to a postbinding deficient mechanism. Cell Immunol 145:1–10

Warren LA, Rossi DJ (2009) Stem cells and aging in the hematopoietic system [Review]. Mech Ageing Dev 130:46–53

Weinberger B, Herndler-Brandstetter D, Schwanninger A, Weiskopf D, Grubeck-Loebenstein B (2008) Biology of immune responses to vaccines in elderly persons. Clin Infect Dis 46:1078–1084

Wenisch C, Patruta S, Daxbock F, Krause R, Horl W (2000) Effect of age on human neutrophil function. J Leukoc Biol 67:40–45

Wessels I, Jansen J, Rink L, Uciechowski P (2010) Immunosenescence of polymorphonuclear neutrophils. Sci World J 10:145–160

Chapter 2
Introduction to Ageing of the Adaptive Immune System

Ludmila Müller and Graham Pawelec

2.1 Introduction

Like other somatic tissues and organs, the vertebrate immune system manifests age-associated alterations to its components and their functions. Unlike in invertebrates, in addition to the innate arm, vertebrates also possess adaptive immunity mediated by both cellular and humoral components. This chapter reviews data on age-associated alterations to adaptive immunity specifically in humans, mostly originating from cross-sectional studies (i.e., comparing young with old people). We summarise what is known about the effects of age on the different components of the adaptive immune system, particularly T cells, which appear most obviously different in the elderly. We consider the serious limitations inherent in cross-sectional studies, and discuss the crucial requirement to perform longitudinal studies (i.e., following the same individuals over time). Despite the logistical and financial constraints, longitudinal follow-up has provided the most biologically meaningful information about which of the many biomarkers apparently changing with age are actually relevant to medical parameters and for late-life health and longevity, and which, in contrast, may change with age but without clinical relevance. Given the lack of consistent data currently available, as a result of performing studies on heterogeneous populations using different analytical techniques, we emphasize the necessity for more numerous, more extensive and more detailed studies including assessments of the impact of psychosocial, nutritional and other thus-far rarely considered parameters on immunological and other biomarkers in longitudinal studies. We consider the mechanisms responsible for the disparate age-associated changes observed, beginning at the level of hematopoiesis, where alterations in the stem cell niches and the stem cells themselves contribute to

G. Pawelec (✉)
Center for Medical Research, University of Tübingen, Waldhörnlestr. 22,
72072 Tübingen, Germany
e-mail: graham.pawelec@uni-tuebingen.de

L. Müller
Max Planck Institute for Human Development, Lentzeallee 94, 14195 Berlin, Germany

J. A. Bosch et al. (eds.), *Immunosenescence*,
DOI 10.1007/978-1-4614-4776-4_2,

age-associated differences in the constantly self-renewing immune system. Thereafter, because T-cell progenitors require further processing in the periphery, thymic involution at puberty also results in a severely decreased production of naïve T cells in later life. This helps to explain why the observed alterations in peripheral T cells seem more extreme than in B cells, which do not require such further processing after their release from the bone marrow. We consider T cells and B cells separately, and describe the constellations of biomarkers, which we term "immune signatures" that may associate with age and/or disease. Determining which of the many potential biomarkers that can be measured nowadays are clinically relevant remains a challenge; thus, we finally introduce the concept of "immune risk profiles" able to predict morbidity and mortality in the elderly, and which may be informative for responses to vaccination, a very important aspect of immunosenescence for public health.

2.2 Overview of the Adaptive Immune System

We begin with a greatly simplified survey of the origins, components and functions of adaptive immunity in humans, and go on to consider what is known concerning the age-associated changes to these. Our adaptive immune system possesses the important function of destroying invading pathogens and neutralising their toxic molecules. Because these responses are potentially destructive to host tissues as well, it is crucial that they act with a high degree of specificity to avoid excessive collateral damage. The ability to distinguish what is "foreign" from what is "self" in this regard is a fundamental feature of the adaptive immune system. This ability relies on clones of immune cells expressing highly heterogeneous clonotypic receptors recognising one single epitope or a limited range of cross-reacting epitopes, with the ability to "remember" the antigen they have encountered. Thus, one hallmark of the adaptive-immune response is its ability to produce memory cells after the infection is cleared, providing long-lasting protection. This allows a more rapid and efficient clearance of the pathogen in case of a re-infection (Rollinghoff 1997). Cellular memory resides in (1) quantitative changes, that is, increase in number of antigen-specific cells and in (2) qualitative changes, that is, altered subset distribution of memory cells with different activation requirements compared to naïve cells. This applies to both the antibody-producing B cells and the cellular effector and helper T cells.

Innate-immune responses, which will be discussed in detail in the next chapter, initially call adaptive-immune responses into play and both work together to eliminate the pathogen. Recently, however, the strict distinction between these two arms of immunity, classically viewed as completely separate, has become less obvious. It has been shown that innate-immune cells can demonstrate memory characteristics under certain conditions, and reciprocally that adaptive-immune cells can express receptor characteristic of innate cells, especially at late stages of differentiation, that is, commonly present in increased numbers later in life and with the appearance of age-associated, potentially senescent, changes (Lanier and Sun 2009).

Adaptive-immune defence is evolutionarily more recent than innate immunity, probably emerging around 500 million years ago with the appearance of clonotypic immunity in both jawed and jawless vertebrates (Cooper and Herrin 2010). The somatic development of a pool of clonally diverse lymphocytes with a myriad of unique antigen-recognition receptors that when ligated under appropriate conditions, trigger cell activation, clonal expansion, and differentiation from naïve cells to effector and memory cells does seem to be a unique feature of an adaptive-immune system. This feature allows a combinatorial generation of an extremely diverse lymphocyte antigen-receptor repertoire by recombining genetic elements at random, increasing the chances of being able to recognize an almost infinite number of potential antigens, and hence elicitation of highly specific responses to them (Litman et al. 2010). In the case of pathogen recognition, this multistep process encompasses (1) recognition of epitopes derived from the pathogen on the surface of antigen presenting cells (APC) and activation of the responding T cells; (2) clonal expansion of initially small numbers of T cells carrying the specific receptor for that antigen in order to produce sufficiently high cell numbers to combat the pathogen, (3) differentiation of some of these to effector cells for elimination of the pathogen by direct cytotoxicity in the case of intracellular pathogens or help B cells to produce antibody against extracellular pathogens, and finally (4) retention of a small fraction of antigen-specific cells as memory cells to respond faster and more vigorously to re-exposure to the same pathogen (see Fig. 2.1 and Fig. 2.3, left).

Adaptive immunity is thus constituted by two separate cell types, depending for its function on B lymphocytes which produce antibody (part of so-called humoral-immune response) and T lymphocytes which mediate cytotoxicity, secrete pro- and anti-inflammatory cytokines, and provide help for B cells (part of so-called cellular immunity, see Fig. 2.1). Immature B cells are continuously formed in the bone marrow, exported to the periphery and there differentiate to antibody-producing cells under appropriate activation by antigen and help by T cells. Several steps of somatic recombination at the receptor (antibody) loci lead to production of a unique variable domain in surface immunoglobulin functioning as the antigen receptor of each individual B cell. Millions of different clonotypes of B cells circulate continuously in the blood and lymphatic system, carrying out immune surveillance. Upon direct recognition of foreign antigen and on receiving additional signals from a helper T cell, they differentiate into plasma cells and produce antibodies specific for the target antigen, and may further differentiate into memory B cells (see Fig. 2.1).

Similarly, T-cell progenitors are also originally produced in the bone marrow from hematopoietic stem cells (see Fig. 2.2), but they must further migrate to the thymus, where they give rise to a large population of CD4- and CD8-double-negative immature thymocytes (which do not express either main coreceptors facilitating binding to MHC class II or class I molecules, respectively) (Schwarz and Bhandoola 2006). These cells then rearrange antigen receptor genes to express the surface receptor for antigen (the TCR) and become CD4- and CD8-double-positive, expressing both CD4 and CD8 and differentiate into single-positive CD4 (MHC class II-restricted "helper") or CD8 (class I-restricted "cytolytic") T cells. These mature cells are then released from the thymus. Within the thymus, two steps of selection are required to

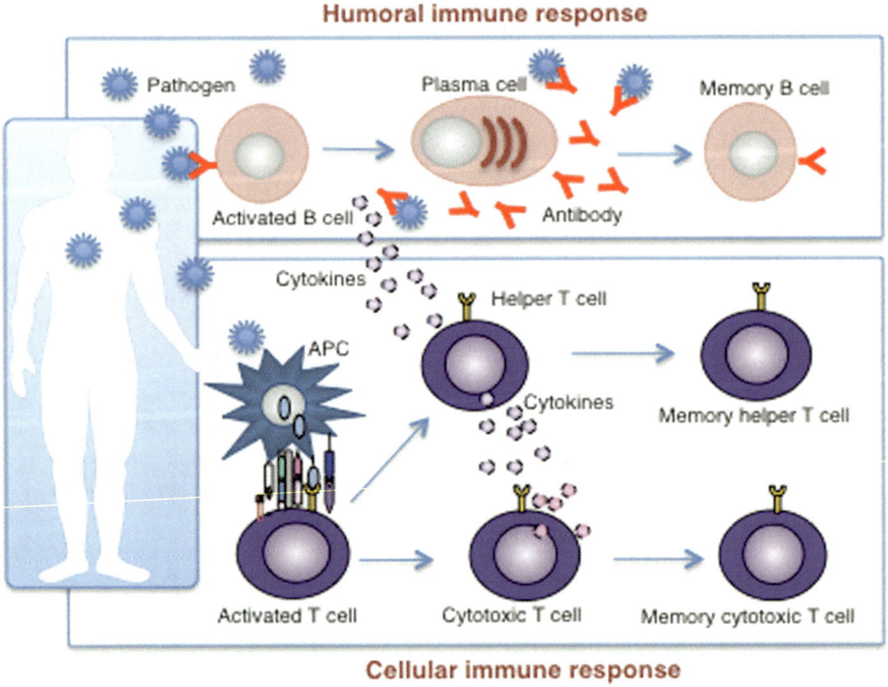

Fig. 2.1 Overview of the components of the adaptive immune system. *APC* Antigen Presenting Cells

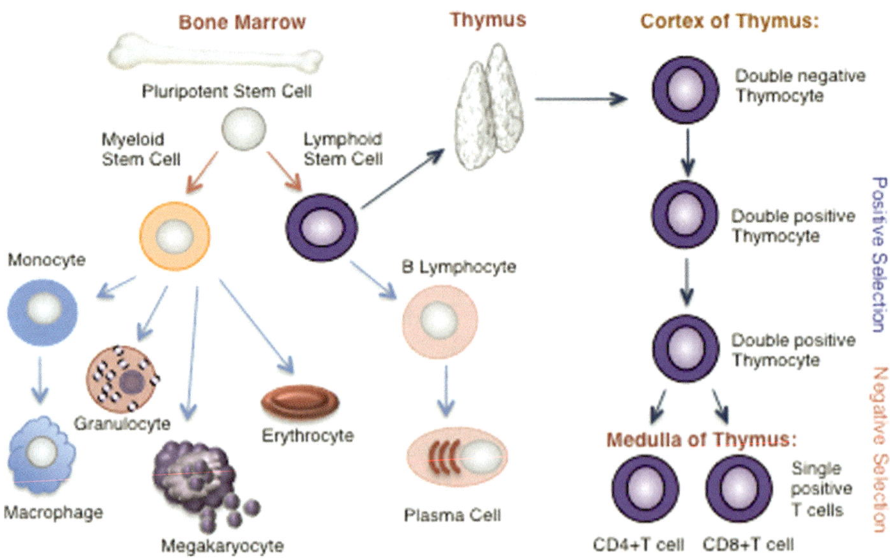

Fig. 2.2 Overview of thymopoiesis and haematopoiesis

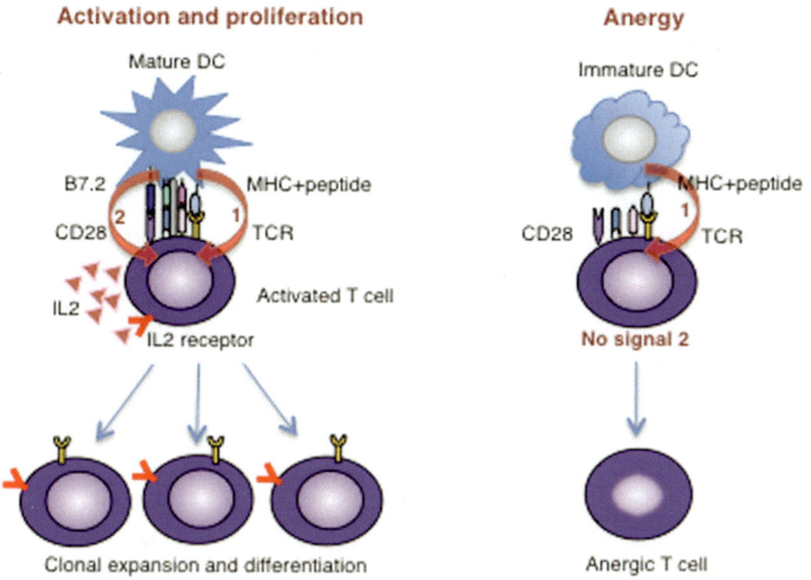

Fig. 2.3 Requirement for co-stimulatory signals for T-cell activation. *DC* Dendritic Cell, *MHC* Major Histocompatibility Complex, *TCR* T-Cell Receptor, *IL2* Interleukin 2

prevent T cells from recognising self antigens and to ensure recognition of a variety of different pathogens: This is known as negative selection of thymocytes binding strongly to self antigen presented by MHC class I or II and then positive selection for thymocytes binding weakly to self in the thymus (see Fig. 2.2, right). The released CD4+ cells mainly act as helper T cells providing signals required by B cells for their activation and antibody production, as well assisting CD8 + T cells to differentiate to cytotoxic effectors. The CD4 subset also contains T-regulatory cells (Tregs), which inhibit responses of other T cells. Most CD8+T cells are cytotoxic and are involved in lysis of infected cells, but some may also function as regulatory cells. Although the majority of mature peripheral T cells carries either CD4 or CD8 as outlined above, and expresses a TCR composed of an α chain and a β chain, a minority of around 5 % remains CD4-CD8-double negative and expresses a genetically distinct TCR composed of a γ- and a δ-chain. These latter cells possess markedly different antigenic specificity, recognising mostly lipid antigens, and commonly remaining CD4- and CD8-double-negative and not MHC-restricted (Rollinghoff 1997).

In the periphery, αβT-cell activation occurs through binding of the TCR to its cognate peptide–MHC complex presented on specialised APC, particularly dendritic cells (Fig. 2.3, left). This interaction between the APC and the T cell provides signal 1 for T-cell activation. In addition to the signal from their receptor, T cells require a second signal in the form of costimulation, provided by the ligation of T-cell surface receptors by their respective ligands on the APC (Fig. 2.3, left), of which CD27 and CD28 recognising CD70 and CD80/86 (B7.1/B.7.2), respectively, are two major examples. Thus, the first signal determines specificity, while the second signal ensures that a response is executed when needed. Provided that appropriate signals are received, the T cells become fully activated and embark on a program of clonal expansion and differentiation, controlled by the availability of antigen and other signals from the APC (mostly prominent cytokines such as IL 12 and IFN-γ), and giving rise to a large pool of effector cells. In the absence of a costimulatory signal, or the presence of negative regulatory signals, however, ligation of the TCR leads to inactivation of the T cell, and development of a state termed anergy, thought to be important in immune tolerance (see Fig. 2.3, right).

We will now briefly consider in what way adaptive-immune parameters differ in younger and older people, in terms of cell generation, function, representation in the periphery and implications for health status. As the immune system is a dynamic self-renewing tissue, depending on sustained production of peripheral cells from the hematopoietic system in the bone marrow, we will start there.

2.3 Ageing and Hematopoietic Stem Cells

The impact of ageing on the immune system is broad (see Fig. 2.4) and ranges from effects on hematopoietic stem cells and lymphoid progenitors in the bone marrow and thymus to mature lymphocytes in secondary lymphoid organs. These changes combine to result in a perceived diminution of immune responsiveness in the elderly (Linton and Dorshkind 2004). Age-associated functional decline of the immune system is likely to be attributable, at least in part, to changes in the hematopoietic stem cell (HSC) compartment. HSCs continuously replenish the blood and immune system throughout life, giving rise to all cellular (lymphoid and myeloid) components (Fig. 2.2, left). Age-related changes within the HSC compartment may be dependent on or independent of "intrinsic" cellular ageing of HSCs themselves. Hallmarks of age-dependent changes in the HSC compartment include an increase in HSC numbers, but decreased DNA repair capacity, a decrease in their homing efficiency, and a myeloid skewing of their differentiation potential (see Fig. 2.4). It remains unclear whether these changes are caused by gradual intrinsic changes within individual HSC and/or by changes in the cellular composition of the HSC compartment (microenvironment). However, an accumulation of DNA damage combined with an increase in intracellular reactive oxygen species has been found to characterize aged HSCs (Dykstra and de Haan 2008; Warren and Rossi 2009). Additional events such as genomic damage and epigenetic instability can also lead to malignant transformation of HSCs, progenitors or effector cells, manifesting as cancer or myeloproliferative disorders.

Age-associated changes in the lymphoid and myeloid lineage (Fig. 2.4) are believed by some investigators to be central to the decline of immune competence and predisposition to myelogenous diseases in the elderly. Their results indicate that ageing does not change individual HSC, but rather changes the clonal composition of the HSC compartment (Cho et al. 2008). Several studies have recently provided

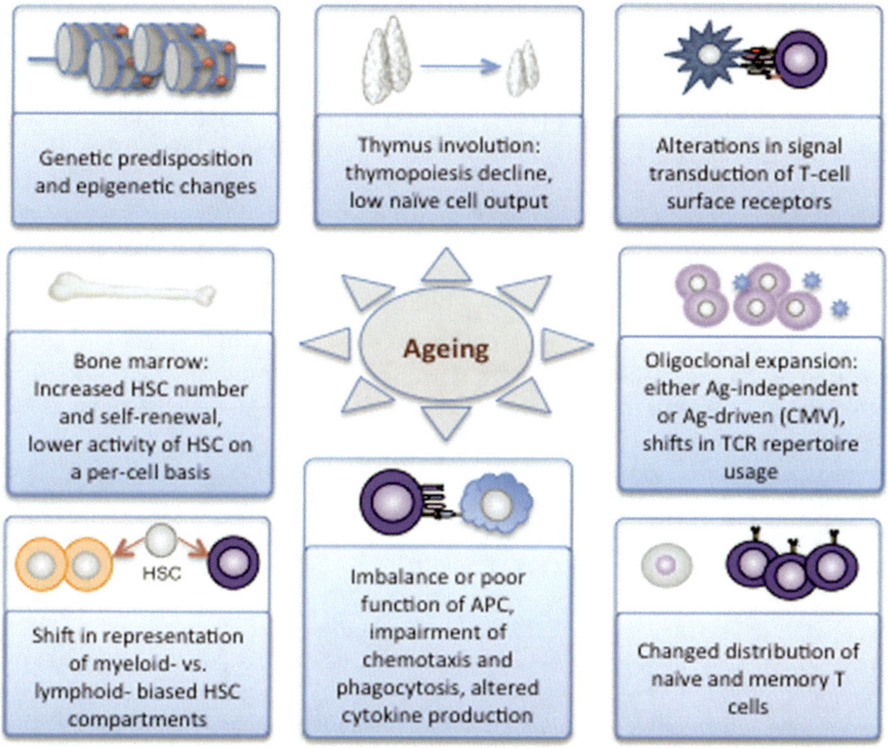

Fig. 2.4 Overview of effects of age on the different components of the immune system. *HSC* Haematopoietic Stem Cell, *APC* Antigen Presenting Cell, *Ag* Antigen, *CMV* Cytomegalovirus, *TCR* T-Cell Receptor

evidence in support of such a model using limiting dilution of whole bone marrow (BM) or purified stem cell assays (Beerman et al. 2010; Dykstra and de Haan 2008). The hematopoietic stem cell pool contains a heterogeneous mixture of HSCs, and ageing is associated with a marked shift in the proportions of these HSC subsets. Additionally, the stromal matrix of the bone marrow compartment, which normally nurtures and drives stem cell production, might play a role (i.e., extrinsic microenvironmental effects). The BM hematopoietic compartment decreases with age and is increasingly composed of fatty adipose tissue, suggesting a very early morphological impact of ageing on immune system ontogeny (Compston 2002; Gruver et al. 2007). Taken together, the results of different studies indicate that age-related alterations in multiple systems are likely to combine to affect the development and function of hematopoietic cells and thus the immune system, but much detail still needs to be filled in.

2.4 Age-related Changes to the Thymus

Thymopoiesis is dependent upon a continuing supply of T-lymphoid progenitors, maintenance of open thymic niches for progenitor engraftment and support of thymocyte migration and productive expansion by the cortical stromal microenvironment. These processes are regulated by reciprocal interactions between the marrow and thymic elements and developing lymphocytes, involving both cytokine signals and direct cell contact-mediated signalling (Hakim and Gress 2005). Age-associated changes at this level of regulation are also considered to have profound consequences for immune function.

The age-dependent involution of the thymus (Fig. 2.4) is associated with subsequent losses in repertoire diversity of T lymphocytes due to decreased production of naïve cells. Loss of thymic function and TCR diversity is thought to contribute to weaker immunosurveillance of the elderly, including increasing incidence of diseases such as cancers, autoimmunity, and new infections (Lynch et al. 2009). The overall physical size of the thymus remains relatively constant after early childhood, but perivascular spaces containing connective tissue expand, and thymic epithelial spaces where thymopoiesis occurs shrink, until thymic medullary and cortical tissue are limited to small islands surrounded by adipose and fibrous tissue (Gruver et al. 2007). The relative levels of thymocytes and stromal cells are reduced proportionately. Despite the quantitative reductions in cortical and medullary tissue, thymopoiesis at some level continues throughout life. Thus, the main effect of thymic involution is quantitative—concomitant with involution, the level of productive thymopoiesis declines (Fig. 2.4). Thymic productivity is determined not only by the immigration into the thymus of progenitor cells (themselves affected by age), but also by the proliferative expansion of thymocytes, the process by which small numbers of progenitors can increase many thousand-fold. Thus, the structural changes in the thymus are associated with a lower intrathymic proliferative expansion of progenitors, contributing to a lower thymic productivity (Hakim and Gress 2005; Hakim et al. 2005). Nonetheless, thymic involution occurs so early in life that it must be viewed as a normal developmental process rather than as a degenerative process associated with senescence. However, the consequences of reduced thymic output of naïve T cells are felt throughout life, perhaps with sequelae not foreseen by evolution but manifesting during the present extended human lifespan.

Several studies suggest that the thymic stroma has a crucial role in driving thymic involution (Alpdogan et al. 2006). It has recently been demonstrated that thymic atrophy can be mediated by a shift from a stimulatory to a suppressive cytokine microenvironment. Thymopoiesis may also be significantly affected by the endocrine system (Shanley et al. 2009) through such systemic hormones as growth hormone-insulin-like growth factor 1 and the stress hormone cortisol (Dykstra and de Haan 2008; Savino 2007). Thus, factors including systemic hormonal shifts and intrinsic cytokine programs within thymocytes and thymic stromal cells (Gruver et al. 2007) can all affect thymopoiesis. These findings indicate a working model in which age-related defects in both T-cell progenitors and thymic stromal cells and their products contribute to involution of the thymus. These changes could occur simultaneously in each compartment or one could precede the other. In either case, changes in one compartment with age could initiate a downward spiral of degenerative events that affect other compartments (Linton and Dorshkind 2004).

2.5 Age-associated Changes to Cell-Mediated Immunity

Immunosenescence is a descriptive term applied to the multiple age-associated changes to immunity observed in all mammals studied so far. These are perceived as deleterious, although under certain circumstances (e.g., acute organ allograft rejection) they might even be beneficial. In most cases, the question of whether age-associated changes are actually deleterious or not is not well-studied and often more an assumption than a fact. However, an important senescent alteration which is definitely clinically significant in the elderly is the decreased responsiveness to vaccination (Chen et al. 2009; Derhovanessian et al. 2008; Fulop et al. 2009; Grubeck-Loebenstein et al. 2009; Joshi et al. 2009; Lang et al. 2011; McElhaney and Effros 2009; Mysliwska et al. 2004; Ongradi et al. 2009; Pawelec and Larbi 2008; Targonski et al. 2007; Vezys et al. 2009), although even this, with its broad public health implications, is surprisingly poorly researched. There is nonetheless widespread acceptance of the notion that the competence of the adaptive-immune system decreases with age, giving rise to the observed increased incidence and severity of infectious disease, increased incidence of myeloid diseases including leukemias, as well as the onset of anemia in the elderly (Hock 2010; Linton and Dorshkind 2004).

Cross-sectional studies reveal a large number of potentially age-associated differences which might correlate with health outcomes, including immune parameters such as the distribution of subpopulations of naïve and memory T cells as well as their functional integrity (Russell et al. 2009; Sallusto et al. 1999; Weng 2006; Weng et al. 1997a, b). Differentiation-associated markers, some of which are listed in Table 2.1, influencing homing, adhesion, and costimulation, as well as differential expression of transcription factors affecting T-cell lineages, are routinely used to define "subsets" of T cells (Okada et al. 2008; Sallusto et al. 1999, 2000).

As T cells approach a putatively senescent state, such characteristics as surface marker expression alter, but distinguishing between differentiation pathways and "ageing" is problematic. The most biologically meaningful information on various factors predictive of longevity and their clinical relevance could be provided by longitudinal studies, which are, however, very rare and logistically challenging. More extensive studies are needed to follow aged populations over time and to use unique opportunities to assess in more detail their intraindividual changes as well as impact of psychosocial, nutritional and other thus-far rarely considered parameters on immunological and other biomarkers.

A consensus on the essential features of immunosenescence would probably include altered innate immunity (e.g., natural killer cell cytotoxicity on a per-cell basis;

reduced number and function of DC in blood and Langerhans cells in skin—see the previous chapter for a discussion of changes to innate immunity); and for adaptive immunity, decreased pools of naïve T and B cells, and increases in the numbers and percentages of memory and effector T and B cells. Altered distributions of naïve and memory T cells in peripheral blood of the elderly are commonly reported for CD8 T cells and to a lesser degree also for CD4 cells. It is often stated that one of the hallmarks of age-associated alterations in the human immune system is that aged T cells do not display CD28, a molecule critical for signal 2 transduction and naïve T-cell activation, leading to accumulation of these T cells in the periphery. In particular, an accumulation of late-differentiated effector T cells, commonly associated with cytomegalovirus (CMV) infection (see below), (Derhovanessian et al. 2009) likely contributes to a decline in the capacity of the adaptive-immune system to respond to novel antigens (Grubeck-Loebenstein et al. 2009).

Thus, low relative levels of naïve T cells, especially CD8 cells, remains the most commonly reported and widely accepted biomarker of immune senescence, usually coupled with reciprocal increase in memory cells with a range of different phenotypes determined by lack of expression of one or both costimulatory receptors CD27 and CD28 (Romero et al. 2007). Cells with such a phenotype may manifest age-related TCR diversity contraction and altered cytokine secretion patterns. Since a large TCR repertoire is generally considered essential for combating new pathogens, such decreases of diversity with age are assumed to have negative implications (Naylor et al. 2005). Additional changes in the distribution of T-cell populations may all conspire with repertoire contraction to negatively impact immune responses to novel infections and to vaccination in older individuals (Goronzy and Weyand 2005; Weng 2006). These could include increased presence of regulatory T cells, as well as accumulations of oligoclonally expanded potentially and functionally incompetent T cells, and a decrease in the cytotoxic activity of NKT cells (covered in Chap 2). However, this conclusion remains speculative and in most instances, we do not know whether this is really the case or not in humans. Thus, many associations have been reported but causation has not been established in any.

Age-related changes have also been repeatedly reported at the level of the individual T cell. Here, more is known about the requirement for signalling and mechanisms of signal transduction, but again, clinical consequences have not been unequivocally demonstrated. Alterations in numbers and efficacy of signal transduction of T cell-surface receptors (Fig. 2.4), both antigen-specific and costimulatory, have been reported which could be partly related to the transit of molecular components through the cell membrane and hence altered formation of the "immunological synapse" (Grakoui et al. 1999; Wetzel et al. 2002) required for initiating signalling to the nucleus. However, the TCR remains unchanged in lymphocytes from the elderly, both in number per cell and structure, but TCR assembly into these functional units ("signalosomes") is compromised due to cell membrane composition and resulting changes to fluidity (Rivnay et al. 1980), which might affect TCR mobility and signalosome assembly. Thus, the increased levels of cholesterol, often found in older adults, are likely to contribute to the age-associated decreases in T-cell activation (Larbi et al. 2005).

Age-associated alterations in APC could also lead to altered T-cell activation because without them T cells cannot be stimulated (Fig. 2.4; see the previous chapter for fuller discussion). As discussed above, in order for T cells to be activated against pathogens, antigens derived from these must first be taken up by professional APCs, usually DCs, processed and presented to T cells. An age-associated imbalance or poor function of APCs could clearly represent a first hurdle for triggering adaptive immunity (Lung et al. 2000). Such effects may be of great clinical significance in elderly people, as illustrated by the findings of van Duin et al. (2007), on success of vaccination, in which they demonstrated that the poor responsiveness to immunisation in the elderly was associated with the decreased expression of the costimulatory molecules CD80 and CD86. Thus, age-associated alterations to basic physiological functioning of DC can have direct clinical effects by reducing the costimulatory signals delivered to the responding T cells, even if the latter are functionally intact. In a preclinical model, age-associated functional impairment of DCs led to decreased T cell-mediated antitumor immunity, suggesting that not only influenza vaccination but cancer immunosurveillance could well be detrimentally affected by age-associated changes to APC (Grolleau-Julius et al. 2008).

2.6 Changes to Humoral Immunity in the Elderly

Immune-mediated protection from infection is attributable to both circulating antibodies to sterilise new infections, and antigen-specific T cells to help antibody production and to lyse cells which have nonetheless become infected (with virus). Such adaptive memory responses are elicited as a result of prior infection or vaccination. It is likely that antibody responses to at least some pathogens generated during youth, before the onset of immunosenescence, persist well into old age (Haynes and Maue 2009). An excellent example of this is the recent realization that survivors of the 1918 influenza pandemic, who are now well into their 90s, still possess highly functional, virus-neutralizing antibodies to this uniquely virulent virus. Hence, we know for certain that humans can sustain circulating B memory cells and antibodies for viruses for many decades after exposure—perhaps indefinitely (Yu et al. 2008). Nonetheless, although age-related changes to B cells are less well studied than T cells, reports suggest that, like T cells, memory B cells accumulate with age, and this may of course have beneficial implications (Colonna-Romano et al. 2003; Veneri et al. 2009). However, conflicting data have also been presented in this context (Breitbart et al. 2002; Chong et al. 2005; Frasca et al. 2005; Shi et al. 2005). One reason for such discrepancies might have been the use of different markers identifying memory cells. As with T cells, B memory cells are heterogeneous (Table 2.2), some being less-differentiated and some late-stage (Sanz et al. 2008).

The former may decline with age in some populations (Shi et al. 2005), but not others (Frasca et al. 2008). Age-associated alterations which may also contribute to the decline of the quality of humoral response in the elderly have been observed in B cells due to the normal processes of class switching and somatic hypermutation in immunoglobulin generation during immune responses which make the BCR so different from the TCR (Frasca et al. 2005). Also some features, which are desired for protective responses, such as long lifespan, prompt and enhanced responses by memory B cells, may be deleterious in terms of avoiding of autoimmunity (Sanz et al. 2008). Additionally, ageing is associated with a higher incidence of B-cell malignancy in older adults with oligoclonally expanded B cells (Weng 2006); this seems not to be the case for T cells where the genetic constitution of the TCR remains stable throughout clonal expansion and contraction. Further data on B-cell alterations are needed, particularly since decreased absolute numbers of B cells are a component of the original immune risk profile (IRP) predicting 2-, 4- and 6-year mortality in very elderly Swedes (see below).

It seems clear from many studies that while all components of adaptive immunity are in some ways different in the elderly compared with the young, the clinical impact of these changes is not always clear, and mechanisms of and biomarkers for immunosenescence remain controversial (Caruso et al. 2009; Derhovanessian et al. 2009. In addition to individual variations of genetic predisposition, epigenetic changes over the full course of human life probably also exert immunomodulating effects (Pawelec and Derhovanessian 2010). Longitudinal studies of different human populations over extended periods, assembling multidisciplinary databases will most likely be required to truly understand the nature and implications of age-associated immune alterations in the elderly.

2.7 Longitudinal Studies Identify an Immune Risk Profile (IRP) and Reveal a Major Impact of Cytomegalovirus on Immunosenescence

Several studies have independently demonstrated, initially quite surprisingly, that many major accepted age-associated changes to parameters used to assess adaptive immune status are markedly influenced by infection with CMV, regardless of infection with other persistent herpesviruses (reviewed in: Derhovanessian et al. 2009; Pawelec and Derhovanessian 2010; Pawelec et al. 2009). Accordingly, as has been mentioned before, the accumulation of memory cells with age may both reflect an adaptive response to the decline of production of naïve lymphocytes through homeostatic cytokine-driven peripheral expansion (i.e., the capacity of the peripheral pool of lymphocytes to undergo cell division without antigenic stimulation in order to fill the available space), as well as the cumulative effect of past and persistent viral infections resulting in the accumulation of virus-specific memory cells (Pawelec et al. 2006; Weng 2006). In the quite rare individuals who remain CMV-seronegative even at advanced age, this loss of naïve CD8 + T cells is much less marked (Chidrawar et al. 2009; Derhovanessian et al. 2010). Whether this may be related to an effect at the level of the thymus is currently unknown, but it is possible that CMV may also decrease residual thymic output of naïve cells in the elderly. The implications of these findings are manifold and emerge unexpectedly from longitudinal studies performed with a very robust clinical endpoint: all-cause mortality. These pioneering studies, designated OCTO-Immune (with subjects selected for very good health) and NONA-Immune (with a representative population of free-living subjects, most of whom were not exceptionally healthy), followed small populations of the very elderly (> 85 years) in Sweden. This resulted in the identification of a set of immune parameters at baseline correlating with mortality at 2-, 4- and 6-year follow-up. Results from the OCTO-immune study performed on the elderly selected for very good health at baseline indicated that around 15 % of them showed a combination of high absolute numbers of CD8 cells and low CD4 cells, and poor mitogen-stimulated T-cell proliferation, together with low B cells, which was associated with a higher 2-year mortality (Olsson et al. 2000; Wikby et al. 1994; Wikby et al. 2002). This cluster of parameters (see Table 3), together with an accumulation of CD8+ CD28-negative T cells, was designated the "immune risk profile" (IRP), (Pawelec et al. 2001).

In follow-up studies at 4 and 6 years, the IRP, together with inflammatory markers such as IL-6, C-reactive protein, transthyretin and albumin, as well as cognitive impairment, were also shown to be predictive of mortality in representative populations, that is, not selected for very good health (Wikby et al. 2006).

The existence of a sequence of stages for IRP individuals that begins with acquisition of CMV infection in earlier life was suggested, followed by generation of CD8+ CD28−cells to control persistent CMV infection, and eventually the development of an IRP associated also with elevated inflammatory status (Olsson et al. 2000; Wikby et al. 1998). Moreover, data at 6-year follow-up suggested that survival to the age of 100 years is associated with selection of individuals with an "inverted" IRP (with high CD4/CD8 ratio and low numbers of CD8-positive and CD28-negative cells), that was stable across time, that is, the oldest cohort of the study who had now become centenarians, which are commonly considered as a paradigm for "successfully ageing", had never entered this at-risk category (Strindhall et al. 2007).

Thus, a hallmark feature of immunosenescence is a cluster of immune parameters including seropositivity for CMV, and characterised by accumulations of clonal expansions of late-differentiated CD8+T cells, many of which are specific for CMV antigens. The latter proliferate poorly on vitro, but are strongly cytotoxic and release factors such as IFN-γ; although their per-cell activity is depressed (many are dysfunctional, perhaps actually "senescent", based on CD28-negativity, possession of short telomeres, refractoriness to apoptosis, and poor proliferative capacity (Effros et al. 2005; Vallejo 2005)). The absolute number of these cells is so much increased that they may themselves cause pathology as a result of their higher ability to produce inflammatory mediators (Pawelec et al. 2010). This may explain the increased low-grade chronic systemic inflammatory status (a state that has been dubbed "inflammaging"), which has been causatively associated with diseases such as cardiovascular disease, diabetes and dementias. Those individuals predisposed in some way (most likely genetic) to "resist" the effects of this increased pro-inflammatory status may be destined to become the long-term survivors in the population (Derhovanessian et al. 2010). Thus, the compromise between levels of immunity required for combating pathogens, and the unavoidable concomitant negative effects of inflammation may become increasingly dysregulated in the ageing individual, possibly amplified by the immunological changes induced by persistent CMV infection.

2.8 Concluding Remarks

Maintaining the integrity of adaptive immunity is crucial for the survival of the organism. This is incontrovertible, given the knowledge that inborn or acquired immunodeficiency results in severe, usually fatal, clinical outcomes. It is also clear that the elderly are often more susceptible to infectious disease and that the outcome is often more severe than in the young. Despite the large number of mostly cross-sectional studies comparing parameters of adaptive immunity, and the resulting identification of many which are different in young and old people, biomarkers of adaptive immunosenescence that actually correlate with clinical outcomes are rare in the extreme. The often dramatically lower number of naïve CD8+ T cells in the elderly has not been conclusively demonstrated to predict mortality or de novo infection. This decline in naïve CD8+ T cells is not seen to anything like the same degree for CD4+ cells in most studies, and seems to be due in large part to infection with CMV and possibly exposure to other persistent antigens than ageing per se. The accumulation of greater numbers of late-differentiated CD8+ T cells has been associated in a limited number of studies with poor responses to influenza vaccination and increased mortality in the very elderly. Lower numbers of B cells are also part of the IRP predicting 2-, 4- and 6-year mortality in the very elderly. Further studies are required in well-characterised diverse human populations, with adequate follow-up, to prove the relevance of the biomarkers studied and to allow some mechanistic insight into causative processes in humans and finally to facilitate intervention to restore and/or maintain appropriate immunity into extreme old age.

Acknowledgments LM and GP cooperate in the BMBF-funded (01UW0808) Berlin Ageing Study II. GP was supported by the European Union-funded Network of Excellence LifeSpan (FP6 036894), the Large Integrated Project IDEAL (FP7 259679), the BMBF project GerontoShield, and by the Deutsche Forschungsgemeinschaft (DFG-PA 361/14-1, DFG-PA 361/11-1).

References

Alpdogan O, Hubbard VM, Smith OM, Patel N, Lu S, Goldberg GL et al (2006) Keratinocyte growth factor (KGF) is required for postnatal thymic regeneration. Blood 107:2453–2460

Beerman I, Bhattacharya D, Zandi S, Sigvardsson M, Weissman IL, Bryder D et al (2010) Functionally distinct hematopoietic stem cells modulate hematopoietic lineage potential during aging by a mechanism of clonal expansion. Proc Natl Acad Sci USA 107:5465–5470

Breitbart E, Wang X, Leka LS, Dallal GE, Meydani SN, Stollar BD (2002) Altered memory B-cell homeostasis in human aging. J Gerontol A Biol Sci Med Sci 57:B304–311

Caruso C, Buffa S, Candore G, Colonna-Romano G, Dunn-Walters D, Kipling D et al (2009) Mechanisms of immunosenescence. Immun Ageing 6:10.

Chen WH, Kozlovsky BF, Effros RB, Grubeck-Loebenstein B, Edelman R, Sztein MB (2009) Vaccination in the elderly: an immunological perspective. Trends Immunol 30:351–359
Chidrawar S, Khan N, Wei W, McLarnon A, Smith N, Nayak L et al (2009) Cytomegalovirus-seropositivity has a profound influence on the magnitude of major lymphoid subsets within healthy individuals. Clin Exp Immunol 155:423–432
Cho RH, Sieburg HB, Muller-Sieburg CE (2008) A new mechanism for the aging of hematopoietic stem cells: aging changes the clonal composition of the stem cell compartment but not individual stem cells. Blood 111:5553–5561
Chong Y, Ikematsu H, Yamaji K, Nishimura M, Nabeshima S, Kashiwagi S et al (2005) CD27(+) (memory) B cell decrease and apoptosis-resistant CD27(−) (naïve) B cell increase in aged humans: implications for age-related peripheral B cell developmental disturbances. Int Immunol 17:383–390
Colonna-Romano G, Bulati M, Aquino A, Scialabba G, Candore G, Lio D et al (2003) B cells in the aged: CD27, CD5, and CD40 expression. Mech Ageing Dev 124:389–393
Compston JE (2002) Bone marrow and bone: a functional unit. J Endocrinol 173:387–394
Cooper MD, Herrin BR (2010) How did our complex immune system evolve? Nat Rev Immunol 10:2–3
Derhovanessian E, Larbi A, Pawelec G (2009) Biomarkers of human immunosenescence: Impact of cytomegalovirus infection. Curr Opin Immunol 21:440–445
Derhovanessian E, Maier AB, Beck R, Jahn G, Hahnel K, Slagboom PE et al (2010) Hallmark features of immunosenescence are absent in familial longevity. J Immunol 185:4618–4624
Derhovanessian E, Solana R, Larbi A, Pawelec G (2008) Immunity, ageing and cancer. Immunity Ageing 5:11.
Dykstra B, de Haan G (2008) Hematopoietic stem cell aging and self-renewal. Cell Tissue Res 331:91–101
Effros RB, Dagarag M, Spaulding C, Man J (2005) The role of CD8+ T-cell replicative senescence in human aging. Immunological Rev 205:147–157
Frasca D, Landin AM, Lechner SC, Ryan JG, Schwartz R, Riley RL et al (2008) Aging down-regulates the transcription factor E2 A, activation-induced cytidine deaminase, and Ig class switch in human B cells. J Immunol 180:5283–5290
Frasca D, Riley RL, Blomberg BB (2005) Humoral immune response and B-cell functions including immunoglobulin class switch are downregulated in aged mice and humans. Semin Immunol 17:378–384
Fulop T, Pawelec G, Castle S, Loeb M (2009) Immunosenescence and vaccination in nursing home residents. [Review]. Clin Infect Dis(An official publication of the Infectious Diseases Society of America) 48:443–448
Goronzy JJ, Weyand CM (2005) T cell development and receptor diversity during aging. Curr Opin Immunol 17:468–475
Grakoui A, Bromley SK, Sumen C, Davis MM, Shaw AS, Allen PM et al (1999) The immunological synapse: A molecular machine controlling T cell activation. Science 285:221–227
Grolleau-Julius A, Harning EK, Abernathy LM, Yung RL (2008) Impaired dendritic cell function in aging leads to defective antitumor immunity. Cancer Res 68:6341–6349
Grubeck-Loebenstein B, Della Bella S, Iorio AM, Michel JP, Pawelec G, Solana R (2009) Immunosenescence and vaccine failure in the elderly. Aging Clin Exp Res 21:201–209
Gruver AL, Hudson LL, Sempowski GD (2007) Immunosenescence of ageing. J Pathol 211:144–156
Hakim FT, Gress RE (2005) Reconstitution of the lymphocyte compartment after lymphocyte depletion: A key issue in clinical immunology. Eur J Immunol 35:3099–3102
Hakim FT, Memon SA, Cepeda R, Jones EC, Chow CK, Kasten-Sportes C et al (2005) Age-dependent incidence, time course, and consequences of thymic renewal in adults. J Clin Invest 115:930–939
Haynes L, Maue AC (2009) Effects of aging on T cell function. Curr Opin Immunol 21:414–417
Hock H (2010) Some hematopoietic stem cells are more equal than others. J Exp Med 207:1127–1130

Joshi SR, Shaw AC, Quagliarello VJ (2009) Pandemic influenza H1N1 2009, innate immunity, and the impact of immunosenescence on influenza vaccine. Yale J Biol Med 82:143–151
Lang PO, Govind S, Michel JP, Aspinall R, Mitchell WA (2011) Immunosenescence: Implications for vaccination programmes in adults. Maturitas 68:322–330
Lanier LL, Sun JC (2009) Do the terms innate and adaptive immunity create conceptual barriers? Nat Rev Immunol 9:302–303
Larbi A, Grenier A, Frisch F, Douziech N, Fortin C, Carpentier AC et al (2005) Acute in vivo elevation of intravascular triacylglycerol lipolysis impairs peripheral T cell activation in humans. Am J Clin Nutr 82:949–956
Linton PJ, Dorshkind K (2004) Age-related changes in lymphocyte development and function. Nat Immunol 5:133–139
Litman GW, Rast JP, Fugmann SD (2010) The origins of vertebrate adaptive immunity. Nat Rev Immunol 10:543–553
Lung TL, Saurwein-Teissl M, Parson W, Schonitzer D, Grubeck-Loebenstein B (2000) Unimpaired dendritic cells can be derived from monocytes in old age and can mobilize residual function in senescent T cells. Vaccine 18:1606–1612
Lynch HE, Goldberg GL, Chidgey A, Van Den Brink MR, Boyd R, Sempowski GD (2009) Thymic involution and immune reconstitution. Trends Immunol 30:366–373
McElhaney JE, Effros RB (2009) Immunosenescence: What does it mean to health outcomes in older adults? Curr Opinion Immunol 21 418–424
Mysliwska J, Trzonkowski P, Szmit E, Brydak LB, Machala M, Mysliwski A (2004) Immunomodulating effect of influenza vaccination in the elderly differing in health status. Exp Gerontol 39:1447–1458
Naylor K, Li G, Vallejo AN, Lee WW, Koetz K, Bryl E et al (2005) The influence of age on T cell generation and TCR diversity. J Immunol 174:7446–7452
Okada R, Kondo T, Matsuki F, Takata H, Takiguchi M (2008) Phenotypic classification of human CD4+ T cell subsets and their differentiation. International Immunol 20:1189–1199
Olsson J, Wikby A, Johansson B, Lofgren S, Nilsson BO, Ferguson FG (2000) Age-related change in peripheral blood T-lymphocyte subpopulations and cytomegalovirus infection in the very old: the Swedish longitudinal OCTO immune study. Mech Ageing Dev 121:187–201
Ongradi J, Stercz B, Kovesdi V, Vertes L (2009) Immunosenescence and vaccination of the elderly II. New strategies to restore age-related immune impairment. Acta Microbiol Immunol Hung 56:301–312
Pawelec G, Derhovanessian E (2010) Role of CMV in immune senescence. Virus Res 157:175–179
Pawelec G, Derhovanessian E, Larbi A, Strindhall J, Wikby A (2009) Cytomegalovirus and human immunosenescence. Rev Med Virol 19:47–56
Pawelec G, Ferguson FG, Wikby A (2001) The SENIEUR protocol after 16 years. Mech Ageing Dev 122:132–134
Pawelec G, Koch S, Franceschi C, Wikby A (2006) Human immunosenescence: does it have an infectious component? Ann N Y Acad Sci 1067:56–65
Pawelec G, Larbi A (2008) Immunity and ageing in man: Annual Review 2006/2007. Exp Gerontol 43:34–38
Pawelec, G, Larbi A, Derhovanessian E (2010) Senescence of the human immune system. J Comp Pathol 142(Suppl 1):S39–44
Rivnay B, Bergman S, Shinitzky M, Globerson A (1980) Correlations between membrane viscosity, serum cholesterol, lymphocyte activation and aging in man. Mech Ageing Dev 12:119–126
Rollinghoff M (1997) Immunity, components of the immune system and immune response. Biologicals 25:165–168
Romero P, Zippelius A, Kurth I, Pittet MJ, Touvrey C, Iancu EM et al (2007) Four functionally distinct populations of human effector-memory CD8 + T lymphocytes. J Immunol 178:4112–4119
Russell DG, Vanderven BC, Glennie S, Mwandumba H, Heyderman RS (2009) The macrophage marches on its phagosome: dynamic assays of phagosome function. Nat Rev Immunol 9:594–600
Sallusto F, Langenkamp A, Geginat J, Lanzavecchia A (2000) Functional subsets of memory T cells identified by CCR7 expression. Curr Top Microbiol Immunol 251:167–171

Sallusto F, Lenig D, Forster R, Lipp M, Lanzavecchia A (1999) Two subsets of memory T lymphocytes with distinct homing potentials and effector functions. Nature 401:708–712
Sanz I, Wei C, Lee FE, Anolik J (2008) Phenotypic and functional heterogeneity of human memory B cells. Sem Immunol 20:67–82
Savino W (2007) Neuroendocrine control of T cell development in mammals: role of growth hormone in modulating thymocyte migration. Exp Physiol 92(5):813–817
Schwarz BA, Bhandoola A (2006) Trafficking from the bone marrow to the thymus: a prerequisite for thymopoiesis. Immunol Rev 209:47–57
Shanley DP, Aw D, Manley NR, Palmer DB (2009) An evolutionary perspective on the mechanisms of immunosenescence. Trends Immunol 30:374–381
Shi Y, Yamazaki T, Okubo Y, Uehara Y, Sugane K, Agematsu K (2005) Regulation of aged humoral immune defense against pneumococcal bacteria by IgM memory B cell J Immunol 175:3262–3267
Strindhall J, Nilsson BO, Lofgren S, Ernerudh J, Pawelec G, Johansson B et al (2007) No immune risk profile among individuals who reach 100 years of age: findings from the Swedish NONA immune longitudinal study. Exp Gerontol 42:753–761
Targonski PV, Jacobson RM, Poland GA (2007) Immunosenescence: role and measurement in influenza vaccine response among the elderly. Vaccine 25:3066–3069
Vallejo AN (2005) CD28 extinction in human T cells: altered functions and the program of T-cell senescence. Immunol Rev 205:158–169
van Duin D, Allore HG, Mohanty S, Ginter S, Newman FK, Belshe RB et al (2007) Prevaccine determination of the expression of costimulatory B7 molecules in activated monocytes predicts influenza vaccine responses in young and older adults. J Infect Dis 195:1590–1597
Veneri D, Ortolani R, Franchini M, Tridente G, Pizzolo G, Vella A (2009) Expression of CD27 and CD23 on peripheral blood B lymphocytes in humans of different ages. Blood Transfus 7:29–34
Vezys V, Yates A, Casey KA, Lanier G, Ahmed R, Antia R et al (2009) Memory CD8 T-cell compartment grows in size with immunological experience. Nature 457:196–199
Warren LA, Rossi DJ (2009) Stem cells and aging in the hematopoietic system. Mech Ageing Dev 130:46–53
Weng NP (2006) Aging of the immune system: how much can the adaptive immune system adapt? Immunity 24:495–499
Weng NP, Granger L, Hodes RJ (1997a) Telomere lengthening and telomerase activation during human B cell differentiation. Proc Natl Acad Sci USA 94:10827–10832
Weng NP, Palmer LD, Levine BL, Lane HC, June CH, Hodes RJ (1997b) Tales of tails: regulation of telomere length and telomerase activity during lymphocyte development, differentiation, activation, and aging. Immunol Rev 160:43–54
Wetzel SA, McKeithan TW, Parker DC (2002) Live-cell dynamics and the role of costimulation in immunological synapse formation. J Immunol 169:6092–6101
Wikby A, Johansson B, Ferguson F, Olsson J (1994) Age-related changes in immune parameters in a very old population of Swedish people: a longitudinal study. Exp Gerontol 29:531–541
Wikby A, Johansson B, Olsson J, Lofgren S, Nilsson BO, Ferguson F (2002) Expansions of peripheral blood CD8 T-lymphocyte subpopulations and an association with cytomegalovirus seropositivity in the elderly: the Swedish NONA immune study. Exp Gerontol 37:445–453
Wikby A, Maxson P, Olsson J, Johansson B, Ferguson FG (1998) Changes in CD8 and CD4 lymphocyte subsets, T cell proliferation responses and non-survival in the very old: the Swedish longitudinal OCTO-immune study. Mech Ageing Dev 102:187–198
Wikby A, Nilsson BO, Forsey R, Thompson J, Strindhall J, Lofgren S et al (2006). The immune risk phenotype is associated with IL-6 in the terminal decline stage: findings from the Swedish NONA immune longitudinal study of very late life functioning. Mech Ageing Dev 127:695–704
Yu X, Tsibane T, McGraw PA, House FS, Keefer CJ, Hicar MD et al (2008) Neutralizing antibodies derived from the B cells of 1918 influenza pandemic survivors. Nature 455:532–536

Chapter 3
The Chronic Stress of Caregiving Accelerates the Natural Aging of the Immune System

Jeanette M. Bennett, Christopher P. Fagundes and Janice K. Kiecolt-Glaser

3.1 Introduction

Caregiving for a spouse with dementia is highly distressing (Taylor Jr et al. 2008). On average, spouses serving as dementia caregivers spend more than 10 h per day providing care for over 5 years (Donelan et al. 2002; Wimo et al. 2002). During this time, they experience a type of "living bereavement" as they watch their partners slowly lose their personality and intellect (Kiecolt-Glaser et al. 1991).

Given this burden, spouses serving as dementia caregivers are at heightened risk for depression with a 46 % prevalence rate (Clyburn et al. 2000; Gallagher et al. 1989), anxiety with a 25 % prevalence rate (Cooper et al. 2007; Mahoney et al. 2005), and poor sleep quality (Brummett et al. 2006). In addition, having lower social support increases the burden perceived by caregivers (Clipp and George 1990; Zarit et al. 1980). These negative effects are particularly detrimental for spousal dementia caregivers because they are typically older adults who are also experiencing a natural decline in immunocompetence (Allen 1994). In turn, they are at increased risk for a number of different illnesses as well as earlier mortality (Schulz and Beach, 1999).

Chronic stressors like caregiving have been linked to many diseases. Stressed individuals are more likely to be depressed and anxious (Cohen and Herbert 1996;

J. M. Bennett (✉)
Division of Oral Biology, College of Dentistry, Institute for Behavioral Medicine Research, College of Medicine, The Ohio State University, OH, USA
e-mail: Jeanette.Bennett@osumc.edu

Department of Psychology, The University of North Carolina at Charlotte, 9201 University City Blvd, Charlotte, NC 28223
e-mail: jeanette.bennett@uncc.edu

C. P. Fagundes
Institute for Behavioral Medicine Research, College of Medicine, The Ohio State University, OH, USA

J. K. Kiecolt-Glaser
Departments of Psychiatry and Psychology, Institute for Behavioral Medicine Research, College of Medicine, The Ohio State University, OH, USA
e-mail: Janice.Kiecolt-Glaser@osumc.edu

J. A. Bosch et al. (eds.), *Immunosenescence*,
DOI 10.1007/978-1-4614-4776-4_3,

Cooper et al. 2007; Gallagher et al. 1989). Stress also has been linked to obesity, cardiovascular disease, cancer, asthma, type 2 diabetes, infectious diseases, and autoimmune diseases (Black and Garbutt 2002; Bloomberg and Chen 2005; Elenkov and Chrousos 2002; Glaser and Kiecolt-Glaser 2005; Godbout and Glaser 2006). The immune system plays a critical role in the relationship between stress and disease (Kiecolt-Glaser 2009; Kiecolt-Glaser and Glaser 1995; Kiecolt-Glaser et al. 1987).

The immune system is a highly complex network of cells and communication mediators that are affected independently by stress and aging (Glaser and Kiecolt-Glaser 2005; Pawelec and Solana 1997). Due to its complexity, investigators commonly study components of the immune system by focusing on a particular facet or process such as cell-mediated immunity or wound healing. Therefore, in this chapter, we review how the chronic stress of caregiving accelerates the natural decline of immune function associated with aging on local immune responses (wound healing), cell-mediated immunity (herpesvirus latency and vaccination responses), systemic inflammation (circulating cytokines), and cellular aging (telomere length).

3.2 Wound Healing

Proper wound healing is critical for maintaining good health by re-establishing the integrity of the skin as a physical barrier and preventing infection. The inflammatory phase, which begins immediately after initial damage, is important for all other phases (Christian et al. 2006). Proinflammatory cytokines protect against infection and prepare injured tissue to be repaired by enhancing phagocytic cell recruitment and activation (Lowry 1993). In addition, proinflammatory cytokines regulate fibroblasts and epithelial cells in order to remodel damaged tissue (Lowry 1993).

Glucocorticoids can suppress production of proinflammatory cytokines such as interleukin (IL)-6, IL-1-beta (1β), and tumor necrosis factor-alpha (TNF-α) (Glaser et al. 1999). Stress facilitates glucocorticoid secretion, which, in turn, delays wound healing. Padgett et al. (1998) demonstrated that mice under stress exhibited slower wound healing compared to non-stressed mice. Yet, when mice under stress were injected with a glucocorticoid receptor antagonist, they healed at the same rate as non-stressed mice. This rodent study confirmed that stress-induced glucocorticoid secretion is one mechanism by which stress retards the wound healing process.

The first human demonstration of stress and wound healing compared female dementia caregivers ($N = 13$; age: 62.3 ± 2.3 years [$\pm$ SEM]) and matched controls ($N = 13$; age: 60.4 ± 2.8 years [$\pm$ SEM]) (Kiecolt-Glaser et al. 1995). Wounds, created via a punch biopsy, took 9 days (24 %) longer to close in the caregiver group than wounds in the control group. Furthermore, caregivers' peripheral blood lymphocytes produced less IL-1β mRNA in response to lipopolysaccharide stimulation than matched controls, suggesting that caregivers' lymphocytes at the wound site also produced less IL-1β, which is responsible for stimulating the wound healing process (Kiecolt-Glaser et al. 1995). These findings have important health implications, because delayed wound healing increases the likelihood of infection and other complications, particularly after surgery.

3.3 Control of Latent Herpesviruses

Infection with herpesviruses such as Epstein–Barr virus (EBV), herpes simplex virus type (HSV)-1, and varicella-zoster virus are very common. For example, 90 % of adults have antibody to EBV (Glaser and Kiecolt-Glaser 1994), and more than 90 % of individuals are HSV-1 seropositive (previously infected) by their 40s (Nahmias and Roizman 1973). Herpesviruses incorporate their genetic material into the host's DNA and establish a latent infection in the host. Latency, or a dormant period, occurs when the cellular immune response controls viral replication. However, when the cellular immune response is compromised, the virus is reactivated (Croen 1991). Viral reactivation stimulates the humoral immune response, increasing antibody production to the latent virus. Therefore, herpesvirus antibody titers serve as a functional marker of cell-mediated immunity, whereby higher herpesvirus titers represent poorer cellular immune system control of latent viruses. For example, older adults have higher herpesvirus antibody titers compared to younger adults; suggesting that as one ages, immunosenescence results in more frequent herpesvirus reactivation (Pawelec et al. 2009).

Caregiver stress can promote reactivation of latent herpesviruses. In one study, investigators examined spousal dementia caregivers ($N = 69$; age: 67.26 ± 0.98 years [$\pm$SEM]) and matched controls ($N = 69$; age: 67.75 ± 0.93 years $\pm$SEM]) over a 1-year period to understand how chronic stress affects EBV latency. Dementia caregivers had higher EBV antibody titers compared to matched controls (Kiecolt-Glaser et al. 1991). In addition, caregivers' peripheral blood lymphocytes proliferated less in response to mitogenic stimulation than those from controls (Kiecolt-Glaser et al. 1991). The fact that caregivers had higher antibody titers to latent EBV as well as lower lymphocyte proliferation than controls suggests that caregivers have a poorer cell-mediated immune response than what might be expected simply as a result of aging.

Another study addressed the effects of chronic stress on HSV-1 latency in family dementia caregivers ($N = 71$; age: 60.55 ± 1.52 years [$\pm$SEM]) and controls ($N = 58$; age: 62.41 ± 1.98 years [$\pm$SEM]) (Glaser and Kiecolt-Glaser 1997). Caregivers had higher HSV-1 antibody titers than controls. However, caregivers' and controls' antibody titers showed the same ability to neutralize the latent virus, suggesting that chronic stress decreases control over viral reactivation, but does not modulate the antibody's ability to function. Furthermore, caregivers had poorer *in vitro* memory T cell proliferation to HSV-1 infected cells compared to controls (Glaser and Kiecolt-Glaser 1997). Again, these findings suggest that chronic stress makes individuals less able to control reactivation of a latent herpesvirus in addition to the natural age-related declines in cell-mediated immunity.

3.4 Vaccine Responses

Among adults aged 65 years or older, influenza and pneumonia infections are the fourth leading cause of death (Thompson et al. 2003). Accordingly, the Center for Disease Control and Prevention (CDC) recommends that individuals 50 or older receive a protective pneumococcal pneumonia vaccination, and yearly influenza vaccinations (Smith et al. 2006). These vaccinations not only reduce infection-related mortality, but also decrease the number of hospitalizations as well as the duration of hospital stays (Smith et al. 2006). Given the importance of vaccine responses, they have been the subject of many studies. In vaccine response studies, seroconversion, a four-fold or greater increase in vaccine specific antibody titers relative to pre-vaccination levels, is considered a protective vaccine response (Levine et al. 1987).

Chronic stress of caregiving can impair vaccine responses. For example, caregivers had a poorer influenza vaccine response than matched controls (Kiecolt-Glaser et al. 1996). Specifically, when analyzing only those who had a four-fold antibody titer increase to at least one of three vaccine specific viral strains, caregivers had lower antibody titers than controls. Furthermore, age intensified the effect of chronic stress on the vaccine response. In individuals over 70-years old, only 26 % of caregivers had a four-fold response compared to 60 % of controls. In addition, caregivers' lymphocytes produced less IL-2 and IL-1β after exposure to vaccine-specific pathogen compared to controls (Kiecolt-Glaser et al. 1996). These immunological differences were not explained by depression, although caregivers were more depressed than controls (Kiecolt-Glaser et al. 1996).

Similar to the findings described above, another study also showed that chronic stress left older caregivers more vulnerable due to the non-protective influenza vaccine responses; 16 % of caregivers and 39 % of controls showed a four-fold rise in antibody to the vaccine (Vedhara et al. 1999). In addition, caregivers' daily salivary cortisol output was greater than controls (Vedhara et al. 1999). As described earlier, cortisol has immunoregulatory properties and can modulate cell-mediated immunity. Therefore, caregivers' higher salivary cortisol is consistent with their poorer vaccine responses.

Further work addressed relationship between caregiving and pneumococcal pneumonia vaccine response (Glaser et al. 2000). Pneumococcal pneumonia is caused by bacteria, and bacterial vaccine responses occur independently of T-cell activation. Therefore, this bacterial vaccine addresses an additional dimension of the immune response because immune responses to viral vaccinations, like influenza, are T-cell dependent.

Current and former caregivers and controls had similar pneumococcal pneumonia antibody titers before, and 2 and 4 weeks following the vaccination (Glaser et al. 2000). However, at 3- and 6-month follow-up visits, current caregivers' overall vaccine specific antibody titers were lower than those of former caregivers and controls. This finding suggests that caregiving reduces the stability of the IgG antibody response to the pneumococcal pneumonia vaccine, which limits the expected long-term protection (Glaser et al. 2000).

In a younger, caregiving sample, similar results were found for both influenza and pneumococcal pneumonia vaccinations. Caregivers for children who were disabled had a reduced antibody response to B/Malaysia strain of the influenza vaccine and the pneumococcal pneumonia vaccine at 1- and 6-months post-vaccination compared to parents of healthy children (Gallagher et al. 2009a, b).

Although caregivers are more likely to follow the CDC vaccination guidelines for older adults (Brown et al. 2009), chronic stress dysregulates cell-mediated immune responses. In addition, cell-mediated immunity declines with age. Thus, caregivers respond more poorly to both viral and bacterial vaccines compared to age-matched controls, suggesting that caregiver stress diminishes cell-mediated immunity above age-related decline in both younger and older adult populations.

3.5 Systemic Inflammation

Cytokines are soluble proteins that act as messenger substances, similar to hormones and neurotransmitters, between cells of the immune system as well as with non-immune cells. Acute local inflammation, cytokine production by lymphocytes at the site of injury, can be beneficial because it clears viral and bacterial pathogens and enhances wound healing (Kiecolt-Glaser 1995). However, chronic low-grade systemic inflammation, defined as a two- to three-fold elevation in cytokine levels compared to healthy subjects, is associated with poorer physical functioning and disease (Pedersen and Febbraio 2008). Low-grade systemic inflammation occurs due to cytokine production from several sources, including chronically activated immune cells in adipose tissue (Mohamed-Ali et al. 1998).

Elevated proinflammatory cytokines in serum or plasma are reliable predictors of morbidity and all-cause mortality in older adults (Bruunsgaard and Pedersen 2003; De Martinis et al. 2006). For example, inflammation is a risk factor for most cancers because proinflammatory cytokines facilitate tumor promotion, survival, proliferation, invasion, angiogenesis, and metastases (Aggarwal et al. 2006). In addition, higher levels of inflammation are associated with many diseases including: type 2 diabetes, arthritis, osteoporosis, Alzheimer's disease, and periodontal disease (Ershler and Keller 2000). Furthermore, inflammation is involved in every stage of atherosclerosis (Libby 2002).

Inflammation increases with age, and chronic stress magnifies age-related changes in proinflammatory cytokines. Indeed, data from several laboratories have shown that caregivers have higher IL-6 than controls. For example, in a 6-year longitudinal community study (age range: 55–89 years old at entry), caregivers' ($N = 116$) rate of IL-6 increase was four-fold greater than matched controls ($N = 109$) (Kiecolt-Glaser et al. 2003). The differences in caregivers' and controls' IL-6 could not be explained by depression, loneliness, or health behaviors (e.g., chronic health conditions, medication use, sleep, obesity, or smoking).

The fact that caregivers had greater IL-6 production than controls is particularly noteworthy in another context. Epidemiological data suggested that serum IL-6 values greater than or equal to 3.19 pg/mL put individuals at two-fold greater risk of death compared to those whose IL-6 levels were less than 1.9 pg/mL (Harris et al. 1999). Following these guidelines, caregivers would cross on average into this greater risk category at age 75, while controls would not cross, on average, into this category until after age 90. Thus, caregiving accelerates age-related IL-6 increases and elevates risk of mortality.

Among former caregivers whose spouses had died, IL-6 production remained elevated much like that of current caregivers (Kiecolt-Glaser et al. 2003). Hence, the chronic stress of caregiving may have long-term inflammatory consequences even after the stressor is over. These comparisons involve spouses who have been bereaved for at least 3 years, well past the usual time for adjustment to loss of their partners.

Elevated levels of C-reactive protein (CRP), IL-6, and D-dimer have been associated with greater frailty in elderly populations (Walston et al. 2002). Through general consensus, "frailty syndrome" has been characterized by the following symptoms: decrease in lean body mass, decline in walking mobility, and poor endurance with exhaustion or fatigue (Ferrucci et al. 2003). In frail individuals, inflammation is associated with muscle wasting, cognitive decline, and increased disease vulnerability (Cannon 1995; Ferrucci et al. 1999).

In a study investigating dementia caregiving and frailty, IL-6 and D-dimer were higher in caregivers than controls, but CRP levels were similar (von Kanel et al. 2006). Caregivers reported greater role overload, being overwhelmed by life's responsibilities, compared to the controls. Interestingly, role overload explained the D-dimer difference between caregivers and controls, suggesting that role overload stress may be driving the elevation in D-dimer levels. Although D-dimer is not an inflammatory marker, it can stimulate IL-6 and IL-1β production from monocytes *in vitro* (Robson et al. 1994).

In a follow-up study, the impact of sleep on inflammation was investigated in dementia caregivers and controls (von Kanel et al. 2010). As expected, caregivers reported poorer sleep and had a 3 % reduction in percent sleep as measured by actigraphy than controls. Among caregivers those who slept less had higher IL-6 and marginally higher CRP after controlling for differences in BMI, gender, and smoking. Among controls sleep was not associated with IL-6 or CRP (von Kanel et al. 2010).

A recent study assessed the consequences of caregiving for a relative with recently diagnosed brain cancer. At study entry, caregivers and controls had similar CRP levels. However, at the 4-month follow-up, caregivers' CRP was higher than controls (Rohleder et al. 2009). Mirroring the increase in CRP levels, expression of inhibitory-kappa B (I-κB), a transcription factor that produces anti-inflammatory effects, was lower in caregivers at the 4-month follow-up compared to controls (Rohleder et al. 2009). Even in a relative young sample, caregivers (N=40; age: 44.9 ± 7.4 years [±SD]) of children with autism or attention deficit hyperactivity disorder had elevated CRP levels compared to controls (N=17; age: 40.3 ± 6.4 years [±SD]) despite having similar diurnal cortisol patterns (Lovell et al. 2012).

Thus, the increase in inflammation associated with caregiving may lead to a greater propensity to frailty above and beyond the age-related increases observed in the control population. Caregivers' poorer sleep and decreased anti-inflammatory control may contribute to the greater frailty and inflammation observed in this group compared to controls.

3.6 Mechanisms Underlying Inflammation and Chronic Stress

There are several possible neuroendocrine mechanisms that may enhance the effects of chronic stress on inflammation. Stress activates the hypothalamic–pituitary–adrenal (HPA)-axis. Thus, one pathway is reduced glucocorticoid regulation of inflammation. Cortisol, a glucocorticoid produced by HPA-axis activation, inhibits immune cell activity by binding to the glucocorticoid receptor and reducing cytokine production (Barnes 1998; Brattsand and Linden 1996). However, chronically elevated cortisol can lead to glucocorticoid insensitivity, such that immune cells downregulate the expression of glucocorticoid receptors (Webster and Cidlowski 1994; Webster et al. 2002). In turn, this downregulation leads to increased inflammation because cortisol no longer has immunoregulatory properties and the immune cells are able to produce cytokines in an unregulated environment (Miller et al. 2002).

Data from one study suggested that caregivers' immune cells were less sensitive to glucocorticoids than controls'. Specifically, dementia caregivers' lymphocytes proliferated at a greater rate during an *in vitro* challenge in the presence of cortisol than did the controls' lymphocytes (Bauer et al. 2000). Caregivers also had higher cortisol levels in the morning and prior to lunchtime compared to the controls (Bauer et al. 2000). In another study, lymphocytes from parents of cancer patients produced higher IL-6 levels *in vitro* in the presence of glucocorticoids compared to lymphocytes from parents of healthy children (Miller et al. 2002). Therefore, caregivers' lymphocytes were less sensitive to cortisol's immunoregulatory effects. This lack of glucocorticoid inhibition on lymphocyte proliferation and cytokine production adds to the ammunition linking chronic stress and inflammation.

In addition to differences in HPA-axis responses between caregivers and controls, elevated sympathetic nervous system activation primes inflammatory responses as well. Specifically, the sympathetic–adrenal–medullary (SAM)-axis stimulates the release of peripheral norepinephrine and epinephrine from the adrenal medulla within seconds of a stressor (Mason 1968). Increased norepinephrine leads to an increase in nuclear factor-kappa B (NF-κB) translocation in mononuclear cells (Bierhaus et al. 2003). NF-κB is a transcription factor that increases inflammatory gene expression, such as IL-6 and IL-8; thus, NF-κB activation results in elevated release of these cytokines (Baldwin 1996; Barnes and Karin 1997). In a recent report, dementia caregivers had a prolonged norepinephrine response to a laboratory stressor compared to controls (Aschbacher et al. 2008). If the extended norepinephrine production is an indicator of chronic sympathetic over-activation, then it could be one of the underlying mechanisms for the elevated IL-6 levels observed in caregivers compared to controls (Kiecolt-Glaser et al. 2003; von Kanel et al. 2006).

Thus, chronic stressors like caregiving may promote inflammation through alterations in neuroendocrine pathways. In addition to the health consequences of inflammation already addressed, inflammation also has implications for cell aging (Kiecolt-Glaser and Glaser 2010). Elevated inflammation promotes T-cell proliferation and differentiation, which is a characteristic of cellular aging (Aviv 2004), as described below.

3.7 Telomere Length and Aging of Immune Cells

A telomere is a group of nucleoprotein complexes that cap chromosomes to protect and stabilize their integrity across the lifespan (Blackburn 2001). Telomere length indicates a cell's ability to replicate; because cell replication naturally shortens the length of a telomere, once a critical length is reached, the cell no longer divides and dies (Blackburn 2001). Therefore, telomere length is a proxy for measuring the biological age of the cell. Telomerase is an enzyme responsible for the production of the telomere cap after cell division (Blackburn 2001). Greater telomerase activity leads to longer telomeres. Thus, both telomerase activity and telomere length can provide insight into how cells age and what factors may expedite or impede the aging process at a molecular level. Further details of the association between telomeres and aging can be found in Chap. 8.

A growing body of literature suggests that chronic stress accelerates telomere shortening (Damjanovic et al. 2007; Epel et al. 2004, 2010; Simon et al. 2006). In a seminal study of mothers (age: 38 ± 6.5 years) who were caregivers for a chronically ill child, women who served as caregivers for a longer period of time had shorter telomeres, lower telomerase activity, and higher oxidative stress than those who were caregivers for a shorter period of time (Epel et al. 2004). These findings suggest that duration of stress affects cell aging and may lead to a decline in immune function and earlier onset of age-related diseases. In addition, although there were no differences between caregivers and controls overall, more stressed women in both groups had shorter telomeres (Epel et al. 2004). Further work that investigated telomere length of T cells and monocytes in dementia caregivers and age-matched controls showed that caregivers' telomeres were shorter than those of controls' (Damjanovic et al. 2007).

The evidence linking chronic stress and telomere shortening provides additional biological pathways connecting stress and aging (Kiecolt-Glaser and Glaser 2010). In addition to shorter telomeres, caregivers had greater inflammation measured by serum TNF-α levels compared to controls (Damjanovic et al. 2007). Higher levels of inflammation activates T-cell proliferation, which leads to shorter telomeres due to increased replication (Aviv 2004). Chronic stress also increases oxidative stress (Epel et al. 2004), which promotes telomere erosion during replication (Aviv 2004). Therefore, the combined effects of stress and inflammation on cellular aging have major implications for the immune system, especially in older individuals.

3.8 Conclusion

The chronic stress of caregiving impairs immune and cellular functions locally and systemically consistent with accelerated aging. Caregivers' wounds heal more slowly than controls' (Kiecolt-Glaser 1995), show more frequent latent herpesviruses reactivation (Glaser and Kiecolt-Glaser 1997; Kiecolt-Glaser et al. 1991), and poor vaccine responses (Gallagher et al. 2009a, b; Glaser et al. 2000; Kiecolt-Glaser et al. 1996;

Vedhara et al. 1999). Furthermore, caregivers have greater inflammation than controls (Damjanovic et al. 2007; Kiecolt-Glaser et al. 2003; Rohleder et al. 2009; von Kanel et al. 2010, 2006), a major predictor for frailty, cognitive decline, and all-cause mortality. Finally, caregivers' immune cells have shorter telomeres, suggesting they age more quickly compared to controls' (Damjanovic et al. 2007; Epel et al. 2004).

Taken together, the accelerated aging of the immune system leaves caregivers open to health problems above and beyond that of their non-caregiving counterparts. For example, caregiving has been categorized as a risk factor for mortality; strained caregivers were 63 % more likely to die over a 4-year period than controls (Schulz and Beach 1999). Another study showed that caregivers were more likely to develop heart disease than controls (Shaw et al. 1999). Furthermore, in the Nurses' Health Study, caring for an ill spouse increased the risk of developing coronary heart disease (Lee et al. 2003). The studies reviewed here clearly demonstrate that caregiving may accelerate the natural decline of the immune system, which may form a key pathway increasing the risk of disease and possibly premature death.

Acknowledgments Work on this chapter was supported in part by the following grants: National Institute on Aging (AG029562), National Institute of Dental & Craniofacial Research (DE014320), and National Cancer Institute (CA126857 and CA131029). The content is solely the responsibility of the authors and does not necessarily represent the official views of the National Institute of Health or any division providing financial support.

References

Aggarwal BB, Shishodia S, Sandur SK, Pandey MK, Sethi G (2006) Inflammation and cancer: How hot is the link? Biochem Pharmacol 72:1605–1621

Allen SM (1994) Gender differences in spousal caregiving and unment need for care. J Gerontol Soc Sci 49:S187–S195

Aschbacher K, Mills PJ, von Kanel R, Hong S, Mausbach BT, Roepke SK et al (2008) Effects of depressive and anxious symptoms on norepinephrine and platelet P-selectin responses to acute psychological stress among elderly caregivers. Brain Behav Immun 22:493–502

Aviv A (2004) Telomeres and human aging: Facts and fibs. Sci Aging Knowledge Environ 51:pe43

Baldwin AS (1996) The NF-kappa B and I kappa B proteins: New discoveries and insights. Ann Rev Immunol 14:649–683

Barnes PJ (1998) Anti-inflammatory actions of glucocorticoids: molecular mechanisms. Clin Sci 94:557–572

Barnes PJ, Karin M (1997) Nuclear factor-kappaB: a pivotal transcription factor in chronic inflammatory diseases. New Engl J Med 336:1066–1071

Bauer ME, Vedhara K, Perks P, Wilcock GK, Lightman SL, Shanks N (2000) Chronic stress in caregivers of dementia patients is associated with reduced lymphocyte sensitivity to glucocorticoids. J Neuroimmunol 103:84–92

Bierhaus A, Wolf J, Andrassy M, Rohleder N, Humpert PM, Petrov D et al (2003) A mechanism converting psychosocial stress into mononuclear cell activation. Proc Natl Acad Sci USA 100:1920–1925

Black PH, Garbutt LD (2002) Stress, inflammation and cardiovascular disease. J Psychosom Res 52:1–23

Blackburn EH (2001) Switching and signaling at the telomere. Cell 106:661–673

Bloomberg GR, Chen E (2005) The relationship of psychologic stress with childhood asthma. Immunol Allergy Clin 25:83–105

Brattsand R, Linden M (1996) Cytokine modulation by glucocorticoids: mechanisms and actions in cellular studies. Aliment Pharm Therap 10:81–90
Brown SL, Smith DM, Schulz R, Kabeto MU, Ubel PA, Poulin M et al (2009) Caregiving behavior is associated with decreased mortality risk. Psychol Sci 20:488–494
Brummett B, Babyak M, Siegler I, Vitaliano P, Ballard E, Gwyther L et al (2006) Associations among perceptions of social support, negative affect, and quality of sleep in caregivers and noncaregivers. Health Psychol 25:220–225
Bruunsgaard H, Pedersen BK (2003) Age-related inflammatory cytokines and disease. Immunol Allergy Clin 23:15–39
Cannon J (1995) Cytokines in aging and muscle homeostasis. J Gerontol: Series A, Biol Sci Med Sci 50:120–123
Christian LM, Graham JE, Padgett DA, Glaser R, Kiecolt-Glaser JK (2006) Stress and wound healing. Neuroimmunomodulation 13:337–346
Clipp EC, George LK (1990) Caregiver needs and patterns of social support. J Gerontol 45:102–222
Clyburn L, Stones M, Hadjistavropoulos T, Tuokko H (2000) Predicting caregiver burden and depression in Alzheimer's disease. J Gerontol Series B: Psychol Sci Soc Sci 55:S2
Cohen S, Herbert TB (1996) Health psychology: psychological factors and physical disease from the perspective of human psychoneuroimmunology. Ann Rev Psychol 47:113–142
Cooper C, Balamurali TBS, Livingston G (2007) A systematic review of the prevalence and covariates of anxiety in caregivers of people with dementia. Intl Psychogeriatrics 19:175–195
Croen KD (1991) Latency of the human herpesviruses. Ann Rev Med 42:61–67
Damjanovic AK, Yang Y, Glaser R, Kiecolt-Glaser JK, Nguyen H, Laskowski B et al (2007) Accelerated telomere erosion is associated with a declining immune function of caregivers of Alzheimer's disease patients. J Immunol 179:4249–4254
De Martinis M, Franceschi C, Monti D, Ginaldi L (2006) Inflammation markers predicting frailty and mortality in the elderly. Exper Molec Pathol 80:219–227
Donelan K, Hill CA, Hoffman C, Scoles K, Feldman PH, Levine C et al (2002) Challenged to care: Informal caregivers in a changing health system. Health Affairs 21:222–231
Elenkov IJ, Chrousos GP (2002) Stress hormones, proinflammatory and anti-inflammatory cytokines, and autoimmunity. Ann NY Acad Sci 966:290–303
Epel ES, Blackburn EH, Lin J, Dhabhar FS, Adler NE, Morrow JD et al (2004) Accelerated telomere shortening in response to life stress. Proc Natl Acad Sci USA 101:17312–17315
Epel ES, Lin J, Dhabhar FS, Wolkowitz OM, Puterman E, Karan L et al (2010) Dynamics of telomerase activity in response to acute psychological stress. Brain Behav Immun 24:531–539
Ershler W, Keller E (2000) Age-associated increased interleukin-6 gene expression, late-life diseases, and frailty. Ann Rev Med 51:245–270
Ferrucci L, Guralnik JM, Cavazzini C, Bandinelli S, Lauretani F, Bartali B et al (2003) The frailty syndrome: a critical issue in geriatric oncology. Crit Rev Oncol/Hematol 46:127–137
Ferrucci L, Harris T, Guralnik J, Tracy R, Corti M, Cohen H et al (1999) Serum IL-6 level and the development of disability in older persons. J Am Geriatr Soc 47:639–646
Gallagher D, Rose J, Rivera P, Lovett S, Thompson LW (1989) Prevalence of depression in family caregivers. Gerontologist 29:449–456
Gallagher S, Phillips AC, Drayson MT, Carroll D (2009a) Caregiving for children with developmental disabilities is associated with a poor antibody response to influenza vaccination. Psychosom Med 71:341–344
Gallagher S, Phillips AC, Drayson MT, Carroll D (2009b) Parental caregivers of children with developmental disabilities mount a poor antibody response to pneumococcal vaccination. Brain Behav Immun 23:338–346
Glaser R, Kiecolt-Glaser JK (1994) Stress-associated immune modulation and its implications for reactivation of latent herpesviruses. In Glaser R, Jones J (eds) Human herpesvirus infections, Dekker, New York, p 245–270
Glaser R, Kiecolt-Glaser JK (1997) Chronic stress modulates the virus-specific immune response to latent herpes simplex virus type 1. Ann Behav Med 19:78–82

Glaser R, Kiecolt-Glaser JK (2005) Stress-induced immune dysfunction: Implications for health. Nature Rev Immunol 5:243–251
Glaser R, Kiecolt-Glaser PK, Marucha PT, MacCallum RC, Laskowski, BF, Malarkey WB (1999) Stress-related changes in proinflammatory cytokine production in wounds. Arch Gen Psychiatr 56:450–456
Glaser R, Sheridan J, Malarkey WB, MacCallum RC, Kiecolt-Glaser JK (2000) Chronic stress modulates the immune response to a pneumococcal pneumonia vaccine. Psychosom Med 62:804–807
Godbout JP, Glaser R (2006) Stress-induced immune dysregulation: Implications for wound healing, infectious disease and cancer. J Neuroimmune Pharmacol 1:421–427
Harris TB, Ferrucci L, Tracy RP, Corti MC, Wacholder S, Ettinger WH, Jr et al (1999) Associations of elevated interleukin-6 and C-reactive protein levels with mortality in the elderly. Am J Med 106:506–512
Kiecolt-Glaser J (2009) Psychoneuroimmunology: Psychology's gateway to the biomedical future. Perspect Psychol Sci 4(367)
Kiecolt-Glaser J, Dura J, Speicher C, Trask O, Glaser R (1991) Spousal caregivers of dementia victims: Longitudinal changes in immunity and health. Psychosom Med 53:345
Kiecolt-Glaser J, Glaser R (1995) Psychoneuroimmunology and health consequences: Data and shared mechanisms. Psychosom Med 57(269)
Kiecolt-Glaser J, Glaser R, Shuttleworth E, Dyer C, Ogrocki P, Speicher C (1987) Chronic stress and immunity in family caregivers of Alzheimer's disease victims. Psychosom Med 49(523)
Kiecolt-Glaser JK, Glaser R (2010) Psychological stress, telomeres, and telomerase. Brain Behav Immun 24:529–530
Kiecolt-Glaser JK, Glaser R, Gravenstein S, Malarkey WB, Sheridan J (1996) Chronic stress alters the immune response to influenza virus vaccine in older adults. Proc Natl Acad Sci USA 93:3043–3047
Kiecolt-Glaser JK, Marucha PT, Malarkey WB, Mercado AM, Glaser R (1995) Slowing of wound healing by psychological stress. Lancet 346:1194–1196
Kiecolt-Glaser JK, Marucha PT, Malarkey WB, Mercado AM, Glaser R (1995) Slowing of wound healing by psychological stress. Lancet 346:1194–1196
Kiecolt-Glaser JK, Preacher KJ, MacCallum RC, Atkinson C, Malarkey WB, Glaser R (2003) Chronic stress and age-related increases in the proinflammatory cytokine IL-6. Proc Natl Acad Sci USA 100:9090–9095
Lee S, Colditz GA, Berkman LF, Kawachi I (2003) Caregiving and risk of coronary heart disease in US women—a prospective study. Am J Prevent Med 24:113–119
Levine M, Beattie BL, McLean DM, Corman D (1987) Characterization of the immune response to trivalent influenza vaccine in elderly men. J Am Geriatr Soc 35:609–615
Libby P (2002) Inflammation in atherosclerosis. Nature 420:868–874
Lovell B, Moss M, Wetherell M (2012) The psychosocial, endocrine and immune consequences of caring for a child with autism or ADHD. Psychoneuroendocrinology 37:534–542
Lowry SF (1993) Cytokine mediators of immunity and inflammation. Arch Surg 28:1235–1241
Mahoney R, Regan C, Katona C, Livingston G (2005) Anxiety and depression in family caregivers of people with Alzheimer disease: the LASER-AD study. Am J Geriatr Psychiatry 13:795
Mason J (1968) A review of psychoendocrine research on the sympathetic-adrenal medullary system. Psychosom Med 30:631
Miller G, Cohen S, Ritchey A (2002) Chronic psychological stress and the regulation of pro-inflammatory cytokines: a glucocorticoid-resistance model. Health Psychol 21:531–541
Mohamed-Ali V, Pinkney J, Coppack S (1998) Adipose tissue as an endocrine and paracrine organ. Intl J Obes 22:1145–1158
Nahmias A, Roizman B (1973) Infection with herpes-simplex viruses 1 and 2. New Engl Jf Med 289:781–789
Padgett DA, Marucha PT, Sheridan JF (1998) Restraint stress slows cutaneous wound healing in mice. Brain Behav Immun 12:64–73

Pawelec G, Derhovanessian E, Larbi A, Strindhall J, Wikby A (2009) Cytomegalovirus and human immunosenescence. Rev Med Virol 19:47–56

Pawelec G, Solana R (1997) Immunosenescence. Immunol Today 18:514–516

Pedersen BK, Febbraio MA (2008) Muscle as an endocrine organ: focus on muscle-derived interleukin-6. Physiol Rev 88:1379–1406

Robson SC, Shephard EG, Kirsch RE (1994) Fibrin degradation product D-dimer induces the synthesis and release of biologically-active IL-1-beta, IL-6 and plasminogen-activator inhibitros from monocytes in-vitro. [Article]. Brit J Haematol 86:322–326

Rohleder N, Marin TJ, Ma R, Miller GE (2009) Biologic cost of caring for a cancer patient: Dysregulation of pro- and anti-inflammatory signaling pathways. J Clin Oncol 27:2909–2915

Schulz R, Beach S (1999) Caregiving as a risk factor for mortality: the caregiver health effects study. J Am Med Assoc 282:2215

Shaw WS, Patterson TL, Ziegler MG, Dimsdale JE, Semple SJ, Grant I (1999) Accelerated risk of hypertensive blood pressure recordings among Alzheimer caregivers. J Psychosom Res 46:215–227

Simon NM, Smoller JW, McNamara KL, Maser RS, Zalta AK, Pollack MH et al (2006) Telomere shortening and mood disorders: preliminary support for a chronic stress model of accelerated aging. Biol Psychiatr 60:432–435

Smith NM, Bresee JS, Shay DK, Uyeki TM, Cox NJ, Strikas RA (2006) Prevention and control of influenza: recommendations of the advisory committee on immunization practices (ACIP). MMWR Recomm Rep 55:1–42

Taylor Jr D, Ezell M, Kuchibhatla M, Østbye T, Clipp E. (2008) Identifying trajectories of depressive symptoms for women caring for their husbands with dementia. J Am Geriatr Soc 56:322–327

Thompson WW, Shay DK, Weintraub E, Brammer L, Cox N, Anderson LJ et al (2003) Mortality associated with influenza and respiratory syncytial virus in the United States. J Am Med Assoc 289:179–186

Vedhara K, Cox NK, Wilcock GK, Perks P, Hunt M, Anderson S et al (1999) Chronic stress in elderly carers of dementia patients and antibody response to influenza vaccination. Lancet 353:627–631

von Kanel R, Ancoli-Israel S, Dimsdale JE, Mills PJ, Mausbach BT, Ziegler MG et al (2010) Sleep and biomarkers of atherosclerosis in elderly Alzheimer caregivers and controls. Gerontology 56:41–50

von Kanel R, Dimsdale JE, Mills PJ, Ancoli-Israel S, Patterson TL, Mausbach BT et al (2006) Effect of Alzheimer caregiving stress and age on frailty markers interleukin-6, C-reactive protein, and D-dimer. J Gerontol A 61:963–969

Walston J, McBurnie MA, Newman A, Tracy RP, Kop WJ, Hirsch CH et al (2002) Frailty and activation of the inflammation and coagulation systems with and without clinical comorbidities: results from the cardiovascular health study. Arch Intern Med 162:2333–2341

Webster JC, Cidlowski JA (1994) Downregulation of the glucocorticoid receptor. A mechanism for physiological adaptation to hormones. Ann NY Acad Sci 746:216–220

Webster JI, Tonelli L, Sternberg EM (2002) Neuroendocrine regulation of immunity. Ann Rev Immunol 20:125–163

Wimo A, von Strauss E, Nordberg G, Sassi F, Johansson L (2002) Time spent on informal and formal care giving for persons with dementia in Sweden. Health Policy 61:255–268

Zarit SH, Reever KE, Bach-Peterson J (1980) Relatives of the impaired elderly: Correlates of feelings of burden. Gerontologist 20:649–655

Chapter 4
Stress and Ageing: Effects on Neutrophil Function

Anna C. Phillips, Riyad Khanfer and Jane Upton

4.1 Introduction

The innate or non-specific immune system consists of soluble components, namely the complement system and cellular elements. The latter includes neutrophils, which make up the majority of the innate immune cells in circulation, and deal with rapidly dividing bacteria. This chapter will focus on the influence of psychological stress and ageing on neutrophil numbers and function.

4.1.1 Neutrophils

Neutrophils are a major component of innate immunity and are the dominant leukocyte in the circulation, making up 60 % of the white cell count. They are also the shortest lived blood cell, dying by apoptosis approximately 24 h after leaving the bone marrow (Savill et al. 1989; Scheel-Toellner et al. 2004). These cells play a crucial role in killing invading pathogens, particularly rapidly dividing bacteria, and are key cellular components of the early phase of inflammatory responses (Nathan 2006). Neutrophils act quickly and without specificity, although their bacterial recognition systems are many and complex. Neutrophils are recruited to sites of infection via chemical homing (chemotactic) signals, such as the chemokine CXCL8 (also known as IL8). Once in contact with the pathogen they uptake the microbe by engulfing (phagocytosis) mediated via opsonic receptors that detect complement proteins C3b and C3Bi or antibody coating the microbe. Once a pathogen is phagocytosed (Smith 1994), neutrophils have the ability to produce a range of cytotoxic and bactericidal molecules such as reactive oxygen species (ROS) (superoxide production), reactive nitrogen species and proteolytic enzymes from cytoplasmic granules (Nauseef 2007). Superoxide production is a means by which neutrophils eliminate pathogens, and

A. C. Phillips (✉) · R. Khanfer · J. Upton
School of Sport and Exercise Sciences, University of Birmingham,
Birmingham, UK B15 2TT
e-mail: A.C.Phillips@bham.ac.uk

J. A. Bosch et al. (eds.), *Immunosenescence*,
DOI 10.1007/978-1-4614-4776-4_4, © Springer Science+Business Media New York 2013

thus is key in combating infections, by, for example, pneumococcal bacteria (Segal 2005, 2006). More recently a third bactericidal mechanism has been described, namely the externalization of DNA decorated with anti-microbial compounds termed neutrophil extracellular traps (NETs), which mediate extracellular killing of bacteria (Brinkmann et al. 2004). NET generation is also triggered by superoxide generation, though superoxide independent NET generation also occurs. Phagocytosis and generation of superoxide trigger the death of the neutrophil, which is then removed by macrophages leading to the resolution of inflammation (Savill et al. 1989). Their function can be enhanced by pro-inflammatory cytokines, such as granulocyte-macrophage colony stimulating factor (GM-CSF), Tumor Necrosis Factor-α (TNFα) and type 1 interferon, which not only amplify their basic bactericidal functions, such as generation of reactive oxygen species, but also extend their lifespan at sites of infection by inhibiting apoptosis (Brach et al. 1992; Scheel-Toellner et al. 1999).

4.1.2 Ageing and Neutrophil Function

With ageing, innate immune responses decline, though not universally. For example, complement activation appears to be unaffected, but macrophage function is modified (reviewed in Chap. 1) including reports of reduced phagocytosis and superoxide function but enhanced basal secretion of IL-6 and TNFα contributing to the raised constitutive level of proinflammatory cytokines seen with age (inflammaging). Natural killer (NK) cells are also affected by ageing; while their numbers increase slightly with age, their killing (cytotoxic) capacity is reduced (Ogata et al. 2001). Neutrophil numbers do not decline with increasing age (Chatta and Dale 1996), and their speed of migration also appears unchanged (Esparza et al. 1996; MacGregor and Shalit 1990), though directional migration is reduced (Wenisch et al. 2000). Moreover, neutrophil bactericidal and phagocytic function is dramatically reduced with age (Butcher et al. 2001; Wenisch et al. 2000). Neutrophils from older adults retain their ability to phagocytose bacterial pathogens coated in antibody (opsonised), such as *E. coli*, per se, but their phagocytic capacity (the number of microbes ingested per cell) is significantly compromised (Butcher et al. 2001). The latter is explained, in part, by a reduction in the number of cell surface opsonic or Fc receptors, specifically CD16, that bind to the antibody coating bacterial pathogens and stimulate their phagocytosis (Butcher et al. 2001).

4.2 Stress and Immunity

Stress can be defined as an environmental stimulus which puts us psychologically or physiologically out of homeostasis (Cannon 1932; Selye and Fortier 1949). Stress can also be perceived psychologically as something which exceeds our capacity to cope (Lazarus and Folkman 1984). When the body is exposed to stress, a fight or

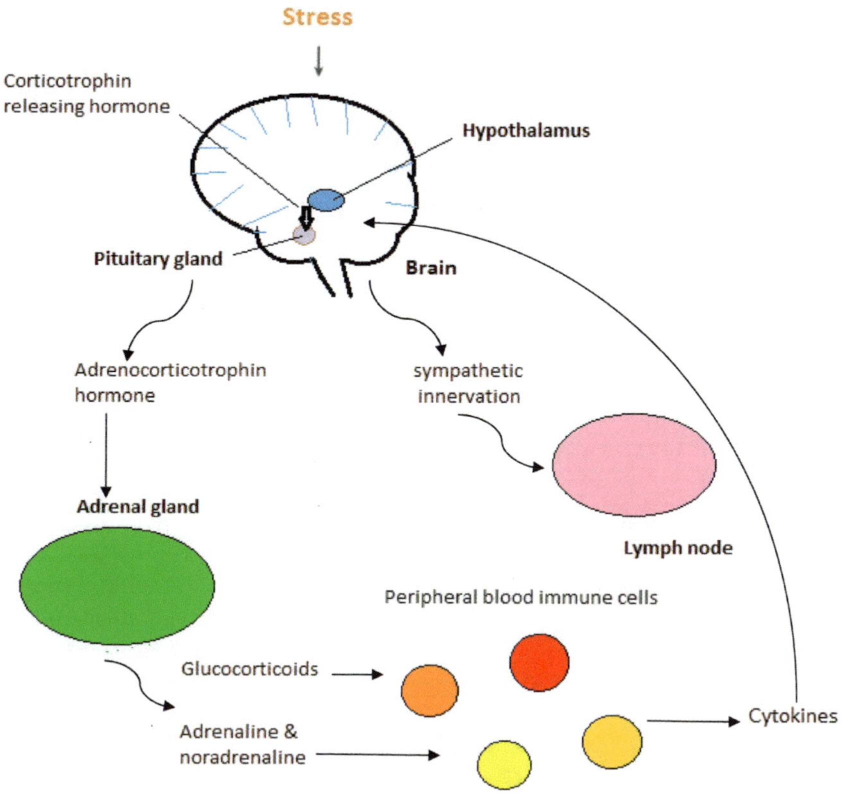

Fig. 4.1 HPA axis and SAM system

flight state is elicited which results in the activation of two key physiological systems within the hypothalamus, the hypothalamic–pituitary–adrenal (HPA) axis and the sympathetic–adrenal–medullary system (Cannon 1932) (see Fig. 4.1). Stress thus induces the release of catecholamines from the adrenal medulla, and both cortisol and dehydroepiandrosterone (DHEA) from the adrenal cortex.

Cortisol is profoundly immunosuppressive and glucocorticoids are used therapeutically to treat acute and chronic inflammation. Of relevance to this chapter, cortisol suppresses neutrophil bactericidal function, notably suppressing superoxide generation (Bekesi et al. 2000; Butcher et al. 2005). DHEA is a precursor to sex hormones which is present in the circulation at micromolar levels in its sulphated form DHEA sulphate (DHEAS) and in contrast to cortisol both DHEA and DHEAS have been shown to be immune enhancing (reviewed in Hazeldine et al. 2010).

The catecholamine hormones exerting an impact on immune function are adrenaline and noradrenaline, sometimes termed epinephrine and norepinephrine. As well as the catecholamines in circulation which can act on immune cells via adrenergic receptors, the main organs of the immune system, also receive direct sympathetic innervation, resulting in the release of adrenaline and noradrenaline from the nerve terminals within these organs (Elenkov et al. 2000). Thus, these hormones are able to produce a variety of effects on immune function. For example, administration of adrenaline in animals has been shown to enhance cell-mediated immunoprotective responses to viruses, bacteria, and fungi in the skin (delayed-type hypersensitivity) and adrenalectomy in animals has also been shown to eliminate this response (Dhabhar 2002, 2003). Adrenergic receptor stimulation generally has also been linked to an inhibition of the production of proinflammatory and stimulation of the production of anti-inflammatory cytokines, thus resulting in a suppression of cellular or Th1 immunity and shift towards humoral or Th2 immunity, although adrenaline can also induce localized inflammatory responses via cytokine release (reviewed in Elenkov et al. 2000).

4.2.1 Acute Versus Chronic Stress

Importantly, acute and chronic psychological stress appears to have a differential impact on the immune system. Generally, chronic stress has been shown to have a negative effect on various aspects of immunity, for example, resulting in poorer antibody responses to vaccination in comparison to matched control participants (Glaser et al. 1992; Phillips et al. 2005a; Vedhara et al. 1999), slower wound healing (Marucha et al. 1998), lower levels of secretory immunoglobulin A in saliva (Phillips et al. 2006b), reduced NK cell cytotoxicity (Esterling et al. 1996) and higher antibody titres against latent viruses such as Cytomegalovirus and Epstein-Barr virus, indicating poorer latent virus control (Glaser et al. 1994). On the other hand, acute stress would seem to be immune enhancing (Segerstrom and Miller 2004). For example, acute stress has been shown to elicit lymphocytosis (Anane et al. 2009; Willemsen et al. 2002), increase NK cell cytotoxicity (Bosch et al. 2005; Sgoutas-Emch et al. 1994) and secretory immunoglobulin A secretion rate (Ring et al. 2002), stimulate aspects of the complement system (Burns et al. 2008) and boost vaccination responses (Edwards et al. 2008).

4.2.2 Stress and Neutrophil Function

Little is known of the effects of chronic or acute stress on human neutrophils. This is perhaps surprising given that neutrophils play a key role in fighting infection, and also in inflammation. Acute stress is usually assessed as the response to a laboratory-based stress test such as mental arithmetic or giving a speech. A 15-min laboratory stress task (Raven's progressive matrices) was used to examine neutrophil activation using the Nitro-blue Tetrazolium assay which indicates oxidative capacity (Ellard et al. 2001). This acute stress was shown to increase neutrophil activation relative to baseline. The effect of examination stress on neutrophils was also examined in a series of studies of individuals with asthma (Kang et al. 1996, 1997, 1998). In these studies, neutrophil function was assessed before, during, and after an examination period, using the superoxide dismutase-inhibitable reduction of ferricytochrome C assay to measure superoxide release. These authors found that examination stress increased superoxide release. However, it has been argued that a prolonged examination period constitutes chronic rather than acute stress, which has different effects, as described above. For example, whereas a single acute examination elicited an increase in secretory immunoglobulin A, a prolonged period of academic examinations was associated with a decrease (Bosch and Carroll 2007). Thus, the increased superoxide release shown in association with this chronic academic stress is somewhat counter intuitive. Further, the assay used in these studies is not an optimally sensitive measure of neutrophil function. A new generation of assays that exploit flow cytometry technology to measure quantitatively the effects of stimulation now offer a more accurate and versatile means of examining different aspects of neutrophil function. These include phagocytosis and superoxide production in response to a range of stimuli (Kampen et al. 2004; Lehmann et al. 2000; Panasiuk et al. 2005). In addition, physiologically relevant agents such as opsonised bacteria can be used to elicit the neutrophil response.

We have recently examined the effects of an acute laboratory psychological stress task, the Paced Auditory Serial Addition Test (PASAT), on neutrophil function, specifically phagocytosis and stimulated superoxide production, using this new generation of assays. Self report and cardiovascular measurements were also taken as a manipulation check, as the effects of this stress task on these variables are well characterised (Phillips et al. 2009; Willemsen et al. 2002). It was hypothesised that acute stress would enhance neutrophil function relative to baseline. In this study of young healthy adults, the self report and cardiovascular data confirmed that the acute stress task provoked emotional and physiological responses. Similarly, as expected, there was an acute stress-induced increase in lymphocyte and granulocyte numbers. The effects of stress on neutrophil function, however, were more complex. For neutrophil phagocytosis, there was an acute increase associated with the stress task (Khanfer et al. 2010). The effects of the acute stress task on neutrophil phagocytosis observed here are consistent with the increase in neutrophil activation to a short mental stressor detailed above (Ellard et al. 2001). This is in contrast with data from studies of chronic stress that reported no effect on neutrophil phagocytosis (Butcher et al. 2005). The acute stress task was also associated with reduction in neutrophil superoxide production when stimulated with *E. coli* (Khanfer et al. 2010). We have recently repeated this study with older adults, and found a similar reduction in neutrophil superoxide production, but no significant increase in phagocytosis (Khanfer et al., in press). Overall comparison of the age groups showed that older adults had lower levels of superoxide production and more of a decline with acute stress. These differences between the studies likely reflect the impact of age, which is discussed below.

Further research is needed to replicate the observed psychological stress-induced changes in neutrophil function but it seems that the stress test used induced a response more akin to a chronic stressor. It should also be noted that studies evaluating neutrophil function and regulation have shown that phagocytosis and intracellular bacterial killing are independent activities of neutrophil-mediated antibacterial defense

(Pechkovsky et al. 1996; Vollebregt et al. 1998). Accordingly, it is feasible that phagocytosis and stimulated superoxide production could be differentially affected by acute stress.

In our study, cardiovascular activity did not appear to be related to neutrophil function (Khanfer et al. 2010). This is surprising, given that neutrophils have been shown to have alpha and beta adrenergic receptors (Abraham et al. 1999; Gosain et al. 2009), and changes in adrenalin and noradrenalin have been shown to relate to circulating neutrophil numbers (Abraham et al. 1999). However, it is possible that the effects observed in the present study are mediated by other stress pathways such as the HPA axis, given that the PASAT has previously been associated with relatively fast increases in cortisol in anticipation and during the stress task (Phillips et al. 2005b). Further, others have shown in an animal model study that dexamethasone (acting as a cortisol-like synthetic stimulant) can significantly decrease neutrophil superoxide production in cats (Hoffmann-Jagielska et al. 2003). *In vitro* studies of human neutrophils have also shown that cortisol is able to suppress neutrophil superoxide generation directly (Bekesi et al. 2000; Butcher et al. 2005). Consequently, the declining superoxide production to *E. coli* observed here might relate to the effects of the acute stress task on cortisol levels, although measurements of serum cortisol would be needed to confirm this. The different phagocytosis and superoxide production results observed may reflect that the regulation of neutrophil function is under the control of many different receptors for a variety of cytokines (Pechkovsky et al. 1996), which are also differentially affected by stress (Steptoe et al. 2001; Yamakawa et al. 2009). These underlying mechanisms of acute stress effects on neutrophil function warrant further investigation.

4.3 Stress Hormones and Ageing

As adrenocortical hormones are generated in response to stress and modulate immune function, any differential change in their production could, therefore, affect immunity. In humans, the production of DHEA and its sulphated form, DHEAS, declines with age, a process termed the adrenopause. The synthesis of DHEA is maximal in humans at age 20–30 and declines gradually thereafter, so that by the seventh decade levels of circulating DHEAS can be as low as 10 % of that seen in young adulthood (Orentreich et al. 1984). Adrenopause occurs at similar rates in both males and females and is a physiological phenomenon unique to the higher primates (Arlt and Hewison 2004). However, although DHEA/S levels fall with age the production of glucocorticoids such as cortisol is remarkably unaltered and, if anything, is slightly increased (Orentreich et al. 1984), resulting in a relative excess of cortisol over DHEA/S and an imbalance of immune suppression over immune enhancement. DHEA is a C19 steroid synthesised from cholesterol in the zona reticularis of the adrenal cortex, a process which requires two enzymes—P450scc and P450c17. A significant proportion of DHEA is converted to DHEAS by the hydroxysteroid sulphotransferase SULT2A1 and DHEAS is the major form of this steroid in serum. As levels of both DHEA and DHEAS fall with age, but cortisol levels do not, a loss of P450c17 function with age has been proposed, though levels of SULT2A1 activity may also decline (Dharia and Parker 2004). The reason for this loss of DHEA synthetic ability is thus not well established, though numbers of zona reticularis cells containing P450c17 are reduced with age and senescence in these cells has been proposed (Hornsby 1997; Staton et al. 2004).

Although sympathetic-adrenal activity is increased with ageing, indexed by increased plasma noradrenaline, adrenaline secretion from the adrenal medulla is reduced by about 40 % in older age (Seals and Esler 2000). However, plasma concentrations of adrenaline do not appear to change due to the reduced clearance rates also observed in ageing resulting from lowered cardiac output or regional blood flow (Seals and Esler 2000). These changes in overall sympathetic nervous system (SNS) activity in ageing can also be observed as reduced adrenaline release in response to range of types of acute stress, particularly in older men (Seals and Esler 2000), likely reflecting elevated baseline SNS activity.

4.4 Synergistic Effects of Ageing and Stress on Immunity

The immunological and endocrinological changes associated with ageing may have implications for resilience to stress in older adults. We have hypothesized previously that the combination of adrenopause, leading to a relative preponderance of cortisol, and an already reduced immune defence against infection, may leave this population particularly vulnerable to the negative effects of stress on immunity (Phillips et al. 2007). Psychosocial stress has been associated with immune system changes in older populations (Graham et al. 2006). For example, relative to age-matched controls, older adults exposed to the chronic stress of being the primary caregiver for a partner with dementia have shown a variety of immunological decrements in comparison to socio-demographically matched control participants, such as: a decreased percentage of T-lymphocytes and helper T-lymphocytes; higher antibody titres to the Epstein-Barr virus; poorer *in vitro* NK cell cytotoxicity, indicating a weakened ability to kill virus-infected cells (Graham et al. 2006); slower wound healing (Kiecolt-Glaser and Newton 2001); and lower antibody titres following both influenza and pneumococcal vaccinations (Glaser et al. 2000; Kiecolt-Glaser et al. 1996). Fewer studies have concentrated on the more mundane stress experienced by older populations, as opposed to the specific chronic stress of caregiving. Our recent research in the United Kingdom focused on a sample of community-based older adults who were attending their General Practice for the National Health Service annual influenza vaccination. Data showed that individuals who reported bereavement and poorer marital satisfaction in the year prior to the vaccination mounted a poorer antibody response to two of the vaccine strains (Phillips et al. 2006a).

Stress management interventions have been studied in the context of ageing, where, following an 8 week cognitive behaviour therapy intervention, more elderly caregivers mounted a 4-fold increase in antibody titre to the influenza vaccine than non-intervention care-givers, although their response was still poorer than non-caregiver controls (Vedhara et al. 2003). Although the mechanisms of the immune enhancement observed are, as yet, not fully understood, these data support the clinical utility of attempts to identify interventions to reduce stress in the elderly.

4.4.1 Chronic Physical Stress and Neutrophil Function in the Elderly

Physical bodily trauma, such as hip fracture resulting from a fall, can also be considered a chronic stressor, and is associated with decrements in immune function. Almost 1 in 3 people over 65 will fall each year, and hip fracture is a frequent consequence of a fall, happening to approximately 86,000 older people in the United Kingdom each year (Donaldson et al. 1990). Fracture is associated with poor outcomes; approximately one quarter of patients are dead 1 year post-fracture (Haentjens et al. 2007) and 1 quarter are institutionalised at discharge rather than returning home (Birge et al. 1994). Few regain pre-fall levels of quality of life (Chaudhry 2007). In a recent study of older (> 65 years) hip fracture patients, who prior to their fall had all been in very good health, it was found that within 4–6 weeks of the fracture 37 % had succumbed to serious infections requiring readmission to hospital (Butcher et al. 2003). These figures for reduced immunity after hip fracture are supported by data from a large Newcastle based study of health outcomes for 531 hip fracture patients, which revealed that post-operative chest infections were the major 6 month mortality risk factor and that pneumonia was the cause of death in 43 % of patients (Wood et al. 1992). Within these patients, the experience of hip fracture was associated with significantly diminished neutrophil function (generation of superoxide) which was greatest in those patients that succumbed to bacterial infection (Butcher et al. 2003). Interestingly, this effect of trauma on neutrophil function and infection rates was not observed in a group of younger (aged < 35 years) fracture patients matched for the level of clinical trauma (Butcher et al. 2003), suggesting that this immune impact of stress is worsened by the presence of already impaired immunity through immunosenescence. Other studies have also reported that age is the most significant risk factor for post-trauma mortality (Connor and Leonard 1998; Park et al. 2006). Physical trauma, such as a fall, is thus a major risk for progression to ill-health and frailty in seniors and reduced immunity is a key underlying frailty factor.

The trauma of hip fracture is associated with the release of stress hormones including cortisol (Butcher et al. 2003), which has potent immunosuppressive properties (Cupps and Fauci 1982), but this does not usually result in increased infections in young individuals. As described above, cortisol production does not decline with age, resulting in a relative preponderance of glucocorticoid-induced immune suppressive effects over immune enhancing effects of DHEA/S, with potentially detrimental consequences for immune regulation in seniors. To test this hypothesis, early morning serum cortisol:DHEAS ratios were measured in older hip fracture patients and a raised cortisol:DHEAS ratio was found to be associated with reduced neutrophil superoxide generation and a higher incidence of post-injury bacterial infections. Moreover, the cortisol:DHEAS ratio was not altered in young participants with limb fractures (Butcher et al. 2003). Adrenocortical hormone balance may thus be a major determinant of immunity in older hip fracture patients.

Adrenal hormones have pleiotropic effects, including a considerable impact on mood. For example, it has been shown that DHEA replacement improves mood and well-being in patients receiving chronic glucocorticoid replacement for adrenal insufficiency (Coles et al. 2005) or in patients on chronic glucocorticoid treatment for systemic lupus erythematosus (Nordmark et al. 2005). Further, DHEA supplementation has been shown to improve minor and major depression in older individuals (Morsink et al. 2007). Raised serum cortisol is strongly associated with melancholic and psychotic depressive subtypes (Duval et al. 2006), and depression is associated with an increased cortisol:DHEAS ratio.

Importantly, the combined immune and endocrine effects of both hip fracture and depression in the context of immunosenescence have not yet been examined. We have recently tested this hypothesis that the addition of psychological distress to the physical stress of hip fracture might work synergistically to amplify effects of stress on immunity in older adults. There is evidence of an increased prevalence of depression and depressive symptomatology in seniors (Mirowsky and Reynolds 2000). Depression is also associated with increased susceptibility to infectious disease and mortality (Connor and Leonard 1998) and with reduced immune cell function. NK T cell function is significantly negatively correlated with depressive symptom scale scores (Park et al. 2006) and resolution of depression was associated with recovery of NK cell activity (Cruess et al. 2005). Further, depression is both a risk factor for (Swantek and Goldstein 2000) and a common co-morbidity of (Holmes and House 2000) hip fracture in seniors. The median prevalence rate for depression in hip fracture patients across 8 United States and United Kingdom studies was 30 % (Miller 1996). Importantly, symptoms of depression in patients with hip fracture hold considerable implications for prognosis; depression has been associated with greater pain persistence (Herrick et al. 2004), poorer rehabilitation participation (Lenze et al. 2004), retarded recovery (Kempen et al. 2003), increased dependence (Magaziner et al. 1990), and higher mortality rates (Nightingale et al. 2001). If immunity is found to be impaired most among hip fracture patients who go on to develop depression, this might be evidence for the treatment of depression in this patient cohort in order to improve immunity and overall outcome. Initial results are promising and we hope to publish these findings soon.

The relative resilience of the younger immune system, described above, has also been observed in non-elderly caregivers of multiple sclerosis patients who, unlike earlier observations in older adults, did not demonstrate reduced antibody responses to vaccination compared to controls (Vedhara et al. 2002). However, this may reflect the severity of the caregiving stress experienced and the nature of the care recipient, as younger parental caregivers for children with developmental disabilities, who report high levels of stress and child problem behaviours, did display poorer antibody titres against pneumonia and influenza than young non-caregivers in our recent study (Gallagher et al. 2009a, b). To our knowledge, this model of the interaction between stress and ageing has not been directly tested in humans with reference to neutrophil function, and is likely to prove fruitful. For example, a meta-analysis revealed that older individuals are more likely to demonstrate negative immune responses to acute naturalistic stressors than younger individuals (Segerstrom and Miller 2004). Further, another review concludes that stress may act to exacerbate the effects of ageing (Graham et al. 2006).

4.4.2 Chronic Psychological Stress and Neutrophil Function in the Elderly

Given our previous findings that older adults who have suffered bereavement show poorer antibody responses to an influenza vaccination, and that recent physical trauma impairs neutrophil immunity, we have recently examined the influence of recent (<2 months) bereavement on neutrophil function among older adults. Further, as the biological mechanisms by which bereavement modulates immune function are unknown, we also measured cortisol and DHEAS among bereaved older adults in comparison to an age- and sex-matched non-bereaved control group. Forty-eight healthy older adults (32 females) were recruited, largely from a local hospice. Analysis of neutrophil phagocytosis and superoxide production were measured by flow cytometry, and cortisol and DHEAS by enzyme linked immunoassay (ELISA). We showed that superoxide production was significantly reduced in the bereaved cohort, whereas there was no effect upon phagocytosis (Khanfer et al. 2011). These data are thus in good agreement with the effects of physical stress (hip fracture) on neutrophil function. This study also showed that the bereaved group had a significantly raised cortisol:DHEAS ratio relative to the non-bereaved controls and this may be one mechanism by which bereavement is associated with higher risk of morbidity and mortality, particularly among older adults (Biondi and Picardi 1996; Clayton 1990; Manor and Eisenbach 2003; Martikainen and Valkonen 1996a, b; Stroebe et al. 2007). A future study will examine neutrophil function among younger bereaved individuals to test our hypothesis that chronic stress is particularly detrimental to immune function among older people, but that younger adults may show resilience to some extent.

4.5 Mechanisms Underlying the Synergism of Chronic Stress and Ageing upon Immunity

A number of mechanistic possibilities underlying the increased immunological vulnerability to stress of older adults have been proposed in another review (Hawkley and Cacioppo 2004). These include evidence that older adults experience more serious and prolonged stressful exposures, have greater stress reactivity, and demonstrate poorer resilience to stress through factors such as reduced social support and poorer sleep patterns. We would also suggest that the increased vulnerability to stress of older adults may be due, at least in part, to the observed age-related imbalance between cortisol and DHEA/S levels (Phillips et al. 2007). This is something we are currently examining within a study comparing older spousal caregivers for those with Alzheimers versus younger caregivers of developmentally disabled children. Neutrophil function is one of the key immune outcomes, and cortisol and DHEAS levels, and psychological factors such as social support and behaviour of the care recipient are being measured as potential mediators.

Although studies on the effects of stress upon neutrophil function *in vivo* are lacking, *in vitro* data suggests that they are key responders to stress hormones. *In vitro* studies have shown that cortisol suppresses neutrophil superoxide generation (Bekesi et al. 2000; Butcher et al. 2005) and importantly this could be overcome by co-incubation with the immune enhancing adrenal steroid DHEAS (Butcher et al. 2005). Further, the effects of DHEAS were shown recently to be direct and mediated via activation of the protein kinase C signaling pathway leading to activation of the enzyme nicotinamide adenine dinucleotide phosphate (NADPH) oxidase to generate superoxide (Radford et al. 2010). Interestingly neutrophils were shown to be the only leucocytes to express a membrane transporter for DHEAS (OATP-D) and thus may be unique in their ability to respond to this steroid (Radford et al. 2010). As mentioned above, SNS stimulation and adrenaline and noradrenaline release systemically can cause a shift towards humoral immunity (Elenkov et al. 2000). However, under certain circumstances, catecholamines can also boost some aspects of cellular immunity and inflammation locally by inducing the release of cytokines, particularly IL-8, which attracts neutrophils to the site of inflammation, potentially to localize and thus limit damage induced by the inflammatory response (Elenkov et al. 2000). These data show that neutrophils are sensitive to stress-responsive systems and are a key target for stress hormones and thus play a key role in mediating the effects of stress upon immunity.

4.6 Conclusions

The literature on stress, ageing and neutrophil function has only begun to emerge in the past decade, showing on the whole that both chronic and acute stress appear to be detrimental to neutrophil function, particularly superoxide production. Our recent and emerging data also suggest that the impact of stress on neutrophil function in the context of older age is much greater, contributing to increased immunosenescence and progression to frailty among older adults. An examination of the potential neuroendocrine and psychological mechanisms will provide routes towards intervention for stress reduction, mood improvement, and boosted neutrophil immunity in this population. The existing literature though sparse suggests that DHEA replacement at times of stress to readjust the exaggerated cortisol:DHEAS ratio may provide significant enhancement of the innate immune response.

Acknowledgments JU is supported by a project grant in the New Dynamics of Ageing initiative funded by research councils UK through the Economic and Social Research Council (ESRC).

References

Abraham E, Kaneko DJ, Shenkar R (1999) Effects of endogenous and exogenous catecholamines on LPS-induced neutrophil trafficking and activation. Am J Physiol 276:L1–8

Anane LH, Edwards KM, Burns VE, Drayson MT, Riddell NE, van Zanten JJ et al (2009) Mobilization of gammadelta T lymphocytes in response to psychological stress, exercise, and beta-agonist infusion. Brain Behav Immun 23:823–829

Arlt W, Hewison M (2004) Hormones and immune function: implications of aging. Aging Cell 3:209–216

Bekesi G, Kakucs R, Varbiro S, Racz K, Sprintz D, Feher J et al (2000) In vitro effects of different steroid hormones on superoxide anion production of human neutrophil granulocytes. Steroids 65:889–894

Biondi M, Picardi A (1996) Clinical and biological aspects of bereavement and loss-induced depression: a reappraisal. Psychother Psychosom 65:229–245

Birge SJ, Morrow-Howell N, Proctor EK (1994) Hip fracture. Clin Geriatr Med 10:589–609

Bosch JA, Berntson GG, Cacioppo JT, Marucha PT (2005) Differential mobilization of functionally distinct natural killer subsets during acute psychologic stress. Psychosom Med 67:366–375

Bosch JA, Carroll D (2007) Stress and mucosal secretory immunity. In: Fink G (ed) Encyclopedia of stress, 2nd ed. Academic Press, Oxford, pp 768–774

Brach MA, deVos S, Gruss HJ, Herrmann F (1992) Prolongation of survival of human polymorphonuclear neutrophils by granulocyte-macrophage colony-stimulating factor is caused by inhibition of programmed cell death. Blood 80:2920–2924

Brinkmann V, Reichard U, Goosmann C, Fauler B, Uhlemann Y, Weiss DS et al (2004) Neutrophil extracellular traps kill bacteria. Science 303:1532–1535

Burns VE, Edwards KM, Ring C, Drayson M, Carroll D (2008) Complement cascade activation after an acute psychological stress task. Psychosom Med 70:387–396

Butcher SK, Chahal H, Nayak L, Sinclair A, Henriquez NV, Sapey E et al (2001) Senescence in innate immune responses: reduced neutrophil phagocytic capacity and CD16 expression in elderly humans. J Leukoc Biol 70:881–886

Butcher SK, Killampalli V, Chahal H, Kaya Alpar E, Lord JM (2003) Effect of age on susceptibility to post-traumatic infection in the elderly. Biochem Soc Trans 31:449–451

Butcher SK, Killampalli V, Lascelles D, Wang K, Alpar EK, Lord JM (2005) Raised cortisol:DHEAS ratios in the elderly after injury: potential impact upon neutrophil function and immunity. Aging Cell 4:319–324

Cannon WB (1932) The wisdom of the body. Norton, New York

Chatta GS, Dale DC (1996) Aging and haemopoiesis. Implications for treatment with haemopoietic growth factors. Drugs Aging 9:37–47

Chaudhry S (2007) The management of subcapital fractures in the elderly—with an emphasis on economic aspects. Trauma 9:119–126

Clayton PJ (1990) Bereavement and depression. J Clin Psychiatry 51(Suppl):34–38; discussion 39–40

Coles AJ, Thompson S, Cox AL, Curran S, Gurnell EM, Chatterjee VK (2005) Dehydroepiandrosterone replacement in patients with Addison's disease has a bimodal effect on regulatory (CD4 + CD25hi and CD4 + FoxP3+) T cells. Eur J Immunol 35:3694–3703

Connor TJ, Leonard BE (1998) Depression, stress and immunological activation: the role of cytokines in depressive disorders. Life Sci 62:583–606

Cruess DG, Douglas SD, Petitto JM, Have TT, Gettes D, Dube B et al (2005) Association of resolution of major depression with increased natural killer cell activity among HIV-seropositive women. Am J Psychiatry 162:2125–2130

Cupps TR, Fauci AS (1982) Corticosteroid-mediated immunoregulation in man. Immunol Rev 65:133–155

Dhabhar FS (2002) Stress-induced augmentation of immune function—the role of stress hormones, leukocyte trafficking, and cytokines. Brain Behav Immun 16:785–798

Dhabhar FS (2003) Stress, leukocyte trafficking, and the augmentation of skin immune function. Ann NY Acad Sci 992:205–217

Dharia S, Parker CR Jr (2004) Adrenal androgens and aging. Semin Reprod Med 22:361–368

Donaldson LJ, Cook A, Thomson RG (1990) Incidence of fractures in a geographically defined population. J Epidemiol Commun Health 44:241–245

Duval F, Mokrani MC, Monreal-Ortiz JA, Fattah S, Champeval C, Schulz P et al (2006) Cortisol hypersecretion in unipolar major depression with melancholic and psychotic features: dopaminergic, noradrenergic and thyroid correlates. Psychoneuroendocrinology 31:876–888

Edwards KM, Burns VE, Adkins AE, Carroll D, Drayson M, Ring C (2008) Meningococcal A vaccination response is enhanced by acute stress in men. Psychosom Med 70:147–151

Elenkov IJ, Wilder RL, Chrousos GP, Vizi ES (2000) The sympathetic nerve—an integrative interface between two supersystems: the brain and the immune system. Pharmacol Rev 52:595–638

Ellard DR, Castle PC, Mian R (2001) The effect of a short-term mental stressor on neutrophil activation. Int J Psychophysiol 41:93–100

Esparza B, Sanchez H, Ruiz M, Barranquero M, Sabino E, Merino F (1996) Neutrophil function in elderly persons assessed by flow cytometry. Immunol Invest 25:185–190

Esterling BA, Kiecolt-Glaser JK, Glaser R (1996) Psychosocial modulation of cytokine-induced natural killer cell activity in older adults. Psychosom Med 58:264–272

Gallagher S, Phillips AC, Drayson MT, Carroll D (2009a) Parental caregivers of children with developmental disabilities mount a poor antibody response to pneumococcal vaccination. Brain Behav Immun 23:338–346

Gallagher S, Phillips AC, Drayson MT, Carroll D (2009b) Care-giving for a child with intellectual disabilities is associated with a poor antibody response to influenza vaccination. Psychosom Med 71:341–344

Glaser R, Kiecolt-Glaser JK, Bonneau RH, Malarkey W, Kennedy S, Hughes J (1992) Stress-induced modulation of the immune response to recombinant hepatitis B vaccine. Psychosom Med 54:22–29

Glaser R, Pearl DK, Kiecolt-Glaser JK, Malarkey WB (1994) Plasma cortisol levels and reactivation of latent Epstein-Barr virus in response to examination stress. Psychoneuroendocrinology 19:765–772

Glaser R, Sheridan J, Malarkey WB, MacCallum RC, Kiecolt-Glaser JK (2000) Chronic stress modulates the immune response to a pneumococcal pneumonia vaccine. Psychosom Med 62:804–807

Gosain A, Gamelli RL, DiPietro LA (2009) Norepinephrine-mediated suppression of phagocytosis by wound neutrophils. J Surg Res 152:311–318

Graham JE, Christian LM, Kiecolt-Glaser JK (2006) Stress, age, and immune function: toward a lifespan approach. J Behav Med 29:389–400

Haentjens P, Autier P, Barette M, Venken K, Vanderschueren D, Boonen S (2007) Survival and functional outcome according to hip fracture type: a one-year prospective cohort study in elderly women with an intertrochanteric or femoral neck fracture. Bone 41:958–964

Hawkley LC, Cacioppo JT (2004) Stress and the aging immune system. Brain Behav Immun 18:114–119

Hazeldine J, Arlt W, Lord JM (2010) Dehydroepiandrosterone as a regulator of immune cell function. J Steroid Biochem Mol Biol 120:127–136

Herrick C, Steger-May K, Sinacore DR, Brown M, Schechtman KB, Binder EF (2004) Persistent pain in frail older adults after hip fracture repair. J Am Geriatr Soc 52:2062–2068

Hoffmann-Jagielska M, Winnicka A, Jagielski D, Lechowski R (2003) Influence of dexamethasone on some cellular aspects of the immune system in cats. Vet Res Commun 27:643–652

Holmes JD, House AO (2000) Psychiatric illness in hip fracture. Age Ageing 29:537–546
Hornsby PJ (1997) DHEA: a biologist's perspective. J Am Geriatr Soc 45:1395–1401
Kampen AH, Tollersrud T, Lund A (2004) Flow cytometric measurement of neutrophil respiratory burst in whole bovine blood using live *Staphylococcus aureus*. J Immunol Methods, 289:47–55
Kang DH, Coe CL, Karaszewski J, McCarthy DO (1998) Relationship of social support to stress responses and immune function in healthy and asthmatic adolescents. Res Nurs Health 21:117–128
Kang DH, Coe CL, McCarthy DO (1996) Academic examinations significantly impact immune responses, but not lung function, in healthy and well-managed asthmatic adolescents. Brain Behav Immun 10:164–181
Kang DH, Coe CL, McCarthy DO, Ershler WB (1997) Immune responses to final exams in healthy and asthmatic adolescents. Nurs Res 46:12–19
Kempen GI, Sanderman R, Scaf-Klomp W, Ormel J (2003) The role of depressive symptoms in recovery from injuries to the extremities in older persons. A prospective study. Int J Geriatr Psychiatry 18:14–22
Khanfer R, Carroll D, Lord JM, Phillips AC (in press) Reduced neutrophil superoxide production among healthy older adults in response to acute psychological stress. Int J Psychophysiol
Khanfer R, Lord JM, Phillips AC (2011) Neutrophil function and cortisol: DHEAS ratio in bereaved older adults. Brain Behav Immun 25:1182–1186
Khanfer R, Phillips AC, Carroll D, Lord JM (2010) Altered human neutrophil function in response to acute psychological stress. Psychosom Med 72:636–640
Kiecolt-Glaser JK, Glaser R, Gravenstein S, Malarkey WB, Sheridan J (1996) Chronic stress alters the immune response to influenza virus vaccine in older adults. Proc Natl Acad Sci USA 93:3043–3047
Kiecolt-Glaser JK, Newton TL (2001) Marriage and health: his and hers. Psychol Bull 127:472–503.
Lazarus RS, Folkman S (1984) Stress, coping and adaptation. Springer, New York
Lehmann TG, Wheeler MD, Schoonhoven R, Bunzendahl H, Samulski RJ, Thurman RG (2000) Delivery of Cu/Zn-superoxide dismutase genes with a viral vector minimizes liver injury and improves survival after liver transplantation in the rat. Transplantation 69:1051–1057
Lenze EJ, Munin MC, Dew MA, Rogers JC, Seligman K, Mulsant BH et al (2004) Adverse effects of depression and cognitive impairment on rehabilitation participation and recovery from hip fracture. Int J Geriatr Psych 19:472–478
MacGregor RR, Shalit M (1990) Neutrophil function in healthy elderly subjects. J Gerontol 45:M55–60
Magaziner J, Simonsick EM, Kashner TM, Hebel JR, Kenzora JE (1990) Predictors of functional recovery one year following hospital discharge for hip fracture: a prospective study. J Gerontol 45:M101–107
Manor O, Eisenbach Z (2003) Mortality after spousal loss: are there socio-demographic differences? Soc Sci Med 56:405–413
Martikainen P, Valkonen T (1996a) Mortality after death of spouse in relation to duration of bereavement in Finland. J Epidemiol Commun Health 50:264–268
Martikainen P, Valkonen T (1996b) Mortality after the death of a spouse: rates and causes of death in a large Finnish cohort. Am J Public Health 86:1087–1093
Marucha PT, Kiecolt-Glaser JK, Favagehi M (1998) Mucosal wound healing is impaired by examination stress. Psychosom Med 60:362–365
Miller RA (1996) The aging immune system: primer and prospectus. Science 273:70–74
Mirowsky J, Reynolds JR (2000) Age, depression, and attrition in the National Survey of Families and Households. Soc Method Res 28:476–504
Morsink LF, Vogelzangs N, Nicklas BJ, Beekman AT, Satterfield S, Rubin SM et al (2007) Associations between sex steroid hormone levels and depressive symptoms in elderly men and women: results from the health ABC study. Psychoneuroendocrinology 32:874–883
Nathan C (2006) Neutrophils and immunity: challenges and opportunities. Nat Rev Immunol 6:173–182

Nauseef WM (2007) How human neutrophils kill and degrade microbes: an integrated view. Immunol Rev 219:88–102

Nightingale S, Holmes J, Mason J, House A (2001) Psychiatric illness and mortality after hip fracture. Lancet 357:1264–1265

Nordmark G, Bengtsson C, Larsson A, Karlsson FA, Sturfelt G, Ronnblom L (2005) Effects of dehydroepiandrosterone supplement on health-related quality of life in glucocorticoid treated female patients with systemic lupus erythematosus. Autoimmunity 38:531–540

Ogata K, An E, Shioi Y, Nakamura K, Luo S, Yokose N et al (2001) Association between natural killer cell activity and infection in immunologically normal elderly people. Clin Exp Immunol 124:392–397

Orentreich N, Brind JL, Rizer RL, Vogelman JH (1984) Age changes and sex differences in serum dehydroepiandrosterone sulfate concentrations throughout adulthood. J Clin Endocrinol Metab 59:551–555

Panasiuk A, Wysocka J, Maciorkowska E, Panasiuk B, Prokopowicz D, Zak J et al (2005) Phagocytic and oxidative burst activity of neutrophils in the end stage of liver cirrhosis. World J Gastroenterol 11:7661–7665

Park EJ, Lee JH, Chae JH, Lee KH, Han SI, Jeon YW (2006) Natural Killer T cells in patients with major depressive disorder. Psychiatry Res 144:237–239

Pechkovsky DV, Potapnev MP, Zalutskaya OM (1996) Different patterns of cytokine regulation of phagocytosis and bacterial killing by human neutrophils. Int J Antimicrob Agents 7:33–40

Phillips AC, Burns VE, Carroll D, Ring C, Drayson M (2005a) The association between life events, social support and antibody status following thymus-dependent and thymus-independent vaccinations in healthy young adults. Brain Behav Immun 19:325–333

Phillips AC, Burns VE, Lord JM (2007) Stress and exercise: getting the balance right for an ageing immune system. Exercise Sport Sci Rev 35:35–39

Phillips AC, Carroll D, Burns VE, Drayson M (2005b) Neuroticism, cortisol reactivity, and antibody response to vaccination. Psychophysiology 42:232–238

Phillips AC, Carroll D, Burns VE, Drayson MT (2009) Cardiovascular activity and the antibody response to vaccination. J Psychosom Res 67:37–43

Phillips AC, Carroll D, Burns VE, Ring C, Macleod J, Drayson M (2006a) Bereavement and marriage are associated with antibody response to influenza vaccination in the elderly. Brain Behav Immun 20:279–289

Phillips AC, Carroll D, Evans P, Bosch JA, Clow A, Hucklebridge F et al (2006b) Stressful life events are associated with low secretion rates of immunoglobulin A in saliva in the middle aged and elderly. Brain Behav Immun 20:191–197

Radford DJ, Wang K, McNelis JC, Taylor AE, Hechenberger G, Hofmann J, Chahal H, Arlt W, Lord JM (2010) Dehydroepiandrosterone sulfate directly activates protein kinase C-beta to increase human neutrophil superoxide generation. Mol Endocrinol 24:813–821

Ring C, Drayson M, Walkey DG, Dale S, Carroll D (2002) Secretory immunoglobulin A reactions to prolonged mental arithmetic stress: inter-session and intra-session reliability. Biol Psychol 59:1–13

Savill JS, Wyllie AH, Henson JE, Walport MJ, Henson PM, Haslett C (1989) Macrophage phagocytosis of aging neutrophils in inflammation. Programmed cell death in the neutrophil leads to its recognition by macrophages. J Clin Invest 83:865–875

Scheel-Toellner D, Pilling D, Akbar AN, Hardie D, Lombardi G, Salmon M et al (1999) Inhibition of T cell apoptosis by IFN-beta rapidly reverses nuclear translocation of protein kinase C-delta. Eur J Immunol 29:2603–2612

Scheel-Toellner D, Wang KQ, Webb PR, Wong SH, Craddock R, Assi LK et al (2004) Early events in spontaneous neutrophil apoptosis. Biochem Soc Trans 32:461–464

Seals DR, Esler MD (2000) Human ageing and the sympathoadrenal system. J Physiol 528:407–417

Segal AW (2005) How neutrophils kill microbes. Annu Rev Immunol 23:197–223

Segal AW (2006) How superoxide production by neutrophil leukocytes kills microbes. Novart Found Symp 279:92–98; discussion 98–100, 216–109

Segerstrom SC, Miller GE (2004) Psychological stress and the human immune system: a meta-analytic study of 30 years of inquiry. Psychol Bull 130:601–630

Selye H, Fortier C (1949). Adaptive reactions to stress. Res Publ Assoc Res Nerv Ment Dis 29:3–18

Sgoutas-Emch SA, Cacioppo JT, Uchino BN, Malarkey W, Pearl D, Kiecolt-Glaser JK et al (1994) The effects of an acute psychological stressor on cardiovascular, endocrine, and cellular immune response: a prospective study of individuals high and low in heart rate reactivity. Psychophysiology 31:264–271

Smith JA (1994) Neutrophils, host defense, and inflammation: a double-edged sword. J Leukoc Biol 56:672–686

Staton BA, Mixon RL, Dharia S, Brissie RM, Parker CR Jr (2004) Is reduced cell size the mechanism for shrinkage of the adrenal zona reticularis in aging? Endocr Res 30:529–534

Steptoe A, Willemsen G, Owen N, Flower L, Mohamed-Ali V (2001) Acute mental stress elicits delayed increases in circulating inflammatory cytokine levels. Clin Sci (Lond) 101:185–192

Stroebe M, Schut H, Stroebe W (2007) Health outcomes of bereavement. Lancet 370:1960–1973

Swantek SS, Goldstein MZ (2000) Practical geriatrics: age and gender differences of patients with hip fracture and depression. Psychiat Ser 51:1501–1503

Vedhara K, Bennett PD, Clark S, Lightman SL, Shaw S, Perks P et al (2003). Enhancement of antibody responses to influenza vaccination in the elderly following a cognitive-behavioural stress management intervention. Psychother Psychosom 72:245–252

Vedhara K, Cox NK, Wilcock GK, Perks P, Hunt M, Anderson S et al (1999) Chronic stress in elderly carers of dementia patients and antibody response to influenza vaccination. Lancet 353:627–631

Vedhara K, McDermott MP, Evans TG, Treanor JJ, Plummer S, Tallon D et al (2002) Chronic stress in non-elderly caregivers: psychological, endocrine and immune implications. J Psychosom Res 53:1153–1161

Vollebregt M, Hampton MB, Winterbourn CC (1998) Activation of NF-kappaB in human neutrophils during phagocytosis of bacteria independently of oxidant generation. FEBS Lett 432:40–44

Wenisch C, Patruta S, Daxbock F, Krause R, Horl W (2000) Effect of age on human neutrophil function. J Leukoc Biol 67:40–45

Willemsen G, Carroll D, Ring C, Drayson M (2002) Cellular and mucosal immune reactions to mental and cold stress: associations with gender and cardiovascular reactivity. Psychophysiology 39:222–228

Wood DJ, Ions GK, Quinby JM, Gale DW, Stevens J (1992) Factors which influence mortality after subcapital hip fracture. J Bone Joint Surg Br 74:199–202

Yamakawa K, Matsunaga M, Isowa T, Kimura K, Kasugai K, Yoneda M et al (2009) Transient responses of inflammatory cytokines in acute stress. Biol Psychol 82:25–32

Chapter 5
Stress, Aging, and Wound Healing

Christopher G. Engeland

5.1 Introduction

The world is rapidly aging. By 2030, older individuals (aged 65+) are predicted to comprise 20 % of the population of the United States (US Census Bureau 2004), up from 12.9 % in 2009 (US Census Bureau 2009). The fastest growing segments are those 85+ years of age (Pittman 2007). Given that the population is aging so quickly, research on age-related health disparities is of increasing importance. An example of this is tissue repair. Healing impairments cost United States health services more than nine billion dollars annually, much of which have been attributed to healing delays in older adults (Ashcroft and Mills 2002). Age-impaired healing is an important and growing issue which meaningfully impacts national health care costs.

This chapter will discuss how aging and stress have an impact upon immunity and wound healing in humans. As demonstrated by the other contributors of this book, stress appears to promote senescence in immune cells in a manner comparable to chronological aging. Thus, some of the mechanisms by which aging and stress affect immunity appear to be overlapping if not identical. In support of this, a recent meta-analysis has shown robust and consistent associations between wound healing and psychological stress (Walburn et al. 2009). As will be reviewed in this chapter, older individuals are more affected by stress than young adults. As a result, stress has particularly strong ramifications for both inflammatory responses and healing outcomes in older adults.

C. G. Engeland (✉)
Center for Wound Healing and Tissue Regeneration, Department of Periodontics,
Department of Women, Child and Family Health Science
College of Dentistry, University of Illinois at Chicago, 801 S Paulina St,
Room 458, M/C 859, Chicago, IL 60612, USA
e-mail: engeland@uic.edu

J. A. Bosch et al. (eds.), *Immunosenescence*,
DOI 10.1007/978-1-4614-4776-4_5, © Springer Science+Business Media New York 2013

5.2 Phases of Wound Healing

Our skin forms an essential and constantly renewing barrier to the external environment. When this barrier is compromised its rapid repair is critical to both health and survival. The cascade of events that comprises tissue repair can be categorized into three phases: the inflammatory phase (lasts hours to days), the proliferative phase (lasts days to weeks), and the remodeling phase (lasts weeks to months). Although these phases are overlapping, the timing and activity of each phase depends upon the resolution of the previous phase. As a result, any dysregulation of this cascade can impair tissue repair at multiple levels. This is particularly true of the inflammatory phase, which starts off the healing cascade. Importantly for this chapter, both aging and stress have been shown to dysregulate immune function at every phase of healing (for reviews, see Engeland and Marucha (2009), Engeland and Gajendrareddy (2011)).

5.2.1 Inflammatory Phase

Immediately following injury, the inflammatory phase begins with blood coagulation and fibrin clot formation. The clot is more than just a barrier to the external environment; it releases numerous important early signaling factors and also serves as a scaffold by which cells can migrate into the wound from the margins. Constitutive reserves of the cytokine IL-1α are also released by epithelial cells shortly after injury. Similarly, in the epidermis, dendritic (Langerhans) cells release inflammatory mediators and connective tissue (mast) cells release histamine and TNF-α. These mediators act to recruit and activate inflammatory cells by inducing the expression of cellular adhesion molecules (e.g., ICAM and E-selectin) on endothelial cells. In addition, chemokines are released which recruit inflammatory cells (e.g., IL-8 for neutrophils; MCP-1, MIP-1α for monocytes/macrophages) to the site of injury (for review, see Eming et al. (2007)).

An appreciable number of neutrophils arrive at the injury site within minutes, and monocytes within hours; these cells clear the site of bacteria and damaged tissue via phagocytosis, and release growth factors to promote the next "proliferative phase" of healing by signaling keratinocytes, fibroblasts, and endothelial cells (e.g., to migrate, proliferate). Monocytes/macrophages also limit tissue damage through the phagocytosis of degenerated neutrophils. It is important to note that under conditions of low bacterial burden and rapid bacterial clearance, the recruitment of neutrophils and macrophages is limited and the inflammatory phase is brief. However, under high bacterial burden (i.e., infection) the resolution of the inflammatory phase takes longer, which delays the subsequent healing phases and increases the incidence of scarring (Stramer et al. 2007).

5.2.2 Proliferative Phase

The proliferative phase, in which lost or damaged tissues are rebuilt, begins within the first few hours of healing. A dramatic proliferation of fibroblasts, epithelial cells, and endothelial cells occurs, all of which aid in reforming the extracellular matrix (ECM), which is the bioscaffold that surrounds and supports the tissue cells. This healing phase also involves the reformation of the epithelial barrier, sensory connections, and blood supply. These processes are strongly modulated by the first phase of healing, as growth factors released by monocytes and tissue cells spur these processes forward, and many of these growth factors are upregulated by inflammation (e.g., vascular endothelial growth factor (VEGF), which is important for angiogenesis) (Werner and Grose 2003).

Shortly after injury, endothelial cells migrate and proliferate to repair and form new vasculature structures, which mature to form functional blood vessels. This revascularization of tissues is guided by signaling from growth factors (e.g., VEGF), maturation factors (e.g., angiopoietins), oxygen balance, and inflammation. Oxygen demands in injured tissue are high, in part, due to oxidative bursts from neutrophils for the purpose of microbial clearance. As a result, injured tissues are typically hypoxic (Chang et al. 1983). Due to the high metabolic demands of the repair process, the degree of revascularization typically exceeds what is needed for tissue maintenance and some of the newly formed vasculature then regresses during the remodeling phase.

Fibroblasts play an essential role in both manufacturing and remodeling connective (granulation) tissue during the proliferative phase. Many of the fibroblasts, under control of a cascade of growth factors, secrete collagen, elastin, and other components of the connective tissue matrix. A subset of fibroblasts migrates to the wound margin and transforms into their contractile phenotype, myofibroblasts. This enables wound contraction to occur, thereby reducing the wound size and minimizing the amount of reconstruction needed.

Tissue re-epithelialization stems from the wound margins and residual intact hair follicles. Keratinocytes, the predominant cell type of the epidermis, form a hyperproliferative advancing front and slowly cover the wound. As this occurs, epithelial cells signal for a reduction of inflammation. Once the wound is covered, the barrier continues to thicken and these cells secrete structural proteins, such as keratins and involucrin (Werner et al. 2007), to stabilize the newly formed tissue.

5.2.3 Remodeling Phase

The remodeling phase occurs over a period of weeks to months, largely after wound closure has occurred. The ECM is relatively weak at this point, and it is now remodeled into a stronger, more organized matrix. Matrix metalloproteinases (MMPs) are the enzymes that are released to degrade the newly formed ECM, allowing tissue remodeling to occur for the purpose of strengthening the healing tissue. This is the phase in which scarring becomes apparent, which closely relates to the restoration of both function and form. Scarring is characterized by smaller diameter collagen fibers which are less organized and cross-linked, and have weaker tensile strength, compared to normal tissue. Greater scarring typically occurs under conditions of higher or more prolonged inflammation, for instance, in wounds that are slow to close. This in turn results in excessive/extended production of growth factors and, hence, more granulation tissue (for a detailed review of these healing phases, see Werner and Grose (2003)).

5.3 Stress Pathways

Psychological stress activates two primary neuroendocrine pathways, both of which can modulate tissue repair. The first is the hypothalamic–pituitary–adrenal (HPA) axis. Following the perception of stress, corticotropin-releasing hormone (CRH) is released into the hypophyseal portal system by the paraventricular nucleus of the hypothalamus. CRH in turn signals the release of adrenocorticotropic hormone (ACTH) from the anterior pituitary gland into the general blood circulation, which in turn causes the release of glucocorticoids (GCs) from the adrenal glands. The primary GC in humans and non-human primates is cortisol, whereas in rodents it is corticosterone. GCs signal back to GC receptors located on various brain regions (e.g., hypothalamus, hippocampus) and on the pituitary, inhibiting further GC release. This negative feedback mechanism helps to prevent the constant elevated release of GCs under conditions of unremitting (chronic) stress.

The second pathway is the sympathetic nervous system (SNS). Under conditions of stress, CRH signals the locus coeruleus, an autonomic brainstem region, to release norepinephrine (NE) from sympathetic nerve endings into the periphery. The adrenal glands are simultaneously stimulated to release epinephrine (EPI). Both GCs and catecholamines (NE, EPI) have well-established and direct effects on immunity and wound healing.

5.3.1 Glucocorticoids and Wound Healing

GC receptors are found on immune cells and GCs have effects on virtually every aspect of the immune system (for reviews, see Chrousos (1998), Flammer and Rogatsky (2011)). As a result, GCs play a critical role in modulating immunity and serve as the body's main brake on inflammation. For example, in the presence of infection, even patients with subclinical adrenal insufficiency succumb to septic shock almost invariably if GC therapy is not given (Berczi 1993). When stress becomes chronic, the HPA axis becomes less responsive to new stressors. In addition, GCs may have reduced immunosuppressive effects due to the formation of GC resistance, in which a reduced responsiveness is developed by immune cells to persistently high levels of these hormones. Through these mechanisms, chronic stress can dysregulate inflammation. This often results in a state of low-grade systemic inflammation accompanied by hyper-inflammatory responses to immune challenge (for review, see Hawkley et al. (2007)).

Altered inflammation can have a negative effect on wound healing, and numerous studies have shown that exogenous GC administration alters tissue repair. Such treatments suppress neutrophil (Clark et al. 1979) and monocyte (Norris et al. 1982) recruitment into the damaged tissue, along with phagocytosis and thereby bacterial clearance (Fauci et al. 1976). For instance, a single administration of corticosteroids in humans reduces circulating lymphocytes by 70 % and monocytes by 90 % (Berczi 1986) and suppresses inflammation at the wound site (Hinz 2007) largely through inhibiting pro-inflammatory cytokine production. Further, GCs inhibit the proliferation of fibroblasts and keratinocytes (Edwards and Dunphy 1958), and collagen production (Beer et al. 2000). Overall, GC administration can result in poor bacterial clearance, slower wound debridement, altered wound contraction, delayed re-epithelialization, slower wound closure, and reduced wound strength.

5.3.2 *Sympathetic Activation and Wound Healing*

Research on stress and wound healing has generally focused on the effects of GCs. However, as indicated above, stress also activates the SNS. The resulting release of NE and EPI alters both blood flow and cell function, and can negatively affect important wound healing parameters such as tissue oxygenation.

Healing tissue has high demands for oxygen and is typically hypoxic (Chang et al. 1983). This stems from both a disruption of the blood supply and a large accumulation of oxygen consuming cells that are involved in the repair process (Chang et al. 1983; Hopf and Rollins 2007). Oxygen tension is also important for normal tissue revascularization and re-epithelialization to occur (Pai and Hunt 1972).

Stress induces the release of NE by sympathetic nerve endings that innervate the vasculature, causing vasoconstriction through its binding with α-adrenergic receptors on endothelial cells. This reduces oxygen and nutrient flow to tissues, and slows the infiltration of immune cells from blood into injured areas. Ultimately, this affects angiogenesis, re-epithelialization, and collagen synthesis, and it reduces the bactericidal actions of neutrophils early in wound healing. NE also contributes to local tissue edema (Koopman 1995) and inhibits the migration of epidermal cells (Donaldson and Mahan 1984), further slowing re-epithelialization.

An infusion of EPI has been shown to reduce wound oxygen levels by 45 % and a discontinuation results in a rebound of these levels. In both humans and animals, wound infection can be predicted by oxygen levels (Hopf et al. 1997; Jonsson et al. 1988). This may explain why exogenous EPI has been associated with increased infection rates during some surgical procedures (England et al. 1983). To sum, stress-induced catecholamines cause vascular changes which limit blood flow and alter a number of important wound healing parameters, ultimately influencing wound closure rates and infection.

5.4 Stress and Wound Healing

It has recently become accepted that psychological stress, even that stemming from everyday events, has the capacity to negatively affect immunity. There is now also strong evidence that stress impairs wound healing. This generally occurs through delays in wound closure which increases the risk of infection and, in the case of surgery, post-surgical complications (Robson 1997). The mechanisms underlying these delays are still being elucidated, but the existing research will be presented below.

5.4.1 Psychological Stress

A recent systematic review of 22 studies, and a meta-analysis of 11 studies, concluded that the relationship between psychological stress and wound healing is intermediate to strong (Walburn et al. 2009). This field of study (stress and healing) is relatively young and the underlying mechanisms are still being delineated. However, it is now generally well accepted that stress can negatively affect tissue repair.

In 1995, it was reported that female Alzheimer's caregivers, who underwent chronic stress, healed experimental skin wounds 24 % more slowly than age-matched controls (Kiecolt-Glaser et al. 1995). When whole blood of these individuals was stimulated with endotoxin (i.e., bacterial cell wall), lower gene expression for IL-1β was observed in the blood of caregivers compared to controls. Hence, in stressed caregivers the ability to mount an early inflammatory response appeared to be impaired, resulting in delayed wound healing.

In a follow-up study, 8 blister wounds were induced by mechanical suction on the forearms of 36 healthy post-menopausal women (mean age 57) and wound fluid was then drawn out at various time points and analyzed. Women with lower levels of IL-1α and IL-8 in the wound site, 24 h after wounding, reported higher stress levels, more negative mood, and had higher levels of salivary cortisol than women with higher cytokine levels (Glaser et al. 1999). Stress was again associated with lower levels of inflammation, and this was now demonstrated directly in the wound site. These effects may have been mediated by the strong anti-inflammatory actions of cortisol. This finding concurs with reports that stress-induced increases in cortisol, through exercise, suppress pro-inflammatory cytokine responses (e.g., IL-1β, TNF-α) to endotoxin in vitro (DeRijk et al. 1997).

In 2004, similar relationships were demonstrated between stress, cortisol, and wound healing in men. Twenty four non-smoking healthy young men received a 4 mm punch biopsy skin wound and healing was assessed using high-resolution ultrasound scanning. This method allows for the base of the wound to be measured, which provides a three dimensional image of the wound that is considered to be more accurate than surface photography (Ebrecht et al. 2004). Slower wound healing correlated with higher perceived stress and lower perceived optimism. In addition, when a median split was performed to divide slower from faster healers, slower healers were found to have significantly higher stress levels, lower trait optimism, and higher cortisol levels upon morning awakening (Ebrecht et al. 2004). These results further demonstrate the associations that exist among stress, cortisol levels, and impaired healing.

Similar relationships between stress, inflammation, and healing have been reported clinically. Prior to undergoing an open incisional hernia operation, 47 patients were assessed for preoperative stress. Higher stress scores predicted lower IL-1 levels in wound fluid, and greater worry about the surgery predicted lower metalloproteinase-9 levels, as well as a slower, more painful recovery (Broadbent et al. 2003). These findings indicate that pre-surgical stress alters both inflammatory and matrix remodeling processes after surgery, resulting in poorer outcomes. Other surgical studies have shown that higher preoperative stress relates to lowered lymphocyte responses, lymphocyte blood counts, and natural killer (NK) cell activity following surgery (Linn et al. 1988; Koga et al. 2001; Starkweather et al. 2006).

The effect of examination stress on healing has been studied in university students. It was reported that examination stress increases the time required to heal from dermal blister wounds. This stressor was also found to have a suppressive effect on neutrophil RNA levels (i.e., transcriptome) (Roy et al. 2005), suggesting that neutrophil function is inhibited by examination stress. Given that neutrophils play a critical role in bacterial clearance and early inflammation, this may explain, at least in part, the slower healing observed during examinations. Similar findings have been reported in mucosal tissues. Dental students ($n = 11$) were found to heal 40 % slower when small mucosal wounds were placed on the mouth's hard palate shortly prior to examinations than during summer vacation. None of the students were found to heal more quickly during examinations than during vacation (Marucha et al. 1998). It is interesting that a stressor, as relatively transient and benign as university examinations, can reliably delay wound healing in young healthy students.

5.4.2 Depression, Anxiety, and Pain

An elevation of depressive symptoms has been shown to negatively correlate with markers of immunity in multiple studies (for review, see Kiecolt-Glaser and Glaser (2002)), suggesting it may also have deleterious effects on wound healing. A study examined the closure rates for a 3.5 mm punch biopsy wound that was placed on the oral hard palate in 183 healthy young adults. Individuals with higher depressive scores (Beck Depression Inventory $\geq$10; $n = 41$) exhibited slower closure of mucosal wounds than subjects with lower scores (Bosch et al. 2007). These results suggest that pre-surgical identification and treatment of depressed individuals, and even non-depressed individuals with high depressive symptom scores, may improve post-surgical outcomes.

The above stress studies have described the healing of experimental or surgical wounds. To study chronic wounds, 53 outpatients with chronic leg ulcers were administered the Hospital Anxiety and Depression Scale (HADS). Higher anxiety and depression scores were associated with slower healing. Moreover, patients with scores in the top 50 % were four times more likely to be categorized as slow healers than the lower 50 % (Cole-King and Harding 2001). Thus, anxiety and depression also relate to impaired healing in chronic wounds.

Post-operative pain is a unique form of stress which can cause or exacerbate healing impairments. Persistent pain after elective gastric bypass surgery has been shown to predict slower closure of 2 mm experimental skin wounds. This occurred independently of pre-existing pain, depressive symptoms, and medical symptoms after discharge (McGuire et al. 2006). Pain can alter immune functioning through both direct (via cytokine activation) and indirect pathways (via stress and behavioral change), and it potentiates the negative effects of stress on immunity (for review, see Engeland and Graham (2011)). In support of this, the immune dysregulating effects of surgery are lessened by the use of post-operative analgesics to control pain (Beilin et al. 2003).

To summarize, it is clear that numerous forms of psychological stress can negatively impact wound healing and surgical outcomes in humans. A recent meta-analysis further supports this finding (Walburn et al. 2009). Higher stress and cortisol levels have been repeatedly associated with altered inflammation and slowed healing, in both clinical and experimental settings. As will be discussed, wound healing is modulated in a similar manner by aging as it is by stress.

5.5 The Aging Skin

As adults get older the human skin undergoes many morphological changes. Aged skin is characterized by atrophy, drying, roughness, altered pigmentation, sagging, wrinkling, and both benign and malignant tumor formation. Furthermore, the skin becomes noticeably thinner after age 70. This is because the skin becomes both less cellular and less vascular (Thomas 2001). This loss in vasculature reduces the ability to adapt to temperature, and may reduce healing capacity by limiting oxygen, nutrient flow, and immune cell infiltration to the tissues (Tsuchida 1993). At this same age the pH of the skin begins to rise above 5.5, becoming less acidic. This increases susceptibility to infection as the skin's acidity inhibits bacterial colonization (Farage et al. 2008).

From 20 to 70 years of age, the turnover rate of the epidermis drops by about 50 % (Cerimele et al. 1990). This is likely due to a loss of responsiveness by keratinocytes to a variety of growth factors (Cook and Dzubow 1997; Tenchini et al. 2001) which limits cell proliferation. In older adult skin there are reductions in vascularization (van de Kerkhof et al. 1994), mast cells (Gilchrest et al. 1982), elastin (Ashcroft et al. 1997c; Thomas 2001), granulation tissue (van de Kerkhof et al. 1994), collagen production and density (van de Kerkhof et al. 1994; Cook and Dzubow 1997; Gosain and DiPietro 2004), and fibroblast numbers and motility (van de Kerkhof et al. 1994; Pienta and Coffey 1990; Reed et al. 2001). It seems sensible that any of these changes could negatively impact wound healing. For instance, reduced vascularity limits oxygen and nutrient availability and reduces immune cell infiltration to tissues (Tsuchida 1993), and a lower number of fibroblasts can hinder wound contraction.

A number of age-associated changes in wound healing parameters have been reported in humans, including increased platelet aggregation (clotting), delays in macrophage infiltration to injured tissue, and reductions in macrophage function, re-epithelialization, angiogenesis, and collagen deposition and remodeling (Thomas 2001; Gosain and DiPietro 2004; Lober and Fenske 1991). These changes result in wounds that have increased rates of infection and reduced tensile strength, the latter largely due to disorganized collagen architecture in the healed tissue (Cook and Dzubow 1997; Gosain and DiPietro 2004). As will be discussed later in this chapter, these morphological changes can alter both healing rates and outcomes.

For largely unknown reasons, older individuals scar less. In older patients incision lines are less red, less hypertrophic scarring occurs, and the skin returns to a "normal" appearance more quickly (Cook and Dzubow 1997). Possibly this occurs due to the reduced collagen deposition or a reduced proliferative response by keratinocytes later in healing, both of which have been reported in the elderly (van de Kerkhof et al. 1994; Cook and Dzubow 1997; Gosain and DiPietro 2004; Tenchini et al. 2001). Alternatively, older individuals produce more elastin and fibrillin during acute wound healing which may reduce scar formation (Ashcroft et al. 1997c). In aged mice this reduction in scar formation has been related to lower levels of transforming growth factor (TGF)-β1 and a late increase in TGF-β3 compared to younger mice (Ashcroft et al. 1997; Ashcroft et al. 1997b). Moreover, administering TGF-β3, or antibodies to TGF-β1, reduces scar formation in mice (Shah et al. 1992). Interestingly, reduced TGF-β1 levels and increased TGF-β3 levels are observed in fetuses, which are known to heal without scarring (Wilgus 2007). Possibly the TGF-β1/TGF-β3 ratio is what determines the degree of scar formation observed in older individuals. Further exploration of this phenomenon is needed.

5.6 Aging and Wound Healing

Most clinical studies which have reported poorer healing in older individuals have not accounted for a variety of other factors, which both impact healing rates and are more common with increasing age. Chief among such factors are concurrent illness/disease (comorbidity) and medication use. As a result, despite the findings that aging alters both immunity and skin morphology, it has been unclear if aging per se impairs or even delays wound healing in humans.

A study, which examined experimental mucosal wounds which were placed on the hard palate of the mouth, reported that older individuals (50+ years) had slower wound closure than young adults (18–35 years) (Engeland et al. 2006). Women in either age group were found to heal more slowly than men. Older women healed the slowest and by day 5 had wounds that were 95 % larger than young men, who were the fastest healing group. When individuals who were taking medication and/or had comorbidity were removed from this analysis, the effect of age-impaired healing remained. Surprisingly, the removal of individuals taking medication increased the healing discrepancy between younger and older adults (Engeland et al. 2006). This suggests that age-associated delays in wound healing are not generally exaggerated by comorbidity and may be partially masked by medication use. It further implies that the deleterious effects of aging on wound healing may be even stronger than previously suspected.

Animal studies strongly indicate that advancing age impairs wound healing (for review, see Engeland and Marucha (2009)). In humans, however, healing may be somewhat delayed but it is essentially normal. The same healing cascade occurs in older as it does in younger individuals, and the same endpoints of healing are eventually met. Moreover, the aesthetic of the healed wound (i.e., scarring) is often better in older individuals (Cook and Dzubow 1997). However, this is in the absence of factors known to impair healing. The major increased risk to older individuals undergoing surgery pertains not to age but to alternate risk factors for impaired healing, including: concomitant disease, malnutrition, inactivity, obesity, self-neglect, and, importantly for this chapter, stress, pain, and depression. These risk factors are more commonly experienced by older individuals, and are more likely to worsen healing than in young adults, increasing the risks for infection and post-surgical complications (van de Kerkhof et al. 1994). Attempts to identify and minimize such factors prior to surgery are strongly encouraged for the clinician.

5.7 Menopause and Wound Healing

Reductions in estrogen following menopause have been associated with atrophy, roughness, dryness, wrinkling, laxity, and poorer healing of the skin (Hall and Phillips 2005). Skin's strength and elasticity are both reduced due to collagen loss and lowered capillary blood flow (Ashcroft et al. 1997, 2003b; Ashcroft and Ashworth 2003; Raine-Fenning et al. 2003). The role of estrogen in wound healing is not fully understood. Overall, estrogen appears to accelerate wound healing (for review, see Hall and Phillips (2005)). Hormone replacement therapy (HRT) in post-menopausal women has been shown to benefit healing in a variety of tissues (Ashcroft et al. 1997; Engeland et al. 2009; Margolis et al. 2002) and also leads to a scarring profile similar to that of younger women (Ashcroft et al. 1997). Studies of ovariectomized animals further suggest that menopause delays, whereas HRT augments, dermal healing (Ashcroft and Ashworth 2003; Ashcroft et al. 2003a).

Using data from a previous study, Engeland et al. (2009) identified a small subset of older women 50–54 years of age who were still naturally cycling. The mucosal healing rates of these women were compared to age-matched women who were post-menopause, and the post-menopausal women healed more slowly. Interestingly, the healing pattern of the older pre-menopausal women was close to identical to the healing pattern of younger pre-menopausal women (Engeland et al. 2009). This suggests that, in women, age may not be a substantial risk for impaired healing until menopause begins. Given that menopause has been related to changes in both stress hormones and multiple health parameters, the effects of stress on wound healing in post-menopausal compared to pre-menopausal women may be substantial but this has not been examined.

5.8 Aging and Cell Senescence

Chromosomes have protective caps on the ends of them called telomeres, which prevent the chromosomal ends from deteriorating or fusing with other chromosomes and becoming unstable (O'Sullivan and Karlseder 2010). Every time a cell replicates its telomeres gets a bit shorter because of limitations of the DNA polymerases in completing the replications of the ends of these molecules (Chan and Blackburn 2004). As a result, a cell's biological age (i.e., its potential for future cell division) can be expressed by the length of its telomeres. Simply put, telomeres shorten as we age. This effect is robust, as telomere length (TL) inversely correlates with chronological age across all adult age groups (Hawkley et al. 2005; Epel et al. 2004). It has been proposed that telomere shortening may have evolved to limit the unregulated cell proliferation that occurs in conditions such as cancer (Harley 1991). As will be discussed, both psychological stress and biological aging have been associated with shortened telomeres (i.e., cell senescence).

Although telomeres appear to be protective against cellular senescence and shorten as cells get older, TL alone cannot determine cellular age. The enzyme telomerase is protective of telomeres and helps to prevent telomere shortening, going so far as to promote telomere lengthening (Chan and Blackburn 2004). Not surprisingly, telomerase activity decreases as cells age. Thus, both TL and telomerase levels are important for assessing cellular senescence.

Dyskeratosis congenita (DKC) is a rare genetic disease which results in a lowered ability to synthesize telomerase, resulting in very short telomeres. Its symptoms resemble those of geriatrics including premature graying, altered skin pigmentation, anemia, and a predisposition to cancer. Interestingly, this condition is associated with numerous epidermal abnormalities such as reticulated skin, hyperkeratosis of the palms and soles, the loss of dermal ridges on fingers and toes (adermatoglyphia), and alopecia affecting the scalp, eyebrows, and eyelashes. This strongly suggests that TL plays a role in epidermal function and homeostasis (Buckingham and Klingelhutz 2011).

Recently, studies have shown that telomeres shorten with age in human skin (for recent review, see Buckingham and Klingelhutz (2011)). In addition, reactive oxygen species (ROS) have been causatively linked with both skin aging (Yaar and Eller 2002) and telomere shortening (von 2002). Interestingly, telomerase is present in epidermal keratinocytes but is close to undetectable in dermal fibroblasts (Boukamp 2005). Less telomere shortening has also been reported in epidermal cells than in dermal fibroblasts (Krunic et al. 2009; Sugimoto et al. 2006), which suggests that telomerase is important for maintaining TL in the epidermis. Indeed, telomerase levels in the epidermis have been shown to increase with UV exposure, suggesting an increased need for this enzyme under conditions of cell damage and repair (Taylor et al. 1996).

5.9 Stress and Cell Senescence

Similar to chronological aging, psychological stress has been related to telomere shortening. Thus, stress appears to contribute to cellular aging. For instance, in mothers caring for a chronically ill child, TLs in peripheral blood mononuclear cells (PBMCs) were inversely related to caregiving duration, even when controlling for maternal age (Epel et al. 2004). Moreover, in both this group and in control mothers without chronic stress, TLs were inversely related with perceived stress levels. Controlling for chronological age, individuals in the highest perceived stress quartile had lymphocytes that were estimated to be 9–17 years older than those of individuals in the lowest quartile (Epel et al. 2004). These data suggest that stress causes biological aging in cells, an effect which can lead to immune senescence.

Another study involved premenopausal women who were the mothers of either a healthy child ($n = 19$) or a chronically ill child ($n = 39$). Although TL was not associated with caregiving per se, it was associated with the chronicity of caregiving stress. Higher levels of both perceived stress and stress chronicity related to higher oxidative stress, lower telomerase activity, and shorter TLs in the PBMCs of these women (Shankar et al. 2011). Moreover, women with the highest perceived stress levels had telomeres that were shorter by the equivalent of one decade of additional aging compared to the low-stress women (Shankar et al. 2011). This study provides evidence for a potential mechanistic pathway by which prolonged psychological stress may impact immune function through cellular senescence.

These results indicate that chronological aging and chronic stress each promote cellular aging, and potentially immunosenescence, through the same mechanisms. Thus, reports of stress-impaired immunity being exacerbated in older adults are not surprising (for reviews, see Bauer et al. (2009); Hawkley et al. (2005); Graham et al. (2006)).

5.10 Health Behaviors

Many of the negative effects of aging on wound healing stem from an increased disposition in older individuals to display certain health behaviors which are not conducive to rapid healing. For instance, poor sleep quality, reduced exercise, malnutrition, pain, inactivity, and self-neglect all relate to slower healing and occur more commonly in older adults (Engeland and Graham 2011). Many of these behaviors promote each other (e.g., pain and poor sleep quality), creating a downward spiral from which it can be hard to break free. Psychological stress promotes many of these same behaviors. As a result, health behaviors provide a separate indirect mechanism through which stress and age can each modulate immunity, and by which the effects of stress may further be potentiated in older adults (for a detailed review of health behaviors and healing see Engeland and Graham (2011)).

5.11 Summary

Both chronological and biological aging, along with psychological stress, can alter immune processes and hinder tissue repair. Due to shared mechanistic pathways, these factors may combine in an additive or synergistic manner to further exacerbate the healing impairments caused by any of these factors alone. In addition, older adults are more prone to experience certain psychosocial stressors than younger adults, such as loneliness and depression. Finally, the anti-inflammatory effects provided by the HPA axis are reduced under conditions of both chronic stress and increasing age. Hence, the effects of stress and age on both immunity and healing are compounded in older adults.

Interestingly, there is little evidence to indicate that older individuals have impaired wound healing compared to younger adults. Although the process of healing is somewhat slower in older individuals, the resulting tissue is similar in quality and often aesthetically superior due to reduced scar formation. It is not aging per se, but aging in the presence of other risk factors that creates a health disparity between the younger and older with respect to wound healing. Such risk factors are more common with increasing age and include comorbidity (e.g., diabetes), stress, depression, pain, malnutrition, inactivity, and self-neglect. The presence of any such factors in older adults should serve as a red flag for the clinician, and stringent efforts should be made to minimize these factors both pre- and post-operatively.

References

Ashcroft GS, Ashworth JJ (2003) Potential role of estrogens in wound healing. Am J Clin Dermatol 4:737–743

Ashcroft GS, Mills SJ (2002) Androgen receptor-mediated inhibition of cutaneous wound healing. J Clin Invest 110(5):615–624

Ashcroft GS, Dodsworth J, van Boxtel E, Tarnuzzer RW, Horan MA, Schultz GS et al (1997a) Estrogen accelerates cutaneous wound healing associated with an increase in TGF-beta1 levels. Nat Med 3:1209–1215

Ashcroft GS, Horan MA, Ferguson MW (1997b) The effects of ageing on wound healing: immunolocalisation of growth factors and their receptors in a murine incisional model. J Anat 190(Pt 3):351–365

Ashcroft GS, Kielty CM, Horan MA, Ferguson MW (1997c) Age-related changes in the temporal and spatial distributions of fibrillin and elastin mRNAs and proteins in acute cutaneous wounds of healthy humans. J Pathol 183:80–89

Ashcroft GS, Mills SJ, Flanders KC, Lyakh LA, Anzano MA, Gilliver SC et al (2003a) Role of Smad3 in the hormonal modulation of in vivo wound healing responses. Wound Repair Regen 11:468–473

Ashcroft GS, Mills SJ, Lei K, Gibbons L, Jeong MJ, Taniguchi M et al (2003b) Estrogen modulates cutaneous wound healing by downregulating macrophage migration inhibitory factor. J Clin Invest 111:1309–1318
Bauer ME, Jeckel CM, Luz C (2009) The role of stress factors during aging of the immune system. Ann NY Acad Sci 1153:139–152
Beer HD, Fassler R, Werner S (2000) Glucocorticoid-regulated gene expression during cutaneous wound repair. Vitam Horm 59:217–239
Beilin B, Shavit Y, Trabekin E, Mordashev B, Mayburd E, Zeidel A et al (2003) The effects of postoperative pain management on immune response to surgery. Anesth Analg 97:822–827
Berczi I (1986) The influence of pituitary-adrenal axis on the immune system. In: Berczi I (ed) Pituitary function and immunity. CRC Press, Boca Raton, p 49
Berczi I (1993) Neuroendocrine defence in endotoxin shock (a review). Acta Microbiol Hung 40:265–302
Bosch JA, Engeland CG, Cacioppo JT, Marucha PT (2007) Depressive symptoms predict mucosal wound healing. Psychosom Med 69:597–605
Boukamp P (2005) Skin aging: a role for telomerase and telomere dynamics? Curr Mol Med 5:171–177
Broadbent E, Petrie KJ, Alley PG, Booth RJ (2003) Psychological stress impairs early wound repair following surgery. Psychosom Med 65:865–869
Buckingham EM, Klingelhutz AJ (2011) The role of telomeres in the ageing of human skin. Exp Dermatol 20:297–302
Cerimele D, Celleno L, Serri F (1990) Physiological changes in ageing skin. Br J Dermatol 122 (Suppl 35):13–20
Chan SR, Blackburn EH (2004) Telomeres and telomerase. Philos Trans R Soc London, Ser. B 359:109–121
Chang N, Goodson WH III, Gottrup F, Hunt TK (1983) Direct measurement of wound and tissue oxygen tension in postoperative patients. Ann Surg 197:470–478
Chrousos GP (1998) Stressors, stress, and neuroendocrine integration of the adaptive response. The 1997 Hans Selye memorial lecture. Ann NY Acad Sci 851:311–335
Clark RA, Gallin JI, Fauci AS (1979) Effects of in vivo prednisone on in vitro eosinophil and neutrophil adherence and chemotaxis. Blood 53:633–641
Cole-King A, Harding KG (2001) Psychological factors and delayed healing in chronic wounds. Psychosom Med 63:216–220
Cook JL, Dzubow LM (1997) Aging of the skin: implications for cutaneous surgery. Arch Dermatol 133:1273–1277
DeRijk R, Michelson D, Karp B, Petrides J, Galliven E, Deuster P et al (1997) Exercise and circadian rhythm-induced variations in plasma cortisol differentially regulate interleukin-1 beta (IL-1 beta), IL-6, and tumor necrosis factor-alpha (TNF alpha) production in humans: high sensitivity of TNF alpha and resistance of IL-6. J Clin Endocrinol Metab 82:2182–2191
Donaldson DJ, Mahan JT (1984) Influence of catecholamines on epidermal cell migration during wound closure in adult newts. Comp Biochem Physiol C 78:267–270
Ebrecht M, Hextall J, Kirtley LG, Taylor A, Dyson M, Weinman J (2004) Perceived stress and cortisol levels predict speed of wound healing in healthy male adults. Psychoneuroendocrinology 29:798–809
Edwards J, Dunphy J (1958) Wound healing II. Injury and abnormal repair. N Eng J Med 259:275–285
Eming SA, Krieg T, Davidson JM (2007) Inflammation in wound repair: molecular and cellular mechanisms. J Invest Dermatol 127:514–525
Engeland CG, Gajendrareddy PK (2011) Wound healing in the elderly. In: Katlic M (ed) Cardiothoracic surgery in the elderly: evidence based practice. Springer, New York, pp 259–270
Engeland CG, Graham JE (2011) Psychoneuroimmunological aspects of wound healing and the role of pain. In: Upton D (ed) Psychological impact of pain in patients with wounds. Wounds UK Limited, London, pp 87–114

Engeland CG, Marucha PT (2009) Wound healing and stress. In: Granstein RD, Luger TA (eds) Neuroimmunology of the skin: basic science to clinical relevance. Springer, Berlin, pp 233–247
Engeland CG, Bosch JA, Cacioppo JT, Marucha PT (2006) Mucosal wound healing: the roles of age and sex. Arch Surg 141:1193–1197
Engeland CG, Sabzehei B, Marucha PT (2009) Sex hormones and mucosal wound healing. Brain Behav Immun 23:629–635
England GT, Randall HW, Graves WL (1983) Impairment of tissue defenses by vasoconstrictors in vaginal hysterectomies. Obstet Gynecol 61:271–274
Epel ES, Blackburn EH, Lin J, Dhabhar FS, Adler NE, Morrow JD et al (2004) Accelerated telomere shortening in response to life stress. Proc Natl Acad Sci USA 101:17312–17315
Farage MA, Miller KW, Elsner P, Maibach HI (2008) Functional and physiological characteristics of the aging skin. Aging Clin Exp Res 20:195–200
Fauci AS, Dale DC, Balow JE (1976) Glucocorticosteroid therapy: mechanisms of action and clinical considerations. Ann Intern Med 84:304–315
Flammer JR, Rogatsky I (2011) Minireview: glucocorticoids in autoimmunity: unexpected targets and mechanisms. Mol Endocrinol 25:1075–1086
Gilchrest BA, Stoff JS, Soter NA (1982) Chronologic aging alters the response to ultraviolet-induced inflammation in human skin. J Invest Dermatol 79:11–15
Glaser R, Kiecolt-Glaser JK, Marucha PT, MacCallum RC, Laskowski BF, Malarkey WB (1999) Stress-related changes in proinflammatory cytokine production in wounds. Arch Gen Psychiatry 56:450–456
Gosain A, DiPietro LA (2004) Aging and wound healing. World J Surg 28:321–326
Graham JE, Christian LM, Kiecolt-Glaser JK (2006) Stress, age, and immune function: toward a lifespan approach. J Behav Med 29:389–400
Hall G, Phillips TJ (2005) Estrogen and skin: the effects of estrogen, menopause, and hormone replacement therapy on the skin. J Am Acad Dermatol 53:555–568
Harley CB (1991) Telomere loss: mitotic clock or genetic time bomb? Mutat Res 256:271–282
Hawkley LC, Berntson GG, Engeland CG, Marucha PT, Masi CM, Cacioppo JT (2005) Stress, aging, and resilience: can accrued wear and tear be slowed? Can Psychol 46:115–125
Hawkley LC, Bosch JA, Engeland CG, Cacioppo JT, Marucha PT (2007) Loneliness, dysphoria, stress, and immunity: a role for cytokines. In: Plotnikoff NP, Faith RE, Murgo AJ (eds) Cytokines: stress and immunity. CRC Press, Boca Raton
Hinz B (2007) Formation and function of the myofibroblast during tissue repair. J Invest Dermatol 127:526–537
Hopf HW, Rollins MD (2007) Wounds: an overview of the role of oxygen. Antioxid Redox Signal 9:1183–1192
Hopf HW, Hunt TK, West JM, Blomquist P, Goodson WH III, Jensen JA et al (1997) Wound tissue oxygen tension predicts the risk of wound infection in surgical patients. Arch Surg 132:997–1004
Jonsson K, Hunt TK, Mathes SJ (1988) Oxygen as an isolated variable influences resistance to infection. Ann Surg 208:783–787
Kiecolt-Glaser JK, Glaser R (2002) Depression and immune function: central pathways to morbidity and mortality. J Psychosom Res 53:873–876
Kiecolt-Glaser JK, Marucha PT, Malarkey WB, Mercado AM, Glaser R (1995) Slowing of wound healing by psychological stress. Lancet 346:1194–1196
Koga C, Itoh K, Aoki M, Suefuji Y, Yoshida M, Asosina S et al (2001) Anxiety and pain suppress the natural killer cell activity in oral surgery outpatients. Oral Surg Oral Med Oral Pathol Oral Radiol Endod 91:654–658
Koopman CF (1995) Cutaneous wound healing, an overview. Otolaryngol Clin N Am 28:835–845
Krunic D, Moshir S, Greulich-Bode KM, Figueroa R, Cerezo A, Stammer H et al (2009) Tissue context-activated telomerase in human epidermis correlates with little age-dependent telomere loss. Biochim Biophys Acta 1792:297–308
Linn BS, Linn MW, Klimas NG (1988) Effects of psychophysical stress on surgical outcome. Psychosom Med 50:230–244

Lober CW, Fenske NA (1991) Cutaneous aging: effect of intrinsic changes on surgical considerations. South Med J 84:1444–1446
Margolis DJ, Knauss J, Bilker W (2002) Hormone replacement therapy and prevention of pressure ulcers and venous leg ulcers. Lancet 359:675–677
Marucha PT, Kiecolt-Glaser JK, Favagehi M (1998) Mucosal wound healing is impaired by examination stress. Psychosom Med 60:362–365
McGuire L, Heffner K, Glaser R, Needleman B, Malarkey W, Dickinson S et al (2006) Pain and wound healing in surgical patients. Ann Behav Med 31:165–172
Norris DA, Capin L, Weston WL (1982) The effect of epicutaneous glucocorticosteroids on human monocyte and neutrophil migration in vivo. J Invest Dermatol 78:386–390
O'Sullivan RJ, Karlseder J (2010) Telomeres: protecting chromosomes against genome instability. Nat Rev Mol Cell Biol 11:171–181
Pai MP, Hunt TK (1972) Effect of varying oxygen tensions on healing of open wounds. Surg Gynecol Obstet 135:756–758
Pienta KJ, Coffey DS (1990) Characterization of the subtypes of cell motility in ageing human skin fibroblasts. Mech Ageing Dev 56:99–105
Pittman J (2007) Effect of aging on wound healing: current concepts. J Wound Ostomy Continence Nurs 34:412–415
Raine-Fenning NJ, Brincat MP, Muscat-Baron Y (2003) Skin aging and menopause: implications for treatment. Am J Clin Dermatol 4:371–378
Reed MJ, Ferara NS, Vernon RB (2001) Impaired migration, integrin function, and actin cytoskeletal organization in dermal fibroblasts from a subset of aged human donors. Mech Ageing Dev 122:1203–1220
Robson MC (1997) Wound infection. A failure of wound healing caused by an imbalance of bacteria. Surg Clin N Am 77:637–650
Roy S, Khanna S, Yeh PE, Rink C, Malarkey WB, Kiecolt-Glaser J et al (2005) Wound site neutrophil transcriptome in response to psychological stress in young men. Gene Expr 12:273–287
Shah M, Foreman DM, Ferguson MW (1992) Control of scarring in adult wounds by neutralising antibody to transforming growth factor beta. Lancet 339:213–214
Shankar A, McMunn A, Banks J, Steptoe A (2011) Loneliness, social isolation, and behavioral and biological health indicators in older adults. Health Psychol
Starkweather AR, Witek-Janusek L, Nockels RP, Peterson J, Mathews HL (2006) Immune function, pain, and psychological stress in patients undergoing spinal surgery. Spine (Phila Pa 1976.) 31:E641–E647
Stramer BM, Mori R, Martin P (2007) The inflammation-fibrosis link? A Jekyll and Hyde role for blood cells during wound repair. J Invest Dermatol 127:1009–1017
Sugimoto M, Yamashita R, Ueda M (2006) Telomere length of the skin in association with chronological aging and photoaging. J Dermatol Sci 43:43–47
Taylor RS, Ramirez RD, Ogoshi M, Chaffins M, Piatyszek MA, Shay JW (1996) Detection of telomerase activity in malignant and nonmalignant skin conditions. J Invest Dermatol 106:759–765
Tenchini ML, Savant F, Paini C, Montefusco MC, Donati V, Malcovati M (2001) Autocrine-acting early secreted mitogenic activity: production and responsiveness in cultures of normal human keratinocytes as a function of in vivo and in vitro age. Cell Biol Int 25:197–204
Thomas DR (2001) Age-related changes in wound healing. Drugs Aging 18:607–620
Tsuchida Y (1993) The effect of aging and arteriosclerosis on human skin blood flow. J Dermatol Sci 5:175–181
US Census Bureau (2004) Interim projections consistent with Census 2000. US Census Bureau, Washington, DC
US Census Bureau (2009) State and county quick facts. US Census Bureau, Washington, DC
van de Kerkhof PC, Van Bergen B, Spruijt K, Kuiper JP (1994) Age-related changes in wound healing. Clin Exp Dermatol 19:369–374
von ZT (2002) Oxidative stress shortens telomeres. Trends Biochem Sci 27:339–344
Walburn J, Vedhara K, Hankins M, Rixon L, Weinman J (2009) Psychological stress and wound healing in humans: a systematic review and meta-analysis. J Psychosom Res 67:253–271

Werner S, Grose R (2003) Regulation of wound healing by growth factors and cytokines. Physiol Rev 83:835–870
Werner S, Krieg T, Smola H (2007) Keratinocyte-fibroblast interactions in wound healing. J Invest Dermatol 127:998–1008
Wilgus TA (2007) Regenerative healing in fetal skin: a review of the literature. Ostomy Wound Manage 53:16–31
Yaar M, Eller MS (2002) Mechanisms of aging. Arch Dermatol 138:1429–1432

Chapter 6
Psychoneuromicrobiology: Cytomegalovirus Infection as a Putative Link Between Stress, Aging, and Immunity

Jos A. Bosch, Jerrald L. Rector, James E. Turner, Natalie E. Riddell, Briain o'Hartaigh and Vikki E. Burns

6.1 Introduction

Epidemiological evidence demonstrates increased morbidity and mortality in populations exposed to adverse psychosocial factors such as low socio-economic status (SES) and protracted psychological distress (Cohen and Herbert 1996; House et al. 1988; Marmot 2005; Schneiderman et al. 2005). While the data are clear, the precise mechanisms underlying these associations are yet to be determined (Antoni et al. 2006; Cacioppo and Hawkley 2003; Glaser and Kiecolt-Glaser 2005; McEwen 1998; Uchino et al. 1996). We, and others, have argued that since increasing age is a major risk factor for a wide range of chronic diseases, the aging process itself may be an important target for such mechanistic research (Bosch et al. 2009; Nilsson 1996).

This chapter discusses immunosenescence as a possible biological pathway linking psychosocial stress and health (Bosch et al. 2009; Pawelec et al. 2012). Immunosenescence refers to a decline in immune competence seen in old age, and it is associated with a dramatic rise in morbidity and mortality from infectious disease (Akbar et al. 2004; Larbi et al. 2008). For example, the elderly exhibit a many-fold higher mortality from otherwise common infections such as gastrointestinal infections, urinary tract infections and influenza (Thompson et al. 2003; Yoshikawa 2000).

J. A. Bosch (✉)
Department of Clinical Psychology, University of Amsterdam, Weesperplein 4, 1018 XA Amsterdam, The Netherlands
e-mail: j.a.bosch@uva.nl

Mannheim Institute of Public Health, Social and Preventive Medicine (MIPH), University of Heidelberg, Mannheim, Germany

J. A. Bosch · J. L. Rector · B. o'Hartaigh · V. E. Burns
School of Sport and Exercise Sciences, University of Birmingham, Birmingham, UK

J. E. Turner
CRUK Institute of Cancer Studies, University of Birmingham, Birmingham, UK

N. E. Riddell
School of Infection and Immunity, University College London, London, UK

J. A. Bosch et al. (eds.), *Immunosenescence*,
DOI 10.1007/978-1-4614-4776-4_6,

This susceptibility also promotes physical frailty and cognitive decline, further increasing mortality risk and reducing quality of life (High et al. 2005; Yoshikawa 2000). Although less well established, immunosenescence may also be a factor in the age-related increase in autoimmune and inflammatory disorders (Franceschi et al. 2000; Prelog 2006).

These clinical complications of immunosenescence mark the terminal phase of a process that already starts in adolescence (Akbar and Fletcher 2005; Nikolich-Zugich 2008a). Around that time the thymus, an organ essential for the maturation of T lymphocytes, reduces in size and function with a concomitant decline in the generation naïve T cells. This process causes immunity to become progressively dependent on the existing pool of memory T cells (Woodland and Blackman 2006). This pool of memory T cells gradually expands compensating for the diminishing influx of naïve T cells via cytokine-induced cell division (denoted 'homeostatic proliferation') and repeated antigen exposure. In particular latent herpes viruses, like cytomegalovirus (CMV) are a major source of continuous antigenic stimulation (Khan et al. 2002; Nikolich-Zugich 2008a). As a consequence the expanding pool of memory cells tends to be oligoclonal, that is, exhibit an inflated antigen receptor diversity that is skewed toward a few immunodominant viral antigens. With an increasing number of cell divisions the memory T cells gradually change their appearance; first losing the surface marker CD28 ('intermediate differentiation') and later also the surface marker CD27 ('late differentiation') (Akbar and Fletcher 2005; Appay et al. 2002). These phenotypic changes are particularly prominent in CD8+ T lymphocytes (CTLs), and the accumulation of CTLs with a late differentiation phenotype is considered a hallmark of immunosenescence (Akbar and Fletcher 2005; Hadrup et al. 2006; Pawelec 2006; Derhovanessian et al. 2011).

Other markers of an aging adaptive immune system include weakened responses to vaccination, a diminished proliferative response to mitogens, reduced IL-2 production, shortening of leukocyte telomeres (a DNA marker of cellular aging), increased serum levels of inflammatory cytokines like IL-6, persistent reactivation of latent herpes viruses (e.g., Epstein–Barr Virus, CMV), and a reduced CD4:CD8 ratio (Effros 2007; Larbi et al. 2008; Pawelec 2006; Moro-Garcia et al. 2012). (See Chaps. 1 and 2 in this book for a comprehensive review.) A reduced CD4:CD8 ratio has been found to be a strong predictor of mortality in very old age (Ferguson et al. 1995; Huppert et al. 2003; Wikby et al. 1998). Significantly, all of the aforementioned immune changes that characterize aging are, in part, a result of the accumulation of T cells with a differentiated phenotype (i.e., CD28−CD27−); that is, these T cells poorly proliferate in response to mitogens, produce little IL-2 but abundant INF-γ, have very short telomeres, expand in response to repeated latent viral reactivation, and the number of these cells correlates with weaker vaccination responses.

Those who are familiar with the psychoneuroimmunology literature may immediately note a striking similarity between the immune changes summarized in the previous paragraph and the immune effects of chronic stress (Bosch et al. 2009). Indeed, meta-analysis has shown that protracted psychosocial stressors are associated with much the same pattern of weakened responses to immunization, diminished T cell proliferation to mitogens, reduced IL-2 production, and reactivation of latent herpes viruses (Segerstrom and Miller 2004). Other distinctive features of immunesenescence, such as elevated serum levels of inflammatory markers and shortened leukocyte telomeres, have also consistently been associated with psychosocial stress (Damjanovic et al. 2007; Epel et al. 2004; Glaser and Kiecolt-Glaser 2005; Simon et al. 2006). (See Chaps. 7 and 11 for extensive reviews on this topic.) The remarkable parallels between the aging and stress literature lead us to hypothesize that psychosocial stress contributes to the process of immunological aging (Bosch et al. 2009). This hypothesis is the topic of the current chapter. Thus, what we propose is that the effects of stress and aging on immunity not only just look the same, but may also share similar pathways. We will argue that one of these shared pathways may be the effect of CMV infection on immunity (for a comparable point of view regarding the link between exercise and immunosenescence, see (Simpson 2010) and Chap. 5 in this volume).

In the sections that follow, we will explain that CMV infection dramatically alters the composition of the blood, causing an accumulation of late-differentiated CD8 T cells (CD27−CD28−); the hallmark of an aging immune system. It is thought that this CMV-driven accumulation of differentiated cells is due to incessant low-grade viral reactivation. Viral reactivation activates CMV-specific T cells, which as a population will both expand (i.e., contributing to a larger part of the total pool of memory T cells) and differentiate (i.e., express surface molecules associated with effector function and T cell senescence) (also see Chap. 2 in this volume). Significantly, there is evidence that physiological and psychosocial stressors, via neuro-endocrine and immunological pathways, may contribute to this viral reactivation, implying that stress exposure may promote this aspect of immunosenescence.

6.2 Cytomegalovirus Infection and the Immunity

6.2.1 Herpes Infections and Latency

Viruses are microscopic pathogens, consisting of a nucleic acid genome inside a protein coat, which can only replicate once inside a living cell. While the immune system is able to eradicate most viral infections, some viruses have developed ways to escape this fate. The herpes viruses are particularly well-known for this ability to evade immune destruction and persist within their human host (Cohrs and Gilden 2001; Croen 1991). A characteristic of herpes viruses is that they remain dormant in the cellular genome, a state denoted as latency, which is interrupted by brief periods of reactivation whereby the virus replicates and infects other cells. Herpes viruses have been extremely successful in colonizing humans. Their typical infection rates range between 30 % and 90 %, depending on the virus, and are in part related to factors such as age, SES, topographical location, sexual experience, and early life exposures (e.g., higher infection rates in children that attended day care) (Arvin et al. 2007). In healthy individuals these viruses typically elicit mild (e.g., cold sores, fatigue) or no disease symptoms at all. However, in immune-suppressed patients (e.g., those with AIDS or taking immune suppressive drugs), these viruses can elicit very severe complications (Nester et al. 2008).

Eight herpes viruses have thus far been identified; herpes simplex virus 1 (HSV-1), which predominantly causes oral lesions (or cold sores); herpes simplex 2 (HSV-2), which predominantly causes genital lesions; *Varizella zoster* virus (VZV), which causes chicken pox and shingles; Epstein–Barr virus (EBV), which causes glandular fever; human herpes virus (HHV)-6 and -7, both T cell-infecting viruses and jointly referred to as 'Roseolovirus' due to the association with the common childhood rash; Karposi's sarcoma-associated herpes virus—a cause of tumors and lymphoma—and finally CMV, which in most individuals does not cause symptoms. Although more often than not the cause of mild infectious symptoms, several of these viruses have also been associated with more serious long-term effects, although these effects are rare. For example, EBV is a primary cause of Burkits lymphoma, a form of leukemia. Research has also implicated CMV in a number of severe maladies and with a reduced life expectancy (see section further below).

6.2.2 CMV: Epidemiology and Pathology

CMV is a highly prevalent herpes virus; approximately 35 % of young children and 90 % of the elderly are seropositive (Staras et al. 2006). Overall sero-prevalence in western societies approaches 60 % (Staras et al. 2006). Individuals free from infection at birth will likely become infected early in life: post-natal or childhood CMV infection is common, and caused by exchange of bodily fluids, including breast-milk (Kenneson and Cannon 2007; van der Meer et al. 1996). An inverse relationship exists between SES and CMV seropositivity. Individuals with less education, lower income, and of non-white race are more likely to become infected, and at an earlier age, than individuals of a higher SES (Dowd and Aiello 2009; Dowd et al. 2009). Studies have also reported differences in human CMV infection between various regions within the United States and Europe. (See Chap. 9 in this book.)

As a seemingly innocent and common infection, CMV infection has consequently not received much public health attention (Wreghitt et al. 2003; Zanghellini et al. 1999). This may be unwarranted. CMV infection during pregnancy, for example, is now recognized as a main cause of infant death and long-term disabilities in the United States, and well exceeds that of other, better known, congenital conditions such as Down syndrome, fetal alcohol syndrome, and spina bifida. Further, immune-compromised individuals, such as HIV patients and transplant recipients, can develop life-threatening complications due to CMV reactivation (Freeman 2009; Sutherland et al. 1992). In aging populations a positive CMV serotatus has been associated with cognitive decline, and is implicated in the pathogenesis and severity of cardiovascular diseases (Aiello et al. 2006; Michaelis et al. 2009; Soderberg-Naucler 2006). In addition, the extent of CMV infection, measured by the concentration of immunoglobulin G (IgG) antibodies to CMV, independently predicts mortality in older adults (Roberts et al. 2010; Strandberg et al. 2009; Pawelec et al. 2012). Thus, this once assumed inconsequential virus has a range of deleterious effects, particularly within the immune system. The next section will outline these effects, which are particularly prominent in CD8+ T lymphocytes.

6.2.3 The Effects of CMV-infection on Immunity

As discussed in Sect. 6.1, one of the hallmarks of an aged immune system is a decline in naïve (CD45RA + CD27 + CD28+) CD8+ T cells and an accumulation of differentiated (CD27−CD28−) CD8+ T cells. These effects are partly caused by the age-associated involution of the thymus. For example, individuals thymectomised in the first few years of life exhibit reduced numbers of naïve T lymphocytes and increased numbers of late-differentiated cells (Eysteinsdottir et al. 2004; Sauce et al. 2009; Torfadottir et al. 2006). However, it is now clear that CMV is the major driving force behind the accumulation of differentiated T cells (Moss 2010; Pawelec et al. 2004; Pawelec and Derhovanessian 2011; Pawelec et al. 2009). In particular, the combination of little or no thymic output, a hallmark of aging (see Chap 2), and a selective expansion of late-differentiated T cell populations leads to a gradual overcrowding by CD8+ T cells that have a limited ('oligoclonal') T cell repertoire, a process denoted as memory inflation (Akbar and Fletcher 2005; Brunner et al. 2010; Sauce et al. 2009; van Lier et al. 2003). Stunningly, in some older adults it has been observed that up to 70 % of the CD8+ T cell memory pool has become specific for CMV epitopes (Appay et al. 2002; Khan et al. 2002), and thus only 30 % of the memory pool of those individuals is available to combat other antigens. The relation between CMV infection and CD8+ T cell differentiation occurs independent of calendar age: young adults infected with CMV similarly exhibit skewing of the T cell repertoire as seen in older adults (Chidrawar et al. 2009; Pita-Lopez et al. 2009; Weinberger et al. 2007). Thus, it seems that infection with this common microorganism accelerates the immunological aging process.

6.2.4 T Cell Phenotypes, Their Development, and Functional Characteristics

Immunologists utilize various analytical strategies based on staining of cell-surface molecules to identify different variants of CD8+ T cell 'subsets'. A commonly used procedure is examining the expression of the cell surface co-stimulatory molecules CD27 and CD28; this approach yields three main sub-populations that represent a continuum of differentiation reflecting early (CD27 + CD28+), intermediate (CD27 +CD28−), and late differentiated cells (CD27−CD28−) (Appay et al. 2002; Appay et al. 2008). Other research groups have used different combinations of surface molecules to identify the same or overlapping subsets. For example, CD45RA (an isoform of the pan-lymphocyte marker CD45) has been used in combination with the surface molecules CD27, CCR7, or CD62 L (Appay et al. 2008; Hamann et al. 1997; Sallusto et al. 2004; Sallusto et al. 1999; van Lier et al. 2003). This approach can be used to discern antigen-inexperienced or naïve cells (CD27+CD45RA+), and three additional subsets of antigen-experienced ('memory') cells; central memory (CD27 +CD45RA−), effector memory (CD27−CD45RA−), and effector memory cells which have re-expressed CD45RA (CD27−CD45RA+). Irrespective of the identification strategy used, the general observation is that repeated contact with the same antigen moves memory cells along a differentiation continuum whereby they gradually change their phenotype and acquire so-called “effector” functions, such as increased cytotoxicity (as determined by granzyme and perforin expression) and attain a preference to migrate to peripheral tissue (e.g., the skin and lungs) rather than the lymphoid tissue. Differentiated CD8+ T cells are also efficient producers of inflammatory cytokines (e.g., TNF-α and IFN-γ,) (Clerici et al. 2001; Sansoni et al. 2008; Zanni et al. 2003), which may explain why CMV infection and the accumulation of late-differentiated T cells has been found to be associated with increased low-grade inflammation in some studies (Markovic-Plese et al. 2001; Schmidt et al. 1996; Sun et al. 2008; Wikby et al. 2006; Zanni et al. 2003). The reason these differentiated CD8+ T cells have been referred to as ‘senescent’ is related to other characteristics, such as having shorter telomeres, a low production of interleukin-2 (IL-2; which stimulates the growth, differentiation and survival of cytotoxic T cells), and the concomitant reduced ability to proliferate response to mitogenic stimulants (Monteiro et al. 1996; Nikolich-Zugich 2008b; van de Berg et al. 2010). Together, this CMV-induced accumulation of T cells displaying the cluster of senescent characteristics listed above, may underlie the heightened inflammation, increased risk for infection, and a reduced ability to respond to novel antigens, as well as accelerated cognitive decline observed in CMV infected adults (Larbi et al. 2009; Saurwein-Teissl et al. 2002; Trzonkowski et al. 2009; Wikby et al. 2005; Moro-Garcia et al. 2012). It is perhaps relevant to reiterate here that this cluster of immunological features is also characteristically found in response to protracted psychological stress (Segerstrom and Miller 2004).

6.2.5 The Number of Late-differentiated T Cells is Related to Viral Activity

One key assumption of the model presented here is that repeated viral reactivation, for example, as a result of stress-induced immune suppression, will promote the accumulation of late-differentiated cells and hereby promote related features of ïmmunosenescence. Virus specific T cells, such as those targeting CMV, are responsible for preventing reactivation and it would therefore be reasonable to assume a direct correlation between the number of those cells and viral load (Ogg et al. 1998; van Baarle et al. 2002). What is the evidence supporting that assumption? Research shows that immune-suppressed individuals (e.g., as a result of medical treatment or immunodeficiency diseases) are unable to maintain CMV in latency, and show larger expansions of late-differentiated T cells relative to and compared to immune-competent individuals (Gamadia et al. 2001). Less data is available for healthy free-living individuals, and recipients of organ transplants have frequently used as a model to study the kinetic effects of infection and reactivation on T cell repertoire. CMV-seronegative patients receiving CMV-infected renal transplants develop primary CMV infections and can therefore be studied longitudinally to examine the effects of CMV infection and incessant viral reactivation. Generally, an increase in viral activity (as assessed by viral load in plasma, or IgG antibodies to CMV) promotes a compensatory increase in the number of CMV-specific T cells which do not express CD27 or CD28 (Cantisan et al. 2009; Gamadia et al. 2003, 2004). Thus, frequent reactivation of CMV, possibly by psychological stress (Coskun et al. 2010; Prösch et al. 2000; Sarid et al. 2001), might further exacerbate immunosenescence. The next section will address the evidence that psychological stressors may promote CMV reactivation.

6.3 Psychosocial Factors and CMV

6.3.1 Stress and CMV Reactivation

Although the literature on stress and CMV reactivation has examined a a wide variety of stressors, including academic exams, caring for spouses with Alzheimer's disease, space flight, and selfreported depressive symptoms, the results of these studies paint a fairly consistent picture whereby higher levels of distress are associated with higher CMV-specific IgG antibodies (used as a marker of viral reactivation). Overall, this literature replicates what has been found for other herpes viruses such as EBV and herpes simplex virus type 1 (HSV-1), which likewise show increased virus-specific IgG antibody titres with higher stress (Esterling et al. 1993; McDade et al. 2000; Shirtcliff et al. 2009).

Most studies that investigated the association between elevated distress and markers of CMV reactivity used plasma levels of virus-specific IgG antibodies as an outcome. This measurement approach is based on the assumption that increased viral load will activate the immune system, thereby stimulating B cells to increase the output of specific antibodies. While there is support for the validity this assumption, mainly from clinical data, it may be relevant to point out that other factors may also determine antibody levels. For example, exposure to different strains of the same virus, denoted as 'super-infection', may generate a more extensive polyclonal antibody response and concomitantly higher virus-specific antibody levels (Novak et al. 2008; Ross et al. 2010). The validity of the studies using antibody levels as an outcome measure are corroborated, by research that used other analytical techniques. CMV-specific IgG antibodies are found to increase in line with the increase in viral load, measured by real-time PCR in leukocytes (Kuo et al. 2008), and viral load in multiple tissues coincide with CMV-attributable clinical symptoms and disease (Boeckh and Boivin 1998). For example, Toro and Ossa assessed CMV viral load bi-weekly by quantitative PCR (qPCR) in blood, throat washings and urine. CMV viral load was then correlated with stressful psychological (e.g., academic tasks, work overload, insomnia) and physical events (e.g., trauma, surgery, X-ray or UV radiation exposure). The results revealed a positive association between CMV viral load and stress-producing events in the workplace (Toro and Ossa 1996).

Academic examinations are commonly employed as a model for acute and protracted stress (Bosch et al. 2001, 2003), and several studies have reported that academic stress causes reactivation of herpes viruses such as EBV, HSV-1, and CMV. Glaser et al. (1985) investigated the effects of academic stress on CMV specific IgG antibody in 20 medical students and found a significant increase in antibody levels of during the first day of exams as compared to a non-stress baseline (the end of summer vacation) (Glaser et al. 1985). Changes in salivary IgG antibodies to CMV were observed in a cohort of 54 first-year nursing and physiotherapy students undergoing two exams, 2 weeks apart. This academic stressor was found to cause an approximate 65 % increase in CMV-specific salivary IgG and a 46 % increase in CMV-specific salivary IgA, as compared to baseline measurements performed at the beginning of the semester (Sarid et al. 2004). Matalka et al. (2000) investigated the effect of examination stress on CMV and EBV reactivation, defined as a 30 % increase in virus-specific IgG levels in 56 female nursing students. In both the summer and winter semester, blood samples were collected during a non-stressful period (start of semester) and a stressful-period of study (immediately after exams). Analysis of sera revealed reactivation in 26 % ($N = 12$) of all CMV-seropositive students during exams in both semesters (Matalka et al. 2000). The results of this study also suggested that there is a seasonal difference (summer vs. fall-winter) in CMV-reactivation due to academic stress, with larger effects during the winter season.

Appels et al. (2000) studied the link between depressive symptoms, vital exhaustion (a constellation of symptoms characterized by listlessness and fatigue), and CMV antibody titers in 30 patients with coronary artery disease. The patients classified as exhausted ($N = 15$) showed higher levels of inflammatory cytokines (i.e., IL-1β, IL-6 and TNF-α) in addition to elevated CMV titers compared to non-exhausted patients. After adjustment for depression the IL-1β and CMV IgM titers differences were still significant (Appels et al. 2000). However, this cross-sectional study was unable to determine if inflammation and viral reactivation caused, or were caused by, symptoms of fatigue. Inflammation might act as a pathway linking feelings of exhaustion and CMV reactivation (see the section below). In a follow-up, van der Ven et al. (2003) did not find associations between inflammatory cytokines (i.e. IL-6, IL-10, and IL-1Ra) and CMV antibody levels in 59 healthy individuals (van der Ven et al. 2003).

Phillips et al. (2008) found no association between CMV serostatus and psychological morbidity (depression and anxiety) in 137 older adults. However, in CMV seropositive individuals higher IgG antibody titers were positively correlated with symptoms of depression and anxiety (Phillips et al. 2008). The cross-sectional nature of this study did not allow for a determination of causality although the findings appear consistent with those of longitudinal studies showing that distress can trigger a reactivation of CMV (Mehta et al. 2000; Miller et al. 2002).

Miller et al. (2002) investigated the association between depression and inflammatory markers (CRP, IL-1β, IL-6, TNF-α and MCP-1) and infectious load (CMV and *Chlamydia pneumoniae*) in a group of 100 adults, of which 50 were clinically depressed. This study showed significant differences in CMV serostatus between depressed (42 % positive) and control (52 % positive) individuals. In a subsequent study, this group of researchers investigated the association between CMV serostatus and depressive symptoms in a sample of 65 patients recovering from an acute coronary syndrome (Miller et al. 2005). After splitting the subjects into three equal-sized groups based on depression scores (using the Beck Depression Inventory), cardiac patients in the highest depression tertile were more likely to be CMV seropositive than the middle and lower tertiles. CMV+ patients also showed higher CRP levels compared to CMV-negative patients (Miller et al. 2005). The authors did not report data on antibody levels. Taken together, these results suggest that CMV seropositivity is related to a higher severity of depressive symptoms and inflammatory activity in patients with acute coronary.

Mehta et al. (2000) studied the effect of spaceflight on CMV reactivation, which was quantified by assessing the number of viral particles in urine (using qPCR) and by measuring CMV specific IgG antibody. It was observed that 10 % of the astronauts showed evidence of viral shedding (i.e., release of viral particles) in urine samples around the time of their mission, whereas this was only observed in 1 % of controls. Also serum CMV IgG levels increased from pre- to post-flight in astronauts, but only in astronauts who showed evidence of shedding (Mehta et al. 2000), suggesting that IgG titres are valid measures of reactivation. Viral shedding did not show a relation with endocrine measures (24 h urinary cortisol, adrenaline and noradrenaline). The authors also did not present any self-report data that could have corroborated a role for psychological stress, and thus a major role of physical factors (e.g., micro-gravity) cannot be excluded.

Uddin et al. (2010) assessed CMV-specific IgG antibody titers as a marker of immune function in 100 individuals, of whom 23 had lifetime post-traumatic stress disorder (PTSD). PTSD is a severe anxiety disorder that can develop after exposure to a psychologically impactful event, like being a victim of crime, accident, or a disaster or being a close witness to such events. Symptoms of PTSD typically include re-experiencing the original trauma(s) through flashbacks or nightmares, avoidance of situations and other stimuli that one associates with the events, and an elevated state of arousal and hypervigilance. PTSD is a chronic stressor that has inflammatory and immunological consequences (Gill et al. 2009). The results of Uddin et al. (2010) showed an approximately 45 % higher level of CMV-specific IgG antibody in those with PTSD versus unaffected individuals (Uddin et al. 2010).

6.3.2 Mechanisms of Stress-induced CMV Reactivation

The dominant hypothesis in psychoneuroimmunology is that latent viral reactivation is due to temporary or persistent stress-induced immune suppression. This immune suppression, in turn, may be related to dysregulation (typically hyperactivity) of various neuroendocrine stress systems (Bauer 2005; Bauer et al. 2000). Best studied in this regard are the hypothalamic–pituitary–adrenal cortex system, which regulates the release of glucocorticoids, and the sympathetic nervous system, which regulates the release of the catecholamines epinephrine and norepinephrine (Glaser and Kiecolt-Glaser 2005). While there is little doubt that this prototypical psycho-neuro-immunological pathway plays a role in viral reactivation, more direct pathways may be involved as well. One example, which we discuss in more detail here, is direct activation of the CMV promoter in the host genome. The CMV promoter is a gene segment that regulates the expression of downstream major immediate-early (IE) genes. Expression of these genes plays an important role in the initial steps leading to reactivation of CMV from latency (Hermiston et al. 1987; Stenberg et al. 1984; Stinski and Isomura 2008). The CMV promoter region contains binding sites for transcription factors that are increased during inflammation and β-adrenergic stimulation, such Activator Protein 1 (AP-1) and cAMP Response Element Binding (CREB). This fact presents a potential pathway which may explain why systemic inflammation has been associated with CMV reactivation in various patient groups, including cardiovascular, transplant patients (Humar et al. 1999; Tong et al. 2001; Widmann et al. 2008) and intensive care unit (ICU) patients (Chilet et al. 2010; Limaye and Boeckh 2010; Limaye et al. 2008). For example, Docke et al. (1994) investigated the in vivo role of TNF-α on the reactivation of CMV in 60 septic patients in intensive care and observed CMV reactivation in 33 (75 %) of the 44 CMV-seropositive septic patients. The authors attributed the high frequency of CMV reactivation in the septic state to enhanced TNF-α levels. In addition, there is good experimental evidence that elevated sympathetic activation, and the concomitant release of the catecholamines epinephrine and norepinephrine, may likewise play a role (Docke et al. 1994; Prosch et al. 1999; Stein et al. 1993).

Stimulation of β-adrenergic receptors by catecholamines lead to a cascade of signaling that increases transcription factors (e.g., CREB/ATF-1) that directly affects CMV transcriptional activity (Montminy 1997). Prosch et al. (2000) studied myocardial infarction patients, who showed strongly elevated catecholamine but normal TNF-α plasma levels and observed viremia in a majority of patients, which peaked 7 days post-infarction. In vitro studies confirmed that stimulation of CMV-infected monocytes with epinephrine, as well as the beta2-adrenergic agonist propanolol, is capable of inducing CMV reactivation in these cells, as measured by expression of the Immediate/Early gene (I/E gene). The authors found little evidence of concomitant immunosuppression, and therefore concluded that catecholamines most likely induce reactivation directly via activation of the CMV IE enhancer/promoter and viral gene expression (Prosch et al. 2000). These experiments thus present an important non-immunological pathway by which stress may cause CMV reaction. A footnote is that although expression of the I/E genes are a first and essential step in reactivation, it is not sufficient for full viral replication, and therefore the definitive evidence for this mechanism still needs to be established.

6.4 CMV Infection Alters Immune System Responses to Acute Stress

The section above reviewed the evidence suggesting that psychosocial stressors may affect reactivation of CMV, which, we propose, may form a mechanism linking stress with accelerated immunological aging. Recent research from our laboratory has also shown that CMV may, in turn, alter psychobiological responses to stress, in particular stress-induced lymphocytosis. Lymphocytosis is probably one of the best documented effects of stress on the immune system: It involves a rapid (within minutes) increase in the absolute number of lymphocytes in the peripheral blood

(Benschop et al. 1996; Bosch et al. 2005; Dhabhar and McEwen 1999). It has been proposed that this stress-induced redeployment of immune cells enhances immunosurveillance in the face of potential threat, that is, when injury and concomitant infection are more likely (Benschop et al. 1996; Bosch et al. 2005; Dhabhar and McEwen 1997; Dopp et al. 2000). Lymphocytosis is driven by activation of β_2-adrenergic receptors (β_2AR) that are expressed by lymphocytes which become activated during the swift upsurge of catecholamines (epinephrine and norepinephrine) during psychological and physical stress. It is not surprising, then, that the strongest mobilisation is observed with lymphocyte subsets that exhibit the highest β2AR density, such as Natural Killer (NK) cells and CD8+ T cells. Via mechanisms that have not been well-characterized yet, these lymphocytes detach from endothelial cells and possibly other reservoirs, and become released into the blood (Benschop et al. 1996; Bosch et al. 2005; Dhabhar 2002; Dimitrov et al. 2010; Kuhlwein et al. 2001; Mills et al. 1995, 1997; Segerstrom and Miller 2004; Zorrilla et al. 2001).

Taking a closer look at the lymphocyte types (or 'subsets') that become mobilized during stress, it appears that this response is largely confined to cells that have a cytotoxic ability, such as the aforementioned NK cells and CD8+ T lymphocytes (CTLs), but also gamma-delta ($\gamma\delta$) T cells and the small subset of cytotoxic CD4+ T cells (Anane et al. 2009; Campbell et al. 2009; Elenkov et al. 2000). Further, even among these lymphocyte subtypes there is a response heterogeneity whereby phenotypes that show an enhanced effector potential (e.g., the ability for cytotoxicity, and inflammatory cytokine production) and exhibit a high-tissue migrating ability (e.g., as evidenced by the elevated expression adhesion molecules such as CD11a+), show the strongest mobilisation during stress (Anane et al. 2010; Bosch et al. 2005; Campbell et al. 2009; Dimitrov et al. 2010). Coincidentally, these cytotoxic T cell subsets are the same types that become strongly enriched in peripheral blood as a result of CMV infection. Thus, acute psychological stress and exercise have been demonstrated to evoke a strong and robust mobilization of the "effector-like" EMRA (CD45RA+ CD27$^-$/CD28$^-$) and EM (CD45RA$^-$CD27$^-$/CD28$^-$) phenotypes. In contrast, the naïve and CM subsets, which have little or no cytotoxic effector functions, are not mobilized by acute stress. The selectivity of the response is likely explained by increased expression of the β_2AR by the stress sensitive EM and, in particular, the EMRA populations (Dimitrov et al. 2009, 2010).

As previously discussed, infection with CMV can dramatically alter the CTL compartment by inducing the accumulation of these β_2AR expressing, stress sensitive, EM and EMRA cells. This enrichment would suggest that CMV infection is associated with an elevated lymphocytosis response to stress and exercise. This is indeed what we found; we observed an increase in lymphocytosis? (how was reactivity measured?) in CMV-seropositive hosts, when compared to CMV-seronegative individuals. Subsequent analyses further showed that CMV-specific CTLs had a greater propensity to mobilize than total CTLs and contained a larger proportion of EM and EMRA cells (Riddell, manuscript in preparation). Further, the same correlation between CMV seropositivity and a greater CTL mobilization is also apparent during exercise (Turner et al. 2010).

While a fascinating physiological phenomenon, it is still unclear if the enhanced stress-induced recruitment of cytotoxic cells in CMV-seropositive individuals has any immunological or clinical impact. However, the inflammatory potential of these effector-like CTLs is consistent with the idea that acute stress may cause exacerbation of inflammatory conditions, such as atherosclerosis, via promoting cell migration into the inflamed tissues (Bosch et al. 2003a, 2003b; Marsland et al. 2002; Nyklicek et al. 2005). Moreover, the enhanced adrenergic sensitivity of CMV-specific CTLs may directly impact viral control as adrenergic stimulation of T cells can alter a variety of effector functions including proliferation, lytic activity, and cytokine production (Bartik et al. 1993; Borger et al. 1998; Glaser and Kiecolt-Glaser 2005; Gratama et al. 2008; Hatfield et al. 1986; Kalinichenko et al. 1999; Leo and Bonneau 2000; Lilleri et al. 2008; Morita-Hoshi et al. 2008; Ozdemir et al. 2002; Sarid et al. 2001, 2004; Sekut et al. 1995).

In summary, while research in psychoneuroimmunology traditionally emphasizes the role of genotype and psychological experiences as key determinants of immune system responses to stress, infection history may be another and thus far underexplored determinant. One may speculate that amplification of cytotoxic lymphocyte mobilization during stressful events in CMV+ individuals may be conducive to systemic inflammation while also decreasing antigen-specific immunity, both of which are hallmarks of immunosenescence.

6.5 Summary and Conclusion

While much research effort has been dedicated to identifying the biological determinants of immune senescence, the potential role of psychosocial and behavioral factors remains ill-considered. This neglect seems unwarranted considering the remarkable similarities between the functional and phenotypical immune changes observed during immunosenescence and those seen in response to protracted stress. These changes include increases in circulating effector memory CTLs, impaired proliferation in response to mitogens, shortening of telomeres, diminished response to vaccination, and a skewed CD4/CD8 ratio. These parallels beg the question of whether the overlapping age- and stress-related immune effects involve similar pathways. This chapter presented a hypothesis whereby infection history is a proposed as a shared mechanism linking psychosocial stress with aging of the immune system. Specifically, we proposed that latent infection with and subsequent reactivation of CMV may act as a mediator linking stress and age-related decline in immunity.

In recent years, it has become increasingly clear that incessant antigenic stimulation by herpes viruses, in particular CMV, is a major driving force in senescence within the T-cell compartment. This aspect of immunosenescence is characterized by a dramatic expansion of CD8+ T lymphocytes that have an effector-memory phenotype (e.g., CD57+/CD27−/CD28−/CCR7−/CD62 L−). The extent of this expansion predicts reduced immune competence and increased morbidity and mortality in epidemiological and experimental studies. The observations from this literature seem to dovetail remarkably well with evidence from human and animal studies showing that stress is associated with reactivation of latent herpes viruses, including CMV. The processes and mechanisms involved in this reactivation have been well-characterized, and include the impairment of cellular immunity by stress hormones such as catecholamines and glucocorticoids, as well as direct genomic viral activation by these hormones. Hence, here we presented a model that combines these different strands of evidence. In brief, what we propose is that psychological and physical stressors facilitate CMV reactivation and thereby accelerate T cell immunosenescence. A limitation of this model is that although its subcomponents have been confirmed (i.e., stress and latent herpes reactivation, and the role of CMV infection in development of immunosenescence), the full model still awaits empirical scrutiny. This scrutiny seems worthwhile as confirmation will have significant implications for our understanding of the links between stress and immunity. One of its main implications would perhaps be that the host's infection history is an important, and thus far overlooked, mediator of these links.

References

Aiello AE, Haan M, Blythe L, Moore K, Gonzalez JM, Jagust W (2006) The influence of latent viral infection on rate of cognitive decline over 4 years. J Am Geriatr Soc 54(7):1046–1054

Akbar AN, Beverley PC, Salmon M (2004) Will telomere erosion lead to a loss of T-cell memory? Nat Rev Immunol 4(9):737–743

Akbar AN, Fletcher JM (2005) Memory T cell homeostasis and senescence during aging. Curr Opin Immunol 17(5):480–485

Anane LH, Edwards KM, Burns VE, Drayson MT, Riddell NE, van Zanten JJ et al (2009) Mobilization of gammadelta T lymphocytes in response to psychological stress, exercise, and beta-agonist infusion. Brain Behav Immun

Anane LH, Edwards KM, Burns VE, Zanten JJ, Drayson MT, Bosch JA (2010) Phenotypic characterization of gammadelta T cells mobilized in response to acute psychological stress. Brain Behav Immun

Antoni MH, Lutgendorf SK, Cole SW, Dhabhar FS, Sephton SE, McDonald PG et al (2006) The influence of bio-behavioural factors on tumour biology: pathways and mechanisms. Nat Rev Cancer 6(3):240–248

Appay V, Dunbar PR, Callan M, Klenerman P, Gillespie GM, Papagno L et al (2002) Memory CD8 + T cells vary in differentiation phenotype in different persistent virus infections. Nat Med 8(4):379–385

Appay V, van Lier RA, Sallusto F, Roederer M (2008). Phenotype and function of human T lymphocyte subsets: consensus and issues. Cytometry A 73(11):975–983

Appels A, Bar FW, Bar J, Bruggeman C, de Baets M (2000) Inflammation, depressive symptomtology, and coronary artery disease. Psychosom Med 62(5):601–605

Arvin AM, Campadelli-Fiume G, Roizman B (2007) Human herpesviruses: biology, therapy, and immunoprophylaxis. Cambridge University Press, Cambridge

Bartik MM, Brooks WH, Roszman TL (1993) Modulation of T cell proliferation by stimulation of the beta-adrenergic receptor: lack of correlation between inhibition of T cell proliferation and cAMP accumulation. Cell Immunol 148(2):408–421

Bauer ME (2005) Stress, glucocorticoids and ageing of the immune system. Stress 8(1):69–83

Bauer ME, Vedhara K, Perks P, Wilcock GK, Lightman SL, Shanks N (2000) Chronic stress in caregivers of dementia patients is associated with reduced lymphocyte sensitivity to glucocorticoids. J Neuroimmunol 103(1):84–92

Benschop RJ, Rodriguez-Feuerhahn M, Schedlowski M (1996) Catecholamine-induced leukocytosis: early observations, current research, and future directions. Brain Behav Immun 10(2):77–91

Boeckh M, Boivin G (1998) Quantitation of cytomegalovirus: methodologic aspects and clinical applications. Clin Microbiol Rev 11(3):533–554

Borger P, Hoekstra Y, Esselink MT, Postma DS, Zaagsma J, Vellenga E et al (1998) Beta-adrenoceptor-mediated inhibition of IFN-gamma, IL-3, and GM-CSF mRNA accumulation in activated human T lymphocytes is solely mediated by the beta2-adrenoceptor subtype. Am J Respir Cell Mol Biol 19(3):400–407

Bosch JA, Berntson GG, Cacioppo JT, Dhabhar FS, Marucha PT (2003a) Acute stress evokes selective mobilization of T cells that differ in chemokine receptor expression: a potential pathway linking immunologic reactivity to cardiovascular disease. Brain Behav Immun 17(4):251–259

Bosch JA, Berntson GG, Cacioppo JT, Marucha PT (2005) Differential mobilization of functionally distinct natural killer subsets during acute psychologic stress. Psychosom Med 67(3):366–375

Bosch JA, de Geus EJ, Kelder A, Veerman EC, Hoogstraten J, Amerongen AV (2001) Differential effects of active versus passive coping on secretory immunity. Psychophysiology 38(5):836–846

Bosch JA, de Geus EJ, Veerman EC, Hoogstraten J, Nieuw Amerongen AV (2003b) Innate secretory immunity in response to laboratory stressors that evoke distinct patterns of cardiac autonomic activity. Psychosom Med 65(2):245–258

Bosch JA, Fischer JE, Fischer JF (2009) Adverse working conditions are associated with CD8 T cell differentiation indicative of immunosenescence. Brain Behav Immun 23(4):527–534

Brunner S, Herndler-Brandstetter D, Weinberger B, Grubeck-Loebenstein B (2010) Persistent viral infections and immune aging. Ageing Res Rev

Cacioppo JT, Hawkley LC (2003) Social isolation and health, with an emphasis on underlying mechanisms. Perspect Biol Med 46(3 Suppl):S39–52

Campbell JP, Riddell NE, Burns VE, Turner M, van Zanten JJ, Drayson MT et al (2009) Acute exercise mobilises CD8 + T lymphocytes exhibiting an effector-memory phenotype. Brain Behav Immun

Cantisan S, Torre-Cisneros J, Lara R, Rodriguez-Benot A, Santos F, Gutierrez-Aroca J et al (2009) Age-dependent association between low frequency of CD27/CD28 expression on pp65 CD8 + T cells and cytomegalovirus replication after transplantation. Clin Vaccine Immunol 16(10):1429–1438

Chidrawar S, Khan N, Wei W, McLarnon A, Smith N, Nayak L et al (2009) Cytomegalovirus-seropositivity has a profound influence on the magnitude of major lymphoid subsets within healthy individuals. Clin Exp Immunol 155(3):423–432

Chilet, M, Aguilar, G, Benet, I, Belda, J, Tormo, N, Carbonell, JA et al (2010) Virological and immunological features of active cytomegalovirus infection in nonimmunosuppressed patients in a surgical and trauma intensive care unit. J Med Virol 82(8):1384–1391

Clerici M, Saresella M, Trabattoni D, Speciale L, Fossati S, Ruzzante S et al (2001) Single-cell analysis of cytokine production shows different immune profiles in multiple sclerosis patients with active or quiescent disease. J Neuroimmunol 121(1–2):88–101

Cohen S, Herbert TB (1996) Health psychology: psychological factors and physical disease from the perspective of human psychoneuroimmunology. Annu Rev Psychol 47:113–142

Cohrs RJ, Gilden DH (2001) Human herpesvirus latency. Brain Pathol 11(4):465–474

Coskun O, Sener K, Kilic S, Erdem H, Yaman H, Besirbellioglu AB et al (2010) Stress-related Epstein-Barr virus reactivation. Clin Exp Med 10(1):15–20

Croen KD (1991) Latency of the human herpesviruses. Annu Rev Med 42:61–67

Damjanovic AK, Yang Y, Glaser R, Kiecolt-Glaser JK, Nguyen H, Laskowski B et al (2007) Accelerated telomere erosion is associated with a declining immune function of caregivers of Alzheimer's disease patients. J Immunol 179(6):4249–4254

Derhovanessian E, Maier AB, Hahnel K, Beck R, de Craen AJ, Slagboom EP, Westendorp RG, Pawelec G (2011) Infection with cytomegalovirus but not herpes simplex virus induces the accumulation of late-differentiated CD4+ and CD8+ T-cells in humans. J Gen Virol 92:2746–2756
Dhabhar FS (2002) Stress-induced augmentation of immune function—the role of stress hormones, leukocyte trafficking, and cytokines. Brain Behav Immun 16(6):785–798
Dhabhar FS, McEwen BS (1997) Acute stress enhances while chronic stress suppresses cell-mediated immunity in vivo: a potential role for leukocyte trafficking. Brain Behav Immun, 11(4):286–306
Dhabhar FS, McEwen BS (1999) Enhancing versus suppressive effects of stress hormones on skin immune function. Proc Natl Acad Sci USA 96(3):1059–1064
Dimitrov S, Benedict C, Heutling D, Westermann J, Born J, Lange T (2009) Cortisol and epinephrine control opposing circadian rhythms in T cell subsets. Blood 113(21):5134–5143
Dimitrov S, Lange T, Born J (2010) Selective mobilization of cytotoxic leukocytes by epinephrine. J Immunol 184(1):503–511
Docke WD, Prosch S, Fietze E, Kimel V, Zuckermann H, Klug C et al (1994) Cytomegalovirus reactivation and tumour necrosis factor. Lancet 343(8892):268–269
Dopp JM, Miller GE, Myers HF, Fahey JL (2000) Increased natural killer-cell mobilization and cytotoxicity during marital conflict. Brain Behav Immun 14(1):10–26
Dowd JB, Aiello AE (2009) Socioeconomic differentials in immune response. Epidemiology 20(6):902–908
Dowd JB, Aiello AE, Alley DE (2009) Socioeconomic disparities in the seroprevalence of cytomegalovirus infection in the US population: NHANES III. Epidemiol Infect 137(1):58–65
Effros RB (2007) Role of T lymphocyte replicative senescence in vaccine efficacy. Vaccine 25(4):599–604
Elenkov IJ, Wilder RL, Chrousos GP, Vizi ES (2000) The sympathetic nerve—an integrative interface between two supersystems: the brain and the immune system. Pharmacol Rev 52(4):595–638
Epel ES, Blackburn EH, Lin J, Dhabhar FS, Adler NE, Morrow JD et al (2004) Accelerated telomere shortening in response to life stress. Proc Natl Acad Sci USA 101(49):17312–17315
Esterling BA, Antoni MH, Kumar M, Schneiderman N (1993) Defensiveness, trait anxiety, and Epstein-Barr viral capsid antigen antibody titers in healthy college students. Health Psychol 12(2):132–139
Eysteinsdottir JH, Freysdottir J, Haraldsson A, Stefansdottir J, Skaftadottir I, Helgason H et al (2004) The influence of partial or total thymectomy during open heart surgery in infants on the immune function later in life. Clin Exp Immunol 136(2):349–355
Ferguson FG, Wikby A, Maxson P, Olsson J, Johansson B (1995) Immune parameters in a longitudinal study of a very old population of Swedish people: a comparison between survivors and nonsurvivors. J Gerontol A 50(6):B378–382
Franceschi C, Bonafe M, Valensin S, Olivieri F, De Luca M, Ottaviani E et al (2000) Inflamm-aging: an evolutionary perspective on immunosenescence. Ann N Y Acad Sci 908:244–254
Freeman RB Jr (2009) The 'indirect' effects of cytomegalovirus infection. Am J Transplant 9(11):2453–2458
Gamadia LE, Remmerswaal EB, Weel JF, Bemelman F, van Lier RA, Ten Berge IJ (2003) Primary immune responses to human CMV: a critical role for IFN-gamma-producing CD4 + T cells in protection against CMV disease. Blood 101(7):2686–2692
Gamadia LE, Rentenaar RJ, Baars PA, Remmerswaal EB, Surachno S, Weel JF et al (2001) Differentiation of cytomegalovirus-specific CD8(+) T cells in healthy and immunosuppressed virus carriers. Blood 98(3):754–761
Gamadia LE, van Leeuwen EM, Remmerswaal EB, Yong SL, Surachno S, Wertheim-van Dillen PM et al (2004) The size and phenotype of virus-specific T cell populations is determined by repetitive antigenic stimulation and environmental cytokines. J Immunol 172(10):6107–6114
Gill JM, Saligan L, Woods S, Page G (2009) PTSD is associated with an excess of inflammatory immune activities. Perspect Psychiatr Care 45(4):262–277

Glaser R, Kiecolt-Glaser JK (2005) Stress-induced immune dysfunction: implications for health. Nat Rev Immunol 5(3):243–251
Glaser R, Kiecolt-Glaser JK, Speicher CE, Holliday JE (1985) Stress, loneliness, and changes in herpesvirus latency. J Behav Med 8(3):249–260
Gratama JW, Brooimans RA, van der Holt B, Sintnicolaas K, van Doornum G, Niesters HG et al (2008) Monitoring cytomegalovirus IE-1 and pp65-specific CD4+ and CD8+T-cell responses after allogeneic stem cell transplantation may identify patients at risk for recurrent CMV reactivations. Cytometry B Clin Cytom 74(4):211–220
Hadrup SR, Strindhall J, Kollgaard T, Seremet T, Johansson B, Pawelec G et al (2006) Longitudinal studies of clonally expanded CD8 T cells reveal a repertoire shrinkage predicting mortality and an increased number of dysfunctional cytomegalovirus-specific T cells in the very elderly. J Immunol 176(4):2645–2653
Hamann D, Baars PA, Rep MH, Hooibrink B, Kerkhof-Garde SR, Klein MR et al (1997) Phenotypic and functional separation of memory and effector human CD8+T cells. J Exp Med 186(9):1407–1418
Hatfield SM, Petersen BH, DiMicco JA (1986) Beta adrenoceptor modulation of the generation of murine cytotoxic T lymphocytes in vitro. J Pharmacol Exp Ther 239(2):460–466
Hermiston TW, Malone CL, Witte PR, Stinski MF (1987) Identification and characterization of the human cytomegalovirus immediate-early region 2 gene that stimulates gene expression from an inducible promoter. J Virol 61(10):3214–3221
High K, Bradley S, Loeb M, Palmer R, Quagliarello V, Yoshikawa T (2005) A new paradigm for clinical investigation of infectious syndromes in older adults: assessing functional status as a risk factor and outcome measure. J Am Geriatr Soc 53(3):528–535
House JS, Landis KR, Umberson D (1988) Social relationships and health. Science 241:540–545
Humar A, St Louis P, Mazzulli T, McGeer A, Lipton J, Messner H et al (1999) Elevated serum cytokines are associated with cytomegalovirus infection and disease in bone marrow transplant recipients. J Infect Dis 179(2):484–488
Huppert FA, Pinto EM, Morgan K, Brayne C (2003) Survival in a population sample is predicted by proportions of lymphocyte subsets. Mech Ageing Dev 124(4):449–451
Kalinichenko VV, Mokyr MB, Graf LH Jr, Cohen RL, Chambers DA (1999) Norepinephrine-mediated inhibition of antitumor cytotoxic T lymphocyte generation involves a beta-adrenergic receptor mechanism and decreased TNF-alpha gene expression. J Immunol 163(5):2492–2499
Kenneson A, Cannon MJ (2007) Review and meta-analysis of the epidemiology of congenital cytomegalovirus (CMV) infection. Rev Med Virol 17(4):253–276
Khan N, Shariff N, Cobbold M, Bruton R, Ainsworth JA, Sinclair AJ et al (2002) Cytomegalovirus seropositivity drives the CD8 T cell repertoire toward greater clonality in healthy elderly individuals. J Immunol 169(4):1984–1992
Kuhlwein EC, Irwin MR, Ziegler MG, Woods VL, Kennedy B, Mills PJ (2001) Propranolol affects stress-induced leukocytosis and cellular adhesion molecule expression. Eur J Appl Physiol 86(2):135–141
Kuo CP, Wu CL, Ho HT, Chen C, Liu SI, Lu YT (2008) Detection of cytomegalovirus reactivation in cancer patients receiving chemotherapy. Clin Microbiol Infect 14(3):221–227
Larbi A, Franceschi C, Mazzatti D, Solana R, Wikby A, Pawelec G (2008) Aging of the immune system as a prognostic factor for human longevity. Physiology (Bethesda) 23:64–74
Larbi A, Pawelec G, Witkowski JM, Schipper HM, Derhovanessian E, Goldeck D et al (2009) Dramatic shifts in circulating CD4 but not CD8 T cell subsets in mild Alzheimer's disease. J Alzheimers Dis 17(1):91–103
Leo NA, Bonneau RH (2000) Mechanisms underlying chemical sympathectomy-induced suppression of herpes simplex virus-specific cytotoxic T lymphocyte activation and function. J Neuroimmunol 110(1–2):45–56
Lilleri D, Fornara C, Chiesa A, Caldera D, Alessandrino EP, Gerna G (2008) Human cytomegalovirus-specific CD4+and CD8+T-cell reconstitution in adult allogeneic hematopoietic stem cell transplant recipients and immune control of viral infection. Haematologica 93(2):248–256

Limaye AP, Boeckh M (2010) CMV in critically ill patients: pathogen or bystander? Rev Med Virol 20(6):372–379

Limaye AP, Kirby KA, Rubenfeld GD, Leisenring WM, Bulger EM, Neff MJ et al (2008) Cytomegalovirus reactivation in critically ill immunocompetent patients. J. Am. Med. Assoc. 300(4):413–422

Markovic-Plese S, Cortese I, Wandinger KP, McFarland HF, Martin R (2001) CD4 + CD28− costimulation-independent T cells in multiple sclerosis. J Clin Invest 108(8):1185–1194

Marmot M (2005) Social determinants of health inequalities. Lancet 365(9464):1099–1104

Marsland AL, Bachen EA, Cohen S, Rabin B, Manuck SB (2002) Stress, immune reactivity and susceptibility to infectious disease. Physiol Behav 77(4–5):711–716

Matalka KZ, Sidki A, Abdul-Malik SM, Thewaini A-J (2000) Academic stress—influence on Epstein-Barr virus and cytomegalovirus reactivation, cortisol, and prolactin. Lab Med 31(3):163–168

McDade TW, Stallings JF, Angold A, Costello EJ, Burleson M, Cacioppo JT et al (2000) Epstein-Barr virus antibodies in whole blood spots: a minimally invasive method for assessing an aspect of cell-mediated immunity. Psychosom Med 62(4):560–567

McEwen BS (1998) Protective and damaging effects of stress mediators. N Engl J Med 338(3):171–179

Mehta SK, Stowe RP, Feiveson AH, Tyring SK, Pierson DL (2000) Reactivation and shedding of cytomegalovirus in astronauts during spaceflight. J Infect Dis 182(6):1761–1764

Michaelis M, Doerr HW, Cinatl J (2009) The story of human cytomegalovirus and cancer: increasing evidence and open questions. Neoplasia 11(1):1–9

Miller GE, Freedland KE, Duntley S, Carney RM (2005) Relation of depressive symptoms to C-reactive protein and pathogen burden (cytomegalovirus, herpes simplex virus, Epstein-Barr virus) in patients with earlier acute coronary syndromes. Am J Cardiol 95(3):317–321

Miller GE, Stetler CA, Carney RM, Freedland KE, Banks WA (2002) Clinical depression and inflammatory risk markers for coronary heart disease. Am J Cardiol 90(12):1279–1283

Mills PJ, Berry CC, Dimsdale JE, Ziegler MG, Nelesen RA, Kennedy BP (1995) Lymphocyte subset redistribution in response to acute experimental stress: effects of gender, ethnicity, hypertension, and the sympathetic nervous system. Brain Behav Immun 9(1):61–69

Mills PJ, Karnik RS, Dillon E (1997) L-selectin expression affects T-cell circulation following isoproterenol infusion in humans. Brain Behav Immun 11(4):333–342

Monteiro J, Batliwalla F, Ostrer H, Gregersen PK (1996) Shortened telomeres in clonally expanded CD28−CD8+T cells imply a replicative history that is distinct from their CD28 + CD8+counterparts. J Immunol 156(10):3587–3590

Montminy M (1997) Transcriptional regulation by cyclic AMP. Annu Rev Biochem 66:807–822

Morita-Hoshi Y, Heike Y, Kawakami M, Sugita T, Miura O, Kim SW et al (2008) Functional analysis of cytomegalovirus-specific T lymphocytes compared to tetramer assay in patients undergoing hematopoietic stem cell transplantation. Bone Marrow Transplant 41(6):515–521

Moro-Garcia MA, Alonso-Arias R, Lopez-Vazquez A, Suarez-Garcia FM, Solano-Jaurrieta JJ, Baltar J, Lopez-Larrea C (2012) Relationship between functional ability in older people, immune system status, and intensity of response to CMV. Age 34:479–495

Moss P (2010) The emerging role of cytomegalovirus in driving immune senescence: a novel therapeutic opportunity for improving health in the elderly. Curr Opin Immunol 22(4):529–534

Nester EW, Anderson DG, Roberts Jr CE, Nester MT (2008) Microbiology: a human perspective, 6th edn. McGraw-Hill, New York

Nikolich-Zugich J (2008a) Ageing and life-long maintenance of T-cell subsets in the face of latent persistent infections. Nat Rev Immunol

Nikolich-Zugich J (2008b) Ageing and life-long maintenance of T-cell subsets in the face of latent persistent infections. Nat Rev Immunol 8(7):512–522

Nilsson PM (1996) Premature ageing: the link between psychosocial risk factors and disease. Med Hypotheses 47(1):39–42

Novak Z, Ross S, Patro R, Pati S, Kumbla R, Brice S et al (2008) Cytomegalovirus strain diversity in seropositive women. J Clin Microbiol 46(3):882–886
Nyklicek I, Bosch JA, Amerongen AV (2005) A generalized physiological hyperreactivity to acute stressors in hypertensives. Biol Psychol 70(1):44–51
Ogg GS, Jin X, Bonhoeffer S, Dunbar PR, Nowak MA, Monard S et al (1998) Quantitation of HIV-1-specific cytotoxic T lymphocytes and plasma load of viral RNA. Science 279(5359):2103–2106
Ozdemir E, St John LS, Gillespie G, Rowland-Jones S, Champlin RE, Molldrem JJ et al (2002) Cytomegalovirus reactivation following allogeneic stem cell transplantation is associated with the presence of dysfunctional antigen-specific CD8+T cells. Blood 100(10):3690–3697
Pawelec G (2006) Immunity and ageing in man. Exp Gerontol 41(12):1239–1242
Pawelec G, Akbar A, Caruso C, Effros R, Grubeck-Loebenstein B, Wikby A (2004) Is immunosenescence infectious? Trends Immunol 25(8):406–410
Pawelec G, Derhovanessian E (2011) Role of CMV in immune senescence. Virus Res 157:175–179
Pawelec G, Derhovanessian E, Larbi A, Strindhall J, Wikby A (2009) Cytomegalovirus and human immunosenescence. Rev Med Virol 19(1):47–56
Pawelec G, McElhaney JE, Aiello AE, Derhovanessian E (2012) The impact of CMV infection on survival in older humans. Curr Opin Immunol 24:507–511
Phillips AC, Carroll D, Khan N, Moss P (2008) Cytomegalovirus is associated with depression and anxiety in older adults. Brain Behav Immun 22(1):52–55
Pita-Lopez ML, Gayoso I, Delarosa O, Casado JG, Alonso C, Munoz-Gomariz E et al (2009) Effect of ageing on CMV-specific CD8 T cells from CMV seropositive healthy donors. Immun Ageing 6:11
Prelog M (2006) Aging of the immune system: a risk factor for autoimmunity? Autoimmun Rev 5(2):136–139
Prosch S, Docke WD, Reinke P, Volk HD, Kruger DH (1999) Human cytomegalovirus reactivation in bone-marrow-derived granulocyte/monocyte progenitor cells and mature monocytes. Intervirology 42(5–6):308–313
Prosch S, Wendt CE, Reinke P, Priemer C, Oppert M, Kruger DH et al (2000) A novel link between stress and human cytomegalovirus (HCMV) infection: sympathetic hyperactivity stimulates HCMV activation. Virology 272(2):357–365
Prösch S, Wendt CEC, Reinke P, Priemer C, Oppert M, Krüger DH et al (2000) A novel link between stress and human cytomegalovirus (HCMV) infection: sympathectic hyperactivity stimulates HCMV activation. Virology 272:357–365
Roberts ET, Haan MN, Dowd JB, Aiello AE (2010) Cytomegalovirus antibody levels, inflammation, and mortality among elderly Latinos over 9 years of follow-up. Am J Epidemiol 172(4):363–371
Ross S, Arora N, Novak Z, Fowler K, Britt W, Boppana S (2010) Cytomegalovirus reinfections in healthy seroimmune women. J Infect Dis 201(3):386–389
Sallusto F, Geginat J, Lanzavecchia A (2004) Central memory and effector memory T cell subsets: function, generation, and maintenance. Annu Rev Immunol 22:745–763
Sallusto F, Lenig D, Forster R, Lipp M, Lanzavecchia A (1999) Two subsets of memory T lymphocytes with distinct homing potentials and effector functions. Nature 401(6754):708–712
Sansoni P, Vescovini R, Fagnoni F, Biasini C, Zanni F, Zanlari L et al (2008) The immune system in extreme longevity. Exp Gerontol 43(2):61–65
Sarid O, Anson O, Yaari A, Margalith M (2001) Epstein-Barr virus specific salivary antibodies as related to stress caused by examinations. J Med Virol 64(2):149–156
Sarid O, Anson O, Yaari A, Margalith M (2004) Academic stress, immunological reaction, and academic performance among students of nursing and physiotherapy. Res Nurs Health 27(5):370–377
Sauce D, Larsen M, Fastenackels S, Duperrier A, Keller M, Grubeck-Loebenstein B et al (2009) Evidence of premature immune aging in patients thymectomized during early childhood. J Clin Invest 119(10):3070–3078
Saurwein-Teissl M, Lung TL, Marx F, Gschosser C, Asch E, Blasko I et al (2002) Lack of antibody production following immunization in old age: association with CD8(+)CD28(−) T cell clonal expansions and an imbalance in the production of Th1 and Th2 cytokines. J Immunol 168(11):5893–5899

Schmidt D, Goronzy JJ, Weyand CM (1996) CD4+CD7-CD28-T cells are expanded in rheumatoid arthritis and are characterized by autoreactivity. J Clin Invest 97(9):2027–2037

Schneiderman N, Ironson G, Siegel SD (2005) Stress and health: psychological, behavioral, and biological determinants. Annu Rev Clin Psychol 1:607–628

Segerstrom SC, Miller GE (2004) Psychological stress and the human immune system: a meta-analytic study of 30 years of inquiry. Psychol Bull 130(4):601–630

Sekut L, Champion BR, Page K, Menius JA Jr, Connolly KM (1995) Anti-inflammatory activity of salmeterol: down-regulation of cytokine production. Clin Exp Immunol 99(3):461–466

Shirtcliff EA, Coe CL, Pollak SD (2009) Early childhood stress is associated with elevated antibody levels to herpes simplex virus type 1. Proc Natl Acad Sci USA 106(8):2963–2967

Simon NM, Smoller JW, McNamara KL, Maser RS, Zalta AK, Pollack MH et al (2006) Telomere shortening and mood disorders: preliminary support for a chronic stress model of accelerated aging. Biol Psychiatry

Simpson RJ. (2010) Aging, persistent viral infections, and immunosenescence: can exercise "make space"? Exercise Sport Sci Rev 39(1):23–33

Soderberg-Naucler C (2006) Does cytomegalovirus play a causative role in the development of various inflammatory diseases and cancer? J Intern Med 259(3):219–246

Staras SA, Dollard SC, Radford KW, Flanders WD, Pass RF, Cannon MJ (2006) Seroprevalence of cytomegalovirus infection in the United States, 1988–1994. Clin Infect Dis 43(9):1143–1151

Stein J, Volk HD, Liebenthal C, Kruger DH, Prosch S (1993) Tumour necrosis factor alpha stimulates the activity of the human cytomegalovirus major immediate early enhancer/promoter in immature monocytic cells. J Gen Virol 74(Pt 11):2333–2338

Stenberg RM, Thomsen DR, Stinski MF (1984) Structural analysis of the major immediate early gene of human cytomegalovirus. J Virol 49(1):190–199

Stinski MF, Isomura H (2008) Role of the cytomegalovirus major immediate early enhancer in acute infection and reactivation from latency. Med Microbiol Immunol 197(2):223–231

Strandberg TE, Pitkala KH, Tilvis RS (2009) Cytomegalovirus antibody level and mortality among community-dwelling older adults with stable cardiovascular disease. J. Am. Med. Assoc. 301(4):380–382

Sun Z, Zhong W, Lu X, Shi B, Zhu Y, Chen L et al (2008) Association of Graves' disease and prevalence of circulating IFN-gamma-producing CD28(−) T cells. J Clin Immunol 28(5):464–472

Sutherland S, Bracken P, Wreghitt TG, O'Grady J, Calne RY, Williams R (1992) Donated organ as a source of cytomegalovirus in orthotopic liver transplantation. J Med Virol 37(3):170–173

Thompson WW, Shay DK, Weintraub E, Brammer L, Cox N, Anderson LJ et al (2003) Mortality associated with influenza and respiratory syncytial virus in the United States. J. Am. Med. Assoc. 289(2):179–186

Tong CY, Bakran A, Williams H, Cuevas LE, Peiris JS, Hart CA (2001) Association of tumour necrosis factor alpha and interleukin 6 levels with cytomegalovirus DNA detection and disease after renal transplantation. J Med Virol 64(1):29–34

Torfadottir H, Freysdottir J, Skaftadottir I, Haraldsson A, Sigfusson G, Ogmundsdottir HM (2006) Evidence for extrathymic T cell maturation after thymectomy in infancy. Clin Exp Immunol 145(3):407–412

Toro AI, Ossa J (1996) PCR activity of CMV in healthy CMV-seropositive individuals: does latency need redefinition? Res Virol 147(4):233–238

Trzonkowski P, Mysliwska J, Pawelec G, Mysliwski A (2009) From bench to bedside and back: the SENIEUR Protocol and the efficacy of influenza vaccination in the elderly. Biogerontology 10(1):83–94

Turner JE, Aldred S, Witard OC, Drayson MT, Moss PM, Bosch JA (2010) Latent cytomegalovirus infection amplifies CD8 T-lymphocyte mobilisation and egress in response to exercise. Brain Behav Immun 24(8):1362–1370

Uchino BN, Cacioppo JT, Kiecolt-Glaser JK (1996) The relationship between social support and physiological processes: a review with emphasis on underlying mechanisms and implications for health. Psychol Bull 119(3):488–531.

Uddin M, Aiello AE, Wildman DE, Koenen KC, Pawelec G, de Los Santos R et al (2010) Epigenetic and immune function profiles associated with posttraumatic stress disorder. Proc Natl Acad Sci USA 107(20):9470–9475

van Baarle D, Kostense S, Hovenkamp E, Ogg G, Nanlohy N, Callan MF et al (2002) Lack of Epstein-Barr virus and HIV-specific CD27− CD8+T cells is associated with progression to viral disease in HIV-infection. AIDS 16(15):2001–2011

van de Berg PJ, Griffiths SJ, Yong SL, Macaulay R, Bemelman FJ, Jackson S et al (2010) Cytomegalovirus infection reduces telomere length of the circulating T cell pool. J Immunol 184(7):3417–3423

van der Meer JT, Drew WL, Bowden RA, Galasso GJ, Griffiths PD, Jabs DA et al (1996) Summary of the international consensus symposium on advances in the diagnosis, treatment and prophylaxis and cytomegalovirus infection. Antiviral Res 32(3):119–140

van Lier RA, ten Berge IJ, Gamadia LE (2003) Human CD8(+) T-cell differentiation in response to viruses. Nat Rev Immunol 3(12):931–939

van der Ven A, van Diest R, Hamulyák K, Maes M, Bruggeman C, Appels A (2003) Herpes viruses, cytokines, and altered hemostasis in vital exhaustion. Psychosom Med 65(2):194–200

Weinberger B, Lazuardi L, Weiskirchner I, Keller M, Neuner C, Fischer KH et al (2007) Healthy aging and latent infection with CMV lead to distinct changes in CD8(+) and CD4(+) T-cell subsets in the elderly. [Article]. Human Immunol 68(2):86–90

Widmann T, Sester U, Gartner BC, Schubert J, Pfreundschuh M, Kohler H et al (2008) Levels of CMV specific CD4 T cells are dynamic and correlate with CMV viremia after allogeneic stem cell transplantation. PLoS One 3(11):e3634

Wikby A, Ferguson F, Forsey R, Thompson J, Strindhall J, Lofgren S et al (2005) An immune risk phenotype, cognitive impairment, and survival in very late life: impact of allostatic load in Swedish octogenarian and nonagenarian humans. J Gerontol A Biol Sci Med Sci 60(5):556–565

Wikby A, Maxson P, Olsson J, Johansson B, Ferguson FG (1998) Changes in CD8 and CD4 lymphocyte subsets, T cell proliferation responses and non-survival in the very old: the Swedish longitudinal OCTO-immune study. Mech Ageing Dev 102(2–3):187–198

Wikby A, Nilsson BO, Forsey R, Thompson J, Strindhall J, Lofgren S et al (2006) The immune risk phenotype is associated with IL-6 in the terminal decline stage: findings from the Swedish NONA immune longitudinal study of very late life functioning. Mech Ageing Dev 127(8):695–704

Woodland DL, Blackman, MA (2006) Immunity and age: living in the past? Trends Immunol 27(7):303–307

Wreghitt TG, Teare EL, Sule O, Devi R, Rice P (2003) Cytomegalovirus infection in immunocompetent patients. Clin Infect Dis 37(12):1603–1606

Yoshikawa TT (2000) Epidemiology and unique aspects of aging and infectious diseases. Clin Infect Dis 30(6):931–933

Zanghellini F, Boppana SB, Emery VC, Griffiths PD, Pass RF (1999) Asymptomatic primary cytomegalovirus infection: virologic and immunologic features. J Infect Dis 180(3):702–707

Zanni F, Vescovini R, Biasini C, Fagnoni F, Zanlari L, Telera A et al (2003) Marked increase with age of type 1 cytokines within memory and effector/cytotoxic CD8+T cells in humans: a contribution to understand the relationship between inflammation and immunosenescence. Exp Gerontol 38(9):981–987

Zorrilla EP, Luborsky L, McKay JR, Rosenthal R, Houldin A, Tax A et al (2001) The relationship of depression and stressors to immunological assays: a meta-analytic review. Brain Behav Immun 15(3):199–226

Chapter 7
Psychosocial Factors and Telomere Length in Peripheral Blood

Christine G. Parks and Lisa A. DeRoo

7.1 Introduction

Telomeres are repetitive DNA sequences that, together with protein complexes, cap and protect eukaryotic chromosomes (Blackburn 2005b). Telomeres are shortened when cells undergo mitosis due to the inability of DNA polymerase to fully extend chromosome ends during cell division (Von Zglinicki 2003). Telomeres prevent the ends of the chromosomes from being recognized as a double stranded DNA break, potentially triggering cell cycle arrest (Jain and Cooper 2010). Critically short telomeres can lead to cellular senescence, when replication is no longer possible, and genomic instability, for example, when chromosomes fuse end to end (Blackburn 2005a; Shay and Wright 2005). Loss of telomere length (TL) can be countered by the enzyme telomerase, which extends TL by adding DNA sequence repeats to the ends of the chromosomes, a process that is carefully regulated throughout the lifespan of the organism (Bekaert et al. 2004; Hathcock et al. 2005). Although telomerase expression is absent or low in most adult somatic cells, its expression in lymphocytes is necessary for the extensive rounds of cell division that characterize the adaptive immune response (Weng et al. 1998). Telomerase expression also helps maintain stem cell function in rapidly dividing tissues, including hematopoetic stem cells and precursors (Lansdorp 2005).

Average leukocyte TL provides a cumulative marker of cellular aging integrating the effects of multiple pathways, including DNA damage due to oxidative stress (Opresko et al. 2005) and the replication history of lymphocytes and hematopoetic stem cells (Aubert and Lansdorp 2008; Lansdorp 2009). Most studies have measured leukocyte TL in DNA obtained from whole blood or peripheral mononuclear cells (PBMC). The development of a high-throughput polymerase chain reaction (PCR)-based assay for average TL (Cawthon 2002) has led to the rapid growth of telomere research. Other TL assays include southern blot of genomic DNA (Kimura et al.

C. G. Parks (✉) · L. A. DeRoo
Epidemiology Branch, A3-05, NIEHS, PO Box 12233, Research Triangle Park, NC 27599, USA
e-mail: Parks1@mail.nih.gov

J. A. Bosch et al. (eds.), *Immunosenescence*,
DOI 10.1007/978-1-4614-4776-4_7,

2010b) and assessment of TL in cell subsets, either directly using fluorescence in situ hybridization (flow FISH) or PCR-based methods on cells sorted by flow cytometry (Lin et al. 2010a; Rufer et al. 1998). Assay type generally does not affect comparability of results of most association studies, at least when using relative measures, but differences in cell types assayed may influence interpretations (discussed later). Although we refer readers to the different source tissues used across studies, not all indicated whether leukocyte DNA was obtained from whole blood or PBMCs. When not specified, we assumed leukocytes or white blood cells were obtained from whole blood.

Recent studies have generated interest in the possibility that leukocyte TL may be a useful biomarker for studying immune aging and associated outcomes. Although a causal role has not been established, shorter TL has been associated with mortality and the risk of chronic diseases and aging-related outcomes including cardiovascular disease, diabetes, and some cancers (Brouilette et al. 2007; Kim et al. 2009; Kimura et al. 2008; McGrath et al. 2007; Tentolouris et al. 2007), many of which involve potential inflammatory or infectious etiologies. Shorter TL has also been associated with increased mortality from infections (Cawthon et al. 2003). Together, findings provide compelling rationale for investigating leukocyte TL as a marker in studies of immune aging and associated outcomes.

A growing body of research also suggests leukocyte TL may be a biologically relevant intermediary between psychological and physiological stress and aging. Here we review the clinical and epidemiologic literature on psychosocial stressors and leukocyte TL, incorporating findings from basic research and experimental studies to provide context. We also discuss considerations in the use of TL as a biomarker in research on stress and immune aging. The wide distribution of TL in the general population generally implies a need for large sample sizes to detect meaningful differences (Aviv et al. 2006); our review highlights the observation that many initial studies on stress and TL derive from clinical studies with relatively small sample sizes and limited generalizability. Furthermore, most studies to date are cross-sectional and so can only be used to infer stress-related TL shortening within individuals.

7.1.1 Chronic and Perceived Stress

Table 7.1 summarizes studies of chronic and perceived stress. Early clinical studies examined TL in the context of caregiving as a chronic major stressor (Damjanovic et al. 2007; Epel et al. 2004). In their small, but carefully designed, landmark study, Epel et al. (2004) found shorter leukocyte TL was associated with years of caregiving in mothers of a chronically ill child, and also with higher perceived stress in both caregivers and control mothers. They hypothesized that TL shortening resulted not only from the more extensive stress in caregivers, but also from stress at normative levels. In the same study sample, shorter TL was also associated with higher nocturnal urinary catecholamines and cortisol.

Table 7.1 Summary of studies on perceived and chronic stress in relation to telomere length

Study	Population/sample (*N*); demographics	Design/methodology* covariates	Findings
Humphreys et al. (2011)	Western United States *N* = 66 women with past history of intimate partner violence (mean age 35, 33 % white) and 46 without (mean age 29, 44 % white) Not including those currently in violent partnership	Cross-sectional clinical study, volunteer sample PBMC; qPCR History of past abuse as adult (age and duration), current depression, lifetime trauma exposure, social stress inventory Age, BMI, education, current financial, employment, marital and parenting status	Formerly abused women had significantly shorter TL; having children and years in abusive relationship both significantly associated with shorter TL (not adjusted for age) Association of BMI and shorter TL only seen in formerly abused women ($r = -0.187$), with significant interaction of BMI-TL association by abuse status ($p = 0.045$)
Parks et al. (2011)	United States, The Sister Study *N* = 647 sisters of women with breast cancer (see Parks et al. 2009) 83 % white, 7 % African-American Ages 35 to 74 (45 % ages 55 and older)	Cross-sectional study, national volunteer sample Leukocytes; qPCR Age, race, BMI, smoking education, childbearing history, cardiovascular disease/diabetes or poor self-reported health First morning urinary cortisol and catecholamines	Significantly shorter TL for current full-time work (versus part-time or not working); long-term fulltime schedule associated with shorter TL adjusting for current work hours Current work schedule associated with shortest TL in women with high perceived stress and epinephrine levels Association not explained by covariates, but differences suggested effects of schedule on TL may be buffered by higher SES or later childbearing
Parks et al. (2009)	United States, The Sister Study *N* = 647 sisters of women with breast cancer sampled enriched for higher than average perceived stress, non-white race, smoking 83 % white, 7 % African-American Ages 35 to 74 (45 % ages 55 and older)	Cross-sectional study, national volunteer sample Leukocytes; qPCR Perceived Stress Scale (4-item), recent major losses (divorce, death of first degree relative), major life losses; catecholamines and cortisol in first morning urines Age, race, BMI, education, marital status, smoking, ever diagnosed with cardiovascular disease or diabetes, clinical depression, self-reported health	Shorter TL with moderately elevated perceived stress; significantly shorter TL seen for higher than average stress levels in women over age 55, those with recent major losses and higher urinary catecholamines Shorter TL also associated with obesity and current smoking, but no significant association of TL with urinary stress biomarkers, major life losses, education, marital status, depression

Table 7.1 (continued)

Study	Population/sample (N); demographics	Design/methodology* covariates	Findings
Damjanovic et al. (2007)	OH, USA $N = 41$ Alzheimer caregivers, 41 age and sex-matched controls 75 % female Mean age 65 ± 1	Cross-sectional clinical study, volunteer sample Leukocytes; Southern blot Depressive symptoms Flow cytometry (cell subsets), T-cell stimulation and proliferation, cytokines, telomerase activity	Caregivers had more depressive symptoms and lower T-cell proliferation, greater TNF and Il-10 Caregivers had significantly shorter TL (6.2 vs. 6.4×10^6), and higher telomerase activity
Epel et al. (2006)	CA, USA $N = 44$ mothers of chronically ill child, 22 mothers of healthy children (see Epel et al. 2004) Ages 20 to 50, premenopausal	Cross-sectional clinical study, volunteer sample PBMC; qPCR Telomerase activity, telomere length, overnight 12-h urinary cortisol and catecholamines ($N = 41$) Perceived stress scale; acute stress arousal (Trier Social Stress Test); adjusted for age, BMI	Shorter TL associated with higher urinary catecholamines and cortisol, but not education, negative mood, or difference in heart rate variability to acute stress arousal Lower telomerase activity associated with higher epinephrine and norepinephrine, and less heart rate variability to stress arousal Lower telomerase also correlated with trait negative mood, BMI, current smoking, and lower education Shorter TL associated with higher perceived stress in caregivers and control mothers; also associated with more years of caregiving
Epel et al. (2004)	CA, USA $N = 39$ mothers of chronically ill child, 19 mothers of healthy children Ages 20 to 50, mean age 38 ± 6.5; premenopausal	Cross-sectional clinical study, volunteer sample PBMC; qPCR Telomerase activity, oxidative stress index (urinary isoprostanes/vitamin E) Perceived stress scale; age, BMI, smoking, vitamin use, child's age	Lower telomerase activity associated with higher perceived stress and more years of caregiving Higher oxidative stress associated with higher perceived stress and more years caregiving

Older caregivers of Alzheimer patients had shorter TL than age and sex-matched controls (Damjanovic et al. 2007). Caregivers were more likely to report depressive symptoms and had lower T-cell proliferation and higher inflammatory cytokine levels in vitro after stimulation with anti-CD3/CD28. These findings suggested a link between caregiver stress and immune aging, although it was not reported whether these factors were also associated with TL. There have been no large, population-based studies examining caregiving stress and TL. Larger studies would provide greater statistical power to evaluate other factors that may increase susceptibility to the adverse physiological or psychological effects of caregiving stress (e.g., childhood adversities (Kiecolt-Glaser et al. 2010)) or buffer the effects of the physiological stress response (e.g., physical activity (Puterman et al. 2010)).

We examined the association between perceived stress and TL in a subsample of a national cohort of women who have a sister with breast cancer (Parks et al. (2009); women reporting higher than average perceived stress (i.e., a median of 2 out of 16 possible points in this sample) had somewhat shorter TL that was not explained by health or lifestyle covariates. Although no dose–response effect was seen at the highest stress levels, stratified analyses showed significant associations of perceived stress and shorter telomeres in some subgroups, including women with higher urinary catecholamines (epinephrine and norepinephrine), those experiencing recent major stressors/losses (death of close relative, divorce, or separation), and women aged 55 years and older.

In the same sample, we evaluated the association of TL with longer work hours, a potential chronic stressor (van der Hulst 2003). Compared to women not currently working, those working full- or over-time had significantly shorter TL, differences were most apparent among women with higher perceived stress and greater overnight urinary epinephrine levels, reflecting interactions of epinephrine levels with both work schedule and perceived stress (Parks et al. (2011). Taking into account current schedule, shorter TL was also seen for more years of full-time work and working in multiple jobs. Schedule-related TL differences were most apparent in working women with more births or births at an early age, and those with moderate to lower education and income levels. These findings provide further evidence of an association of chronic stress with TL and suggest a link to research on socio-economic health disparities and the role of employment in women's health.

Although few in number, these studies suggest that exposure to perceived and chronic stress may contribute to shorter TL. Larger, population-based studies are needed, including a diverse range of chronic stressors. Shorter TL has been described in women with a history of intimate partner violence (Humphreys et al. 2011), a substantial and widespread chronic stressor. Clinical studies often have limited statistical power to disentangle the effects of age, years of stress exposures, and other participant characteristics. However, replicating the findings of small clinical studies may be challenging in larger epidemiologic studies due to the heterogeneity of chronic stressors experienced across populations, difficulties in assessing past stress exposures, and variability in the individual stress response.

7.1.2 *Psychosocial, Demographic, and Lifestyle Factors*

7.1.2.1 Emotional Distress, Personality, and Developmental Exposures

Individual variation response to external stressors may be influenced by heritable traits and developmental exposures (Hopwood et al. 2011). Table 7.2 summarizes studies on TL in relation to mental health, personality, and early childhood exposures to adversity or abuse. In several studies, shorter TL was associated with emotional distress phenotypes such as depression or anxiety, which may be a consequence or marker of increased susceptibility to acute or chronic stress. Evidence primarily comes from clinical studies describing shorter TL in patients with major depression (Hoen et al. 2011; Lung et al. 2007; Simon et al. 2006; Wolkowitz et al. 2011a). The effects of depression on TL may require long-term exposure or more severe symptoms (Wolkowitz et al. 2010). One small clinical study showed no significant effect of the duration or severity of major depression on TL (Hartmann et al. 2010), while another indicated longer duration of symptoms was associated with shorter TL in patients with major depression (Wolkowitz et al. 2011). In a large study of coronary heart disease patients, depression was associated with shorter TL at baseline but did not predict 5-year TL shortening after adjusting for baseline TL (Hoen et al. 2011); interpretation of these longitudinal findings is limited, given that past history, duration, and subsequent 5-year course of depression and anxiety symptoms were not described. A large, population-based study also examined TL in relation to current diagnosis of anxiety, which may be triggered by stress or co-occur with depression (Kananen et al. 2010). Although there was no overall association, shorter TL was seen in older individuals with current anxiety, suggesting duration or timing of symptoms may be relevant.

Genetic variants or personality traits may underlie some pathways relating emotional distress and TL. For example, a population-based study reported shorter TL associated with a specific genetic variant for monoamine oxidase (MAO-A) (Lung et al. 2005, 2007) associated with risk of depression in some studies (Rivera et al. 2009; Berry et al. 1994); in a later analysis the authors noted the possibility that this variant might mediate the relationship between TL and major depression (Lung et al. 2007). In healthy post-menopausal women, primarily caregivers, pessimism was strongly correlated with shorter TL and higher levels of the pro-inflammatory cytokine, interleukin-6 (O'Donovan et al. 2009), while optimism was not independently associated with either measure. Shorter TL was associated with lower perceived mental health scores but not depressive symptoms or type D personality in a study of heart failure patients (Huzen et al. 2010).

Response to stress in adulthood also may be influenced by experiences in early life. Childhood abuse and adversity has been associated with lasting alterations in the physiological stress response, with evidence from numerous studies of a "biologic" imprint, for example, abuse-related epigenetic modifications in the promoter of glucocorticoid or serotonin receptor genes (Beach et al. 2010; McGowan et al. 2009). Early life exposures may also directly affect TL attrition during childhood,

Table 7.2 Summary of studies on emotional distress, personality, and developmental exposures

Study	Population/sample (*N*); demographics	Design/methodology* Covariates	Findings
Hoen et al. (2011)	Bay Area, CA, USA Patients with stable coronary heart disease; $N = 206$ with current depression (mean age 62, SD 11, 69 % male); $N = 746$ without depression (mean age = 68, SD 11, 85 % male) 60 % white	Clinic-based prospective cohort, $N = 608$ with 5-year follow-up data Leukocytes; qPCR Computerized diagnostic interview schedule IV (major depressive disorder—past month), patient health questionnaire (symptom severity) Age, sex, ethnicity, education, smoking, alcohol, physical activity, BMI, comorbidities, blood pressure, cholesterol medication, anxiety (hospital anxiety scale)	Depressed patients had significantly shorter TL, adjusting for age, sex, BMI, physical inactivity, smoking, statin/antidepressant use; association similar but no longer significant after also adjusting for anxiety Loss to follow-up not associated with age or TL Depression associated with decreased odds of 5-year TL shortening (>10 %), but association not significant after adjusting for covariates and baseline TL
Wolkowitz et al. (2011)	Bay Area, CA, USA $N = 18$ outpatients with major depression $N = 17$ age-, sex-, and ethnicity-matched controls 65 % female, 71 % Caucasian Mean age 36.6 (±11.8) years	Cross-sectional clinical study, volunteer sample Leukocytes; qPCR Structured clinical Interview for DSM-IV-TR (excluded patients with bipolar, psychosis, post-traumatic stress disorder; controls excluded for depression, alcohol, or substance abuse) Hamilton depression rating scale, lifetime duration depression, anti-depressant treatment history form Covariates: age, sex, BMI, smoking	Leukocyte TL did not differ overall in depressed cases vs. controls; but depression duration was inversely associated with TL in cases. Cases with longer than average (9 years) duration had shorter TL vs. controls Depression severity not related to TL, but years of untreated depression correlated with shorter TL Shorter TL was associated with oxidative stress ratio (F2-isoprostane/Vit C) in the overall sample (cases and controls combined), and with IL-6 levels in cases
Kiecolt-Glaser et al. (2011)	United States Overall sample: $N = 132$ (58 dementia family caretakers, 74 demographically similar controls) Telomere sample: $N = 82$ 72 % women; 7.5 % non-white median age 65.9 years	Cross-sectional PBMCs, Southern blot Childhood Abuse: emotional, physical, sexual, and emotional and physical neglect Childhood adversity: death of parent, severe parental marital problems, severe mental illness in family, alcohol abuse in family, lack of at least one close relationship with an adult Age, sex, BMI, caregiving status	Childhood adversity was significantly associated with TL; association persisted after adjustment for current weekly physical activity, BMI, and sleep Childhood abuse was not significantly associated with TL

Table 7.2 (continued)

Study	Population/sample (N); demographics	Design/methodology* Covariates	Findings
DeRoo et al. (2010)	United States, The Sister Study $N = 576$ women (see Parks et al. (2009) 85 % non-Hispanic whites, 6 % black, 9 % other	Cross-sectional, volunteer national sample Whole blood leukocytes, qPCR Traumatic experiences before age 18: physical abuse, sexual abuse, emotional abuse, witnessing an attack, witnessing someone close attack a family member, death of parent, family substance abuse, family mental illness, feeling unsafe some or all of the time in childhood neighborhood Age, race/ethnicity, father's age at birth childhood family income	Shorter average TL was observed in women reporting physical abuse, emotional abuse or feeling unsafe in neighborhood Significant trend in shorter TL with increasing numbers of experiences reported
Glass et al. (2010)	Twins UK Cohort $N = 123$ exposed to childhood abuse and 1,751 controls	Cross-sectional, twin registry Leukocytes; southern blot Questionnaires on childhood maltreatment (sexual, physical or emotional) and abuse at any time	No significant difference in TL for persons experiencing abuse in childhood or at any time
Hartmann et al. (2010)	Germany $N = 54$ patients with major depression (61 % female), 20 healthy age-matched controls (45 % female) Ages 19–75 years; average 49.5 (sd 14)	Cross-sectional Age, gender, smoking, BMI clinical study, hospital-based sampling Leukocytes; Southern blot Antidepressant medications (total antidepressant dose), electroconvulsive therapy Age, sex, smoking, duration, history of hospitalizations, Hamilton depression score	Patients had significantly shorter TL than controls No association of treatment dose/type, duration of illness, hospitalizations Similar age-related TL differences in patients and controls, and no gender or smoking TL differences observed
Kananen et al. (2010)	Finland $N = 321$ anxiety disorder diagnosis (62 % female), 653 matched controls (63 % female) Ages 30–87 (mean age 49.8, SD 12.7)	Cross-sectional, population health survey Leukocytes; qPCR Current anxiety disorder and subthreshold cases identified by standardized diagnostic interview Age, sex, district, employment, depression, alcohol use, smoking, BMI, physical activity, various cardiovascular risk factors/diabetes [childhood adversity, current psychological distress—general health questionnaire, GHQ12]	No significant overall TL differences in cases vs. non-cases; but older cases (ages > 47) had significantly shorter TL than controls Significant associations of TL with smoking, physical activity, number of childhood adversities No TL association with demographic factors, current stress, depression, alcohol use disorder, recent psychiatric medication use, BMI and cardiovascular risk factors/diabetes

Table 7.2 (continued)

Study	Population/sample (N); demographics	Design/methodology* Covariates	Findings
Huzen et al. (2010)	Netherlands $N = 890$ patients with chronic heart failure (39 % female) Mean age 73 (inter-quartile range 64–79)	Cross-sectional, hospital based sampling Leukocytes; qPCR Perceived mental health (RAND-36); depression symptoms (CES-D), type D personality (DS13) Age, sex, severity of heart failure, COPD, diabetes, history of stroke	Mean TL shorter in patients in lowest tertile of mental health domain (RAND-36) than second and first tertile Not confounded by covariates examined No differences in TL by depression symptom score, or type D personality
O'Donovan et al. (2009)	Bay Area, CA, USA $N = 36$ women (23 dementia caregivers, 13 controls) Post-menopausal Ages 51–79 years (mean 60.7, SD 6.7)	Cross-sectional clinical study, volunteer sample Interleukin-6 Leukocytes; qPCR Optimism/pessimism scale (LOT-R), Perceived stress scale (10 item), Neuroticism (Big Five Inventory) Physical activity, insomnia, BMI	Among caregivers and controls combined, pessimism was significantly associated with shorter TL Shorter TL associated with higher IL-6 Associations persisted adjusting for age, caregiver status, and optimism score
Tyrka et al. (2009)	USA, RI $N = 31$ (21 participants with no history of child maltreatment and 10 history of maltreatment) 22 men, 9 women 68 % Caucasian, 10 % black, 10 % Asian, 12 % other Ages 18 to 64 years Exclusions: current/past major AXIS I psychiatric disorder	Cross-sectional, volunteer sample from study of stress reactivity and psychiatric symptoms Leukocytes; qPCR Childhood Trauma Questionnaire (28-items): physical, emotional and sexual abuse, emotional and physical neglect	Maltreatment group had significantly shorter TL
Lung et al. (2007)	Southern Taiwan $N = 253$ patients with major depression (65 % female), and 411 population-based controls (57 % female) Mean age 45 (SD 15)	Cross-sectional survey, hospital-based case sampling Leukocytes; Southern Blot Major depression screened by in-person interview (DSM-IIIR, Chinese version) MAO genotype, age, and sex	Significantly shorter TL in depression patients versus controls; and significantly sorter TL associated with low activity MAOA genotypes among both patients and controls Model including both patient status and MAOA genotype suggested MAOA/TL association was mediated or confounded by depression diagnosis
Simon et al. (2006)	MA, USA $N = 44$ mood disorder patients (48 % female), 44 age-matched healthy volunteer controls (43 % female) Mean age 50.8 (SD 8)	Cross-sectional, clinical study Leukocytes; Southern blot Major depression ($n = 15$), bipolar disorder ± anxiety ($n = 14$ and $n = 15$) by DSM-IV Age, sex, smoking	TL was significantly shorter in mood disorder patients, and this difference was not changed in models adjusting for covariates

which may be a period of greater vulnerability due to greater leukocyte turnover and TL shortening resulting from rapid growth and development (Baerlocher et al. 2007; Rufer et al. 1999). Several cross-sectional studies in adults have examined self-reported childhood abuse or adversity and TL with mixed findings. One small study found shorter TL among 10 adults reporting moderate-to-severe maltreatment in childhood compared with 21 reporting none (Tyrka et al. 2010), while a larger study found no significant differences in TL for persons reporting abuse in childhood or any time (Glass et al. 2010). These studies used very different participant samples and methods to assess abuse. Tyrka and colleagues used the 28-item version of the Childhood Trauma Questionnaire, whereas Glass and colleagues assessed maltreatment on two occasions and considered subjects exposed to abuse only if they answered consistently in both questionnaires. A small study in caregivers suggested early childhood adversity, but not abuse, was associated with shorter adult TL (Kiecolt-Glaser et al. 2010). We examined abuse and other traumatic experiences in early life (before 18) in relation to adult TL in participants who completed a trauma questionnaire in the Sister Study (DeRoo et al. 2010). Shorter TL was observed in women reporting physical abuse, emotional abuse, or feeling unsafe where they lived, and there was a significant trend in shorter TL with increasing numbers of types of experiences. Contrary to some earlier studies, our results suggest that trauma, abuse, and chronic stressors in childhood may influence leukocyte aging.

In sum, research suggests shorter TL may be associated with emotional distress, pessimism, and early childhood experiences of adversity or abuse. Whether the association of TL with early life exposures are mediated by later life experiences or stress response phenotypes has not been investigated. Larger studies are needed that consider past stress exposures and mental health history, ongoing chronic and acute stress exposures, medication use, health behaviors, as well as the social and demographic context.

7.1.2.2 Socio-economic, Racial/Ethnic and Demographic Factors

Measuring TL may also provide a way to explore the contribution of psychosocial stress to socio-economic and racial/ethnic health disparities. Lower socio-economic status (SES) and racial/ethnic minority status have been independently associated with poorer health and mortality, and could be associated with increased exposure or susceptibility to external stressors (Fig. 7.1) (Lantz et al. 2005; Williams and Jackson 2005). Studying the effects of SES on health is challenging because social position is complex and multidimensional, encompassing economic resources, occupational status, and education level, and may operate at the individual, family or community levels across the lifespan (Braveman et al. 2005). Stress and strain may also arise from social and internal experiences of racism or discrimination, associated community-level exposures, or lack of access to support and resources (Myers 2009). Similar to other forms of adversity, childhood exposures may be the most relevant for examining effects of socio-economic factors on adult health and longevity (Schwartz et al. 1995).

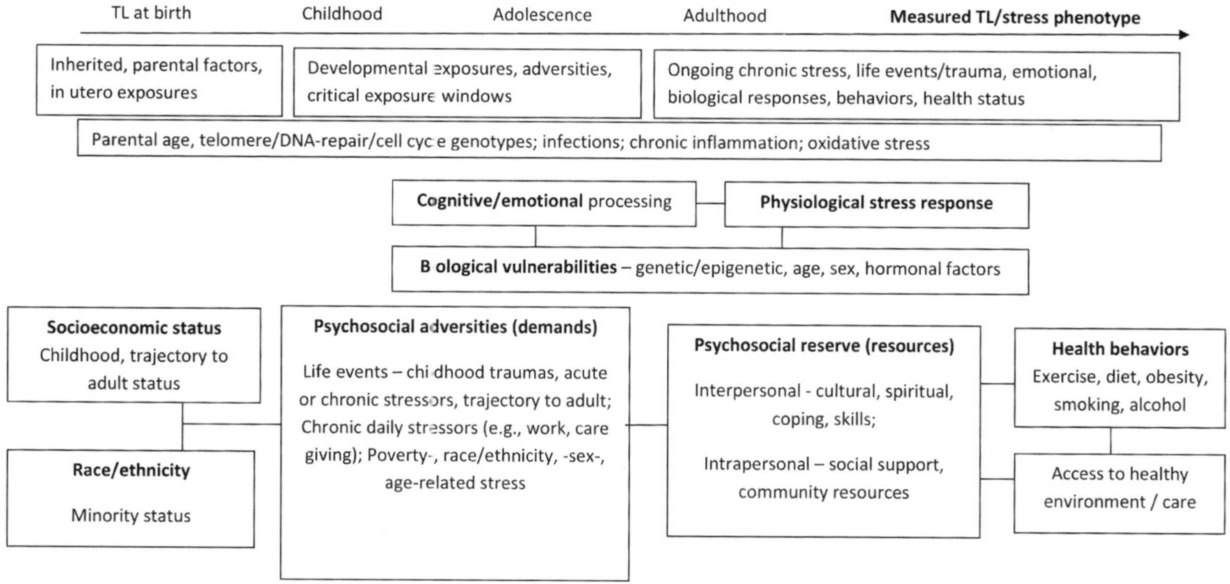

Fig. 7.1 Assessment of psychosocial stressors TL throughout the life course. SES and Race/ethnicity independently and synergistically contribute to psychosocial adversities and reserve; response to stressors can be mitigated by psychosocial reserve. Other modifying factors may include health behaviors or access to a healthy environment, both of which are impacted by SES and minority status. Resulting stress phenotype can be viewed as a product of these external factors in the context of biological vulnerabilities, cognitive/emotional factors, and the physiological stress response; other less direct influences may derive from stress-related genetic variants, acute and chronic infections, chronic inflammation, and oxidative stress

There is a growing literature on socio-economic factors and TL (Table 7.3). An early study (Cherkas et al. 2006) reported shorter TL in twins with manual compared to non-manual occupations. The association was attenuated after adjusting for body mass index (BMI), smoking, and exercise, suggesting that socio-economic factors and lifestyle characteristics may operate along a shared pathway. Unlike the enthusiasm generated by initial perceived stress and TL research, this paper prompted multiple responses urging cautious interpretation. Concerns included the lack of causal mechanisms and unknown direction of causality (Hornsby 2006), the heterogeneity of leukocyte populations studied (Lansdorp 2006), and the need for a life course perspective (Kuh 2006). Several subsequent studies using various measures of adult social position did not replicate the earlier finding (Harris et al. 2006; Kananen et al. 2010; Woo et al. 2010). TL was positively associated with greater educational attainment, but not with current household income or grade of employment, in a UK study of healthy adults aged 53–76 years (Steptoe et al. 2011). The authors proposed that educational attainment is a better measure of long-term socio-economic trajectory than income or employment and that education stimulates the development of problem-solving and coping skills that benefit health over and above increased income, resources or social status. In a study of men at high risk of coronary heart disease (Batty et al. 2009), researchers found unemployment was associated with shorter TL; interpretation was complicated by the lack of trend of TL with employment level and the heterogeneity of the unemployed group, which also included non-workers who were financially secure.

Few studies have examined childhood SES and TL. In a sample of older men and women from the United Kingdom, TL was unrelated to socio-economic trajectory, based on social status measured at birth, age 25, and age 50 (Adams et al. 2007). A large, population-based Finnish study observed parental unemployment in childhood was associated with shorter TL, but adult education and employment were not (Kananen et al. 2010). In a sample of Sister Study participants, we found shorter TL among women reporting childhood food deprivation (times when their family did not have enough to eat) and lower adult education in the household at age 13, with a significant trend in shorter TL for the number of lower childhood SES factors (DeRoo et al. 2009). Although the evidence is scant, these studies suggest that lower SES in childhood may have more effect on TL than adult exposures. Further research taking into account more comprehensive social position characteristics across the lifespan is needed to better understand this relationship.

Several studies have examined racial/ethnic differences in TL, mainly comparing minorities to whites in majority white populations, making it difficult to understand the possible interplay between genetic differences and environmental factors such as increased stress and strain related to minority status. Three studies described longer leukocyte TL in African Americans compared to whites (Aviv et al. 2009; Hunt et al. 2008; Zhu et al. 2010). A possible explanation is "ethnic neutropenia" (lower neutrophil count in persons of African descent and ethnic groups in the Middle East), perhaps due to fewer replications of hematopoetic stem cells and progenitor cells; however, no association of TL with race-related differences in neutrophils was seen in the other studies to consider this possibility (Hunt et al. 2008). Two studies

Table 7.3 Summary of studies on TL and social position

Study	Population/sample (n); demographics	Design/Methodology*Covariates	Findings
Steptoe et al. (2011)	Whitehall II epidemiological cohort, UK $N = 277$ men and 229 women Ages 53–76	Cross-sectional Peripheral blood mononuclear cells, qPCR Educational attainment, current household income, current or most recent employment grade attained in the British civil service Age, sex, current paid employment, systolic BP, glycated hemoglobin, HDL-cholesterol, smoking, BMI, physical activity	Educational attainment was positively associated with TL in both unadjusted and adjusted models No TL association with either current household income or employment grade
Batty et al. (2010)	West of Scotland Primary Prevention Study, Scotland $N = 1{,}542$ men Ages 45–64	Cross-sectional case-control study, population-based screening of men at risk of coronary heart disease based on lipid profiles Leukocytes, qPCR Four SES measures: education; employment status; height (measure of childhood deprivation); carstairs' index (at neighborhood level) Age, smoking, BMI, alcohol, existing illness, statin treatment, case-control status	Employment status was related to TL in both unadjusted and adjusted models Unemployed men had significantly shorter TL than employed
Kananen et al. (2010)	Finland $N = 321$ anxiety disorder cases and 653 controls matched on sex, age, and hospital district 64 % women Ages 30–87 years	Cross-sectional, population health survey Leukocytes; qPCR Adult sociodemographic factors: age, sex, hospital district, education, employment Childhood adversity: longer term financial difficulties, parental unemployment, parental disease/disability, mother/father alcohol problem, mother/father mental health problem, serious family conflict, parental divorce, personal serious/chronic illness, bullied at school (11 items examined individually and as a score (0, 1, 2, 3, 4+) Age, sex, and covariates	When the adult sociodemographic variables were modeled together, age was the only one significantly associated with TL Number of childhood adverse events was significantly associated with TL Individually, childhood adverse events associated with TL were parental unemployment and personal chronic or serious illness

Table 7.3 (continued)

Study	Population/sample (*n*); demographics	Design/Methodology*Covariates	Findings
DeRoo et al. (2009)	United States, The Sister Study $N = 647$ women 83 % non-Hispanic white, 7 % black, 9 % other Ages 35–74 years	Cross-sectional, volunteer national sample Leukocytes; qPCR Self-reported childhood social and economic factors: food insecurity (times when their family did not have enough to eat), highest education level in household, low income or poor, single parent household Age, race/ethnicity	Significantly shorter average TL observed in women reporting childhood food deprivation (times when their family did not have enough to eat) and highest education level in household $\leq$ high school vs. $\geq$ bachelor's degree Significant trend in shorter TL with increasing numbers of lower SES conditions reported
Woo et al. (2009)	Hong Kong $N = 958$ men, 978 women Age $\geq$65 years, community-dwelling elderly Chinese	Cross-sectional, community based volunteer sample Leukocytes; qPCR Self-rated SES: Community ladder (10 rungs, status in community) and Hong Kong ladder (money, education, respected job) BMI, history of chronic diseases, physical activity, smoking, alcohol, diet	For men, higher rank in self-rated social standing was associated with shorter TL
Cherkas et al. (2006)	United Kingdom St. Thomas' Adult Twin Registry (Twins UK) $N = 1552$ Caucasian female twins, 749 dizygotic and 27 monozygotic pairs Ages 18–75	Cross-sectional, volunteer twin registry, whole cohort and discordant twin pair analysis Leukocytes, Southern blot Education and occupation level based on UK National Statistics socio-economic classification (NS-SEC) BMI, smoking, exercise	Shorter TL among manual compared to nonmanual SES groups, not completely accounted for by covariates
Harris et al. (2006)	Scotland Lothian Birth Cohort $N = 82$ men and 108 women out of 550 surviving participants Age 79	Cross-sectional, population sample Leukocytes, qPCR Occupational social class using the standard UK Classification of Occupations from the UK Office of Population and Census Studies (classification published in 1951 was used)	Social class level not associated with TL

reported shorter TL in African Americans versus whites (Geronimus et al. 2010; Roux et al. 2009). In one of these (Geronimus et al. 2010), adjustment for perceived stress partially attenuated racial TL differences, and the authors posited differences might be partly explained by psychosocial stress. Steeper age-related decline in TL among African Americans has also been described in both cross-sectional and longitudinal studies (Aviv et al. 2009; Hunt et al. 2008; Roux et al. 2009), and also among Hispanics (Roux et al. 2009), suggesting higher attrition rates could be more generally related to minority status. A steeper age-related TL decline may be due to an intrinsically faster rate of decline in those with longer baseline TL (Aviv et al. 2009) or increased opportunity for oxidative damage for longer telomeres; the finding could also be due to measurement error or issues related to mathematical dependency in statistical modeling of change within individuals (Giltay et al. 2010). Arguing against inherent racial differences in TL, there were no significant differences in cord blood TL in one study of white, African American and Hispanic newborns (Okuda et al. 2002). The question of racial/ethnic TL differences may be further complicated by other sources of geographic variability; controlling for genetic variation, significant TL differences were observed across fourteen populations across Europe (Eisenberg et al. 2011).

7.1.2.3 Age and Sex-related TL Differences

The rate of TL attrition varies throughout the lifespan, with faster attrition in young children during growth and development, a leveling off, and more gradual shortening in older individuals (Frenck et al. 1998; Rufer et al. 1999). The rate of TL attrition may be influenced by TL at birth. Cross-sectional differences age may also reflect survival factors. The apparent slower attrition in older individuals may be explained by healthier individuals with longer TL surviving to older age (Nordfjall et al. 2009). Some studies have reported stress–TL associations primarily in older individuals (Kananen et al. 2010; Parks et al. (2009), so it would be informative to examine possible age-related differences in perceived and chronic stress, and neuroendocrine responses.

There is also consistent evidence of longer TL in women (Aviv et al. 2005), though whether this is due to endogenous or exogenous hormonal exposure(s), or other factors such as body size or lifestyle, is not known. Evidence supporting the role of hormones includes the absence of sex-related TL differences in newborns (Okuda et al. 2002), longer TL in women with increasing duration of reproductive years (Lin et al. 2010b) and those who used post-menopausal hormones (Lee et al. 2005). Estrodial and androgens can enhance telomerase expression in vitro in hematopoetic cells (Calado 2009). Gonadal steroids and adrenal hormone pathways may interact, with sex-related differences in stress-response (Kajantie and Phillips 2006; Young and Korszun 2010), and so it seems plausible that stress effects on leukocyte TL may be modified or mediated by sex. Notably most research on perceived and chronic stress has been conducted in women.

In sum, evidence pointing toward TL differences by social position and demographic factors (race/ethnicity, age, and gender) is inconclusive. Lower SES and race/ethnicity or minority status are not equivalent constructs, but may interact in determining susceptibility along multiple pathways contributing to health disparities (Fig. 7.1). Longitudinal studies with repeat TL measures across the lifespan are needed to better understand telomere dynamics within individuals and the origins of demographic differences. Further studies of TL in subjects of different racial/ethnic and socio-economic backgrounds, and careful assessment of stress exposure and susceptibility related to minority status and SES, may shed light on these relationships. Studies using a life course perspective with assessment of SES, external stressors (demands), potential modifying characteristics (resources), and lifestyle behaviors across the lifespan would be particularly illuminating (Myers 2009). Research on the stress–TL association also needs to consider the possible role of age- and gender-related differences.

7.1.2.4 Behavioral and Lifestyle Factors

Several behavioral and lifestyle risk factors have been studied with respect to TL, though findings are somewhat inconsistent and a comprehensive review is beyond the scope of this chapter. Shorter TL has been associated with smoking and obesity (Kim et al. 2009; Parks et al. (2009; Valdes et al. 2005; Buxton et al. 2011; Zhu et al. 2010) and longer TL with greater physical activity (LaRocca et al. 2010), multivitamin use (Xu et al. 2009), dietary omega-6 and omega-3 fatty acids and cereal fiber (Cassidy et al. 2010; Farzaneh-Far et al. 2010a), and serum vitamin D levels (Cassidy et al. 2010; Farzaneh-Far et al. 2010a; LaRocca et al. 2010; Richards et al. 2007; Xu et al. 2009). Some associations have been reported only once, while several studies have shown that obesity and abdominal fatness are associated with shorter TL (Farzaneh-Far et al. 2010b; Gardner et al. 2005; Kim et al. 2009). Still, the association between obesity and shorter TL is not consistent across studies (especially among those examining obesity only as a covariate), and it is unclear if the relationship arises in childhood or later: A recent study reported shorter TL associated with obesity in children (Buxton et al. 2011), but another showed no associations between TL and measures of adiposity in adolescents (Zhu et al. 2010).

Lifestyle factors may substantially influence the relationship between stress and TL. For example, the association between perceived stress and shorter TL was observed only among post-menopausal women who were not engaging in vigorous physical activity (Puterman et al. 2010). We are unaware of published studies on other factors such as coping, social support, or sleep, which may also moderate the stress–TL association. Marriage was associated with longer TL in one study (Mainous et al. 2011), but is not necessarily a proxy for social support. Further research is needed on the interrelationship of lifestyle and psychosocial factors in specific pathways related to TL and immune aging.

7.2 Considerations in Telomere Research

Several lines of evidence indicate telomerase enzyme inactivity acts together with shorter TL to mediate immunosenescence. Rare genetic mutations affecting telomerase function can result in aging-related phenotypes such as dyskeratosis congenita, characterized by defects in hematopoetic stem cells resulting in anemia and bone marrow dysfunction (Calado and Young 2009; Savage and Alter 2008). Experimental studies also show that shorter TL is a heritable trait associated with immune defects and, in telomerase insufficient conditions, may be associated with bone marrow failure, mucosal defects, and death from opportunistic infections (Armanios et al. 2009; Chiang et al. 2010). TL dynamics across the human lifespan are not well understood. Growing evidence suggests shorter leukocyte TL may predispose toward subsequent lengthening (Aviv et al. 2009), maybe due to preferential telomerase activation in cells with shorter TL, although the exact mechanisms are unknown.

Some studies on stress and TL have also focused on the effects of stress exposure and related behaviors on telomerase expression. Clinical research initially described inconsistent associations of telomerase activity associated with perceived stress and caregiving (Damjanovic et al. 2007; Epel et al. 2004), though in one study lower telomerase levels were associated with higher urinary epinephrine and autonomic reactivity to an acute stressor (Epel et al. 2006). Dementia patient caregivers with higher stress had lower overall telomerase expression than controls but increased telomerase activity in response to an acute stressor (Epel et al. 2010). The effects of short-term interventions, such as meditation (Jacobs et al. 2010) or lifestyle intervention (Ornish et al. 2008), on TL and telomerase activity suggest that long-term exposure to stress-reducing methods may reduce rate of TL shortening. Elevated telomerase activity was seen in non-medicated patients with major depression, and was directly associated with depression ratings and response to treatment (Wolkowitz et al. 2011b). A large study of depression in heart disease patients showed depressed patients experienced less TL shortening over a 5-year period, however telomerase activity was not assessed (Hoen et al. 2011). More research is needed on the effect of long-term and acute stress exposures on telomerase activity and TL attrition, including longitudinal data and measures of other neuroendocrine and immune effects.

Most studies have measured leukocyte TL in DNA from whole blood or PBMC. Because of the short-turnover of granulocytes/neutrophiles, whole blood leukokcyte TL plausibly reflects underlying hematopoetic stem or progenitor cell TL. TL measured in PBMC may be more influenced by stress-effects on lymphocytes. However, studies have shown high correlation between TL measured in different cell types (Kimura et al. 2010a; Lin et al. 2010a). In a study investigating the influence of varying leukocyte cell types in healthy women, greater TL and telomerase activity were seen in B-cells, while the shortest TL and lower telomerase activity were seen in senescent (CD28−) phenotype CD8+ T-cells (Lin et al. 2010a). Total PBMC-TL was weakly associated with B-cell and CD4+ T cell TL, and more strongly associated with CD8+ TL. Current leukocyte TL therefore reflects developmental, chronic, and acute exposure effects on the immune system, adding to heritable factors determining individual TL. Such variation limits the utility of leukocyte TL as a simple stress biomarker, but underscores its potential role as an indicator of cumulative stress effects on the immune system.

Given the large population variation in TL and correlations between cell types, stress-related variation in cell subtype distributions might have limited influence on the observed association of stress and leukocyte TL. Stress effects on T-cell aging and subset distributions may be difficult to disentangle from stress-related differences in overall leukocyte TL. One study has shown PBMC-TL to be modestly associated with having a higher proportion of senescent CD8+ cells (Lin et al. 2010a). Because of high population TL variation and within-individual TL correlations, studies of etiologic pathways relating stress, TL, and immune aging will require appropriate study design (e.g., longitudinal studies within individuals) with additional biological data to allow examination of specific mechanisms. Compared to cross-sectional studies that require large sample sizes to study inter-individual TL differences, longitudinal studies provide greater power for studying within-individual changes in TL, so smaller numbers of participants are needed (Aviv et al. 2006). Furthermore, assays of TL in cell subtypes or telomerase activity in unstimulated lymphocytes require fresh specimens and/or a large number of cells (Lin et al. 2010a), and thus may be most feasible in smaller clinical studies.

Several biological mechanisms may link stress to immune aging and leukocyte TL; a full discussion is beyond the scope of this review. Observed TL differences in shorter lived granulocytes/neutrophils may reflect the effects of stress on the repeated need for expansion hematopoetic progenitor and stem cells, perhaps mediated by chronic stress-related increased inflammation (Black 2003; Kiecolt-Glaser et al. 2003). TL in PBMC may also indicate differences and telomerase activity in longer lived B and T lymphocyte clones. Stress may have either stimulatory or suppressive effects on lymphocytes depending on the nature of the stressor (acute or chronic) (Glaser and Kiecolt-Glaser 2005) and specific mechanisms examined (Lucas et al. 2007); so observed stress-related differences in PBMC-TL might vary from effects on overall leukocyte TL. One intriguing hypothesis relating stress and TL suggests a role for chronic viral infections driving lymphocyte replication (Effros 2011), and it has been shown that viral infections (e.g., Epstein-Barr virus or Cytomegalovirus) and chronic antigenic exposure may contribute to TL shortening in T-cells (van Baarle et al. 2008; van de Berg et al. 2010). These pathways may be related, as age-related loss of latency of chronic viral infections and decreased adaptive immunity may contribute to shifts in cytokine profiles and greater activation of the innate immune system associated with "inflame-aging" (Almanzar et al. 2005).

7.3 Summary

The literature suggests psychosocial and biological stress may be related to shorter TL, but substantial gaps still exist in understanding this relationship. Findings from small clinical studies need replication in large and diverse samples, with adequate stress assessment and careful attention to the role sociodemographic and lifestyle covariates. In light of emerging evidence on telomere dynamics, stress, and telomerase activity, it seems plausible that intense or prolonged stressors could affect TL. However, just as cross-sectional measures on individual stressors or perceived stress are an inadequate proxy for the lifetime burden of acute and chronic stress, cross-sectional assessment of TL and telomerase activity provide only a brief snapshot of a dynamic system. Lack of understanding of normal TL dynamics throughout the lifespan limits interpretation of cross-sectional associations. Longitudinal studies with repeated measurements in the same individuals are needed to examine determinants of telomere attrition and resilience, with a focus on exposures throughout the lifespan to allow investigation of latent, pathway, and cumulative effects of psychosocial stressors and TL. As long-term studies are often infeasible, there is also a need for more research on stress-related TL determinants at different ages, especially in early childhood when TL attrition rates are greatest or in middle age and older populations with hormonal and aging-related changes in neuroendocrine and immunological responses (Heffner 2011). Mechanistic studies are also needed to provide links to other markers and outcomes related to immunosenescence.

7.4 Author's Note

This review includes studies published through June, 2011. Since that time, several notable studies been published supporting the idea that psychological stressors may contribute to leukocyte telomere length. These include large cross-sectional studies showing shorter telomeres associated with work-related exhaustion, phobic anxiety, and adverse childhood experiences (Ahola et al. 2012; Okereke et al. 2012; Surtees et al. 2011). Added support for the importance of early life stressors also includes a longitudinal study of telomere shortening and exposure to violence in children (Shalev et al. 2012), and a smaller study showing prenatal stress exposure associated with telomere length in young adults (Entringer et al. 2011).

References

Adams J, Martin-Ruiz C, Pearce MS, White M, Parker L, von Zglinicki T (2007) No association between socio-economic status and white blood cell telomere length. Aging Cell 6:125–128

Ahola K, Sirén I, Kivimäki M, Ripatti S, Aromaa A, Lönnqvist J, Hovatta I (2012) Work-related exhaustion and telomere length: a population-based study. PloS One 7(7):e40186

Almanzar G, Schwaiger S, Jenewein B, Keller M, Herndler-Brandstetter D, Wurzner R, Grubeck-Loebenstein B (2005) Long-term cytomegalovirus infection leads to significant changes in the composition of the CD8+T-cell repertoire, which may be the basis for an imbalance in the cytokine production profile in elderly persons. J Virol 79:3675–3683

Armanios M, Alder JK, Parry EM, Karim B, Strong MA, Greider CW (2009) Short telomeres are sufficient to cause the degenerative defects associated with aging. Am J Hum Genet 85:823–832

Aubert G, Lansdorp PM (2008) Telomeres and aging. Physiol Rev 88:557–579

Aviv A, Chen W, Gardner JP, Kimura M, Brimacombe M, Cao X, Berenson GS (2009) Leukocyte telomere dynamics: longitudinal findings among young adults in the Bogalusa Heart Study. Am J Epidemiol 169:323–329
Aviv A, Shay J, Christensen K, Wright W (2005) The longevity gender gap: are telomeres the explanation? Sci Aging Knowledge Environ 23:16
Aviv A, Valdes AM, Spector TD (2006) Human telomere biology: pitfalls of moving from the laboratory to epidemiology. Int J Epidemiol 35:1424–1429
Baerlocher GM, Rice K, Vulto I, Lansdorp PM (2007) Longitudinal data on telomere length in leukocytes from newborn baboons support a marked drop in stem cell turnover around 1 year of age. Aging Cell 6:121–123
Batty GD, Wang Y, Brouilette SW, Shiels P, Packard C, Moore J, Ford I (2009) Socioeconomic status and telomere length: the west of Scotland coronary prevention study. J Epidemiol Commun Health 63:839–841
Beach SR, Brody GH, Todorov AA, Gunter TD, Philibert RA (2010) Methylation at 5HTT mediates the impact of child sex abuse on women's antisocial behavior: an examination of the Iowa adoptee sample. Psychosom Med 73:83–87
Bekaert S, Derradji H, Baatout S (2004) Telomere biology in mammalian germ cells and during development. Dev Biol 274:15–30
Berry MD, Juorio AV, Paterson IA (1994) The functional role of monoamine oxidases A and B in the mammalian central nervous system. Prog Neurobiol 42(3):375–391
Black PH (2003) The inflammatory response is an integral part of the stress response: implications for atherosclerosis, insulin resistance, type II diabetes and metabolic syndrome X. Brain Behav Immun 17:350–364
Blackburn EH (2005a) Telomerase and cancer: Kirk A. Landon-AACR prize for basic cancer research lecture. Mol Cancer Res 3:477–482
Blackburn EH (2005b) Telomeres and telomerase: their mechanisms of action and the effects of altering their functions. FEBS Lett 579:859–862
Braveman PA, Cubbin C, Egerter S, Chideya S, Marchi KS, Metzler M, Posner S (2005) Socioeconomic status in health research: one size does not fit all. JAMA 294:2879–2888
Brouilette SW, Moore JS, McMahon AD, Thompson JR, Ford I, Shepherd J, Samani NJ (2007) Telomere length, risk of coronary heart disease, and statin treatment in the west of Scotland primary prevention study: a nested case-control study. Lancet 369:107–114
Buxton JL, Walters RG, Visvikis-Siest S, Meyre D, Froguel P, Blakemore AI (2011) Childhood obesity is associated with shorter leukocyte telomere length. J Clin Endocrinol Metab 96:1500–1505
Calado RT (2009) Telomeres and marrow failure. Hematol Am Soc Hematol Educ Prog:338–343
Calado RT, Young NS (2009) Telomere diseases. N Engl J Med 361:2353–2365
Cassidy A, De Vivo I, Liu Y, Han J, Prescott J, Hunter DJ, Rimm EB (2010) Associations between diet, lifestyle factors, and telomere length in women. Am J Clin Nutr 91:1273–1280
Cawthon RM (2002) Telomere measurement by quantitative PCR. Nucl Acids Res 30:e47
Cawthon RM, Smith KR, O'Brien E, Sivatchenko A, Kerber RA (2003) association between telomere length in blood and mortality in people aged 60 years or older. Lancet 361:393–395
Cherkas LF, Aviv A, Valdes AM, Hunkin JL, Gardner JP, Surdulescu GL, Spector TD (2006) The effects of social status on biological aging as measured by white-blood-cell telomere length. Aging Cell 5:361–365
Chiang YJ, Calado RT, Hathcock KS, Lansdorp PM, Young NS, Hodes RJ (2010) Telomere length is inherited with resetting of the telomere set-point. Proc Natl Acad Sci USA 107:10148–10153
Damjanovic AK, Yang Y, Glaser R, Kiecolt-Glaser JK, Nguyen H, Laskowski B, Weng NP (2007) Accelerated telomere erosion is associated with a declining immune function of caregivers of Alzheimer's disease patients. J Immunol 179:4249–4254
DeRoo LA, Parks CG, Weinberg CR, Sandler DP (2010) Childhood trauma, abuse, and other stressors in relation to telomere length. Am J Epidemiol 171:S463

DeRoo LA, Parks CG, Kim S, Cawthon RM, Weinberg C, Sandler DP (2009) Childhood socioeconomic factors and telomere length. Am J Epidemiol 169:S88

Effros RB (2011) Telomere/telomerase dynamics within the human immune system: effect of chronic infection and stress. Exp Gerontol 46:135–140

Eisenberg DT, Salpea KD, Kuzawa CW, Hayes MG, Humphries SE (2011) Substantial variation in qPCR measured mean blood telomere lengths in young men from eleven European countries. Am J Hum Biol Jan 10 (Epub ahead of print)

Entringer S, Epel ES, Kumsta R, Lin J, Hellhammer DH, Blackburn EH, Wüst S, Wadhwa PD (2011) Stress exposurein intrauterine life is associated with shorter telomere length in young adulthood. Proc Natl Acad Sci USA 108(33):E513–518

Epel ES, Blackburn EH, Lin J, Dhabhar FS, Adler NE, Morrow JD, Cawthon RM (2004) Accelerated telomere shortening in response to life stress. [see comment]. Proc Natl Acad Sci USA 101:17312–17315

Epel ES, Lin J, Dhabhar FS, Wolkowitz OM, Puterman E, Karan L, Blackburn EH (2010) Dynamics of telomerase activity in response to acute psychological stress. Brain Behav Immun 24:531–539

Epel ES, Lin J, Wilhelm FH, Wolkowitz OM, Cawthon R, Adler NE, Blackburn EH (2006) Cell aging in relation to stress arousal and cardiovascular disease risk factors. Psychoneuroendocrinology 31:277–287

Farzaneh-Far R, Lin J, Epel ES, Harris WS, Blackburn EH, Whooley MA (2010a) Association of marine omega-3 fatty acid levels with telomeric aging in patients with coronary heart disease. JAMA 303(3):250–257

Farzaneh-Far R, Lin J, Epel E, Lapham K, Blackburn E, Whooley MA (2010b) Telomere length trajectory and its determinants in persons with coronary artery disease: longitudinal findings from the heart and soul study. PLoS One 5:e8612

Frenck RW Jr, Blackburn EH, Shannon KM (1998) The rate of telomere sequence loss in human leukocytes varies with age. Proc Natl Acad Sci USA 95:5607–5610

Gardner JP, Li S, Srinivasan SR, Chen W, Kimura M, Lu X, Aviv A (2005) Rise in insulin resistance is associated with escalated telomere attrition. Circulation 111:2171–2177

Geronimus AT, Hicken MT, Pearson JA, Seashols SJ, Brown KL, Cruz TD (2010) Do US black women experience stress-related accelerated biological aging?: a novel theory and first population-based test of black-white differences in telomere length. Hum Nat 21:19–38

Giltay EJ, Hageman GJ, Kromhout D (2010) Spurious association between telomere length reduction over time and baseline telomere length. Int J Epidemiol Dec 12 (Epub ahead of print)

Glaser R, Kiecolt-Glaser JK (2005) Stress-induced immune dysfunction: implications for health. Nat Rev Immunol 5:243–251

Glass D, Parts L, Knowles D, Aviv A, Spector TD (2010) No correlation between childhood maltreatment and telomere length. Biol Psychiatr 68:e21–22; author reply e23–24

Harris SE, Deary IJ, MacIntyre A, Lamb KJ, Radhakrishnan K, Starr JM, Shiels PG (2006) The association between telomere length, physical health, cognitive ageing, and mortality in non-demented older people. Neurosci Lett 406:260–264

Hartmann N, Boehner M, Groenen F, Kalb R (2010) Telomere length of patients with major depression is shortened but independent from therapy and severity of the disease. Depress Anxiety 27:1111–1116

Hathcock KS, Jeffrey Chiang Y, Hodes RJ (2005) In vivo regulation of telomerase activity and telomere length. Immunol Rev 205:104–113

Heffner KL (2011) Neuroendocrine effects of stress on immunity in the elderly: implications for inflammatory disease. Immunol Allergy Clin North Am 31(1):95–108

Hoen PW, de Jonge P, Na BY, Farzaneh-Far R, Epel E, Lin J, Whooley MA (2011) Depression and leukocyte telomere length in patients with coronary heart disease: data from the heart and soul study. Psychosom Med May 9 (Epub ahead of print)

Hopwood CJ, Donnellan MB, Blonigen DM, Krueger RF, McGue M, Iacono WG, Burt SA (2011) Genetic and environmental influences on personality trait stability and growth during the transition to adulthood: a three-wave longitudinal study. J Pers Soc Psychol 100:545–556
Hornsby PJ (2006) Short telomeres: cause or consequence of aging? Aging Cell 5:577–578
Humphreys J, Epel ES, Cooper BA, Lin J, Blackburn EH, Lee KA (2011) Telomere shortening in formerly abused and never abused women. Biol Res Nurs Mar 8 (Epub ahead of print)
Hunt SC, Chen W, Gardner JP, Kimura M, Srinivasan SR, Eckfeldt JH, Aviv A (2008) Leukocyte telomeres are longer in African Americans than in whites: the National Heart, Lung, and Blood Institute Family Heart Study and the Bogalusa Heart Study. Aging Cell 7:451–458
Huzen J, van Der Harst P, de Boer RA, Lesman-Leegte I, Voors AA, van Gilst WH, van Veldhuisen DJ (2010) Telomere length and psychological well-being in patients with chronic heart failure. Age Ageing 39:223–227
Jacobs TL, Epel ES, Lin J, Blackburn EH, Wolkowitz OM, Bridwell DA, Saron CD (2010) Intensive meditation training, immune cell telomerase activity, and psychological mediators. Psychoneuroendocrinology 36:664–681
Jain D, Cooper JP (2010) Telomeric strategies: means to an end. Annu Rev Genet 44:243–269
Kajantie E, Phillips DI (2006) The effects of sex and hormonal status on the physiological response to acute psychosocial stress. Psychoneuroendocrinology 31:151–178
Kananen L, Surakka I, Pirkola S, Suvisaari J, Lonnqvist J, Peltonen L, Hovatta I (2010) Childhood adversities are associated with shorter telomere length at adult age both in individuals with an anxiety disorder and controls. PLoS One 5:e10826
Kiecolt-Glaser JK, Preacher KJ, MacCallum RC, Atkinson C, Malarkey WB, Glaser R (2003) Chronic stress and age-related increases in the proinflammatory cytokine IL-6. Proc Natl Acad Sci USA 100:9090–9095
Kiecolt-Glaser JK, Gouin JP, Weng NP, Malarkey WB, Beversdorf DQ, Glaser R (2010) Childhood adversity heightens the impact of later-life caregiving stress on telomere length and inflammation. Psychosom Med 73:16–22
Kim S, Parks CG, DeRoo LA, Chen H, Taylor JA, Cawthon RM, Sandler DP (2009) Obesity and weight gain in adulthood and telomere length. Cancer Epidemiol Biomarkers Prev 18:816–820
Kimura M, Gazitt Y, Cao X, Zhao X, Lansdorp PM, Aviv A (2010a) Synchrony of telomere length among hematopoietic cells. Exp Hematol 38:854–859
Kimura M, Hjelmborg JV, Gardner JP, Bathum L, Brimacombe M, Lu X, Christensen K (2008) Telomere length and mortality: a study of leukocytes in elderly Danish twins. Am J Epidemiol 167:799–806
Kimura M, Stone RC, Hunt SC, Skurnick J, Lu X, Cao X, Aviv A (2010b) Measurement of telomere length by the southern blot analysis of terminal restriction fragment lengths. Nat Protoc 5:1596–1607
Kuh D (2006) A life course perspective on telomere length and social inequalities in aging. Aging Cell 5:579–580
Lansdorp PM (2005) Role of telomerase in hematopoietic stem cells. Ann N Y Acad Sci 1044:220–227
Lansdorp PM (2006) Stress, social rank and leukocyte telomere length. Aging Cell 5:583–584
Lansdorp PM (2009) Telomeres and disease. Embo J 28:2532–2540
Lantz PM, House JS, Mero RP, Williams DR (2005) Stress, life events, and socioeconomic disparities in health: results from the Americans' Changing Lives Study. J Health Soc Behav 46:274–288
LaRocca TJ, Seals DR, Pierce GL (2010) Leukocyte telomere length is preserved with aging in endurance exercise-trained adults and related to maximal aerobic capacity. Mech Ageing Dev 131:165–167
Lee DC, Im JA, Kim JH, Lee HR, Shim JY (2005) Effect of long-term hormone therapy on telomere length in postmenopausal women. Yonsei Med J 46:471–479
Lin J, Epel E, Cheon J, Kroenke C, Sinclair E, Bigos M, Blackburn E (2010a) Analyses and comparisons of telomerase activity and telomere length in human T and B cells: insights for epidemiology of telomere maintenance. J Immunol Methods 352:71–80

Lin J, Kroenke CH, Epel E, Kenna HA, Wolkowitz OM, Blackburn E, Rasgon NL (2010b) Greater endogenous estrogen exposure is associated with longer telomeres in postmenopausal women at risk for cognitive decline. Brain Res 352:71–80

Lucas RM, Ponsonby AL, Dear K (2007) Mid-life stress is associated with both up- and down-regulation of markers of humoral and cellular immunity. Stress 10:351–361

Lung FW, Chen NC, Shu BC (2007) Genetic pathway of major depressive disorder in shortening telomeric length. Psychiatr Genet 17:195–199

Lung FW, Fan PL, Chen NC, Shu BC (2005) Telomeric length varies with age and polymorphisms of the MAOA gene promoter in peripheral blood cells obtained from a community in Taiwan. Psychiatr Genet 15:31–35

Mainous AG 3rd, Everett CJ, Diaz VA, Baker R, Mangino M, Codd V, Samani NJ (2011) Leukocyte telomere length and marital status among middle-aged adults. Age Ageing 40:73–78

McGowan PO, Sasaki A, D'Alessio AC, Dymov S, Labonte B, Szyf M, Meaney MJ (2009) Epigenetic regulation of the glucocorticoid receptor in human brain associates with childhood abuse. Nat Neurosci 12:342–348

McGrath M, Wong JY, Michaud D, Hunter DJ, De Vivo I (2007) Telomere length, cigarette smoking, and bladder cancer risk in men and women. Cancer Epidemiol Biomarkers Prev 16:815–819

Myers HF (2009) Ethnicity- and socio-economic status-related stresses in context: an integrative review and conceptual model. J Behav Med 32:9–19

Nordfjall K, Svenson U, Norrback KF, Adolfsson R, Lenner P, Roos G (2009) The individual blood cell telomere attrition rate is telomere length dependent. PLoS Genet 5:e1000375

O'Donovan A, Lin J, Dhabhar FS, Wolkowitz O, Tillie JM, Blackburn E, Epel E (2009) Pessimism correlates with leukocyte telomere shortness and elevated interleukin-6 in post-menopausal women. Brain Behav Immun 23:446–449

Okereke OI, Prescott J, Wong JY, Han J, Rexrode KM, De Vivo I (2012) High phobic anxiety is related to lower leukocyte telomere length in women. PloS One 7(7):e40516 (July 11)

Okuda K, Bardeguez A, Gardner JP, Rodriguez P, Ganesh V, Kimura M, Aviv A (2002) Telomere length in the newborn. Pediatr Res 52:377–381

Opresko PL, Fan J, Danzy S, Wilson DM 3rd, Bohr VA (2005) Oxidative damage in telomeric DNA disrupts recognition by TRF1 and TRF2. Nucl Acids Res 33:1230–1239

Ornish D, Lin J, Daubenmier J, Weidner G, Epel E, Kemp C, Blackburn EH (2008) Increased telomerase activity and comprehensive lifestyle changes: a pilot study. Lancet Oncol 9:1048–1057

Parks CG, Deroo LA, Miller DB, McCanlies EC, Cawthon RM, Sandler DP (2011) Employment and work schedule are related to telomere length in women. Occup Environ Med May 2 (Epub ahead of print)

Parks CG, Miller DB, McCanlies EC, Cawthon RM, Andrew ME, DeRoo LA, Sandler DP (2009) Telomere length, current perceived stress, and urinary stress hormones in women. Cancer Epidemiol Biomarkers Prev 18:551–560

Puterman E, Lin J, Blackburn E, O'Donovan A, Adler N, Epel E (2010) The power of exercise: buffering the effect of chronic stress on telomere length. PLoS One 5:e10837

Rhee DB, Ghosh A, Lu J, Bohr VA, Liu Y (2011) Factors that influence telomeric oxidative base damage and repair by DNA glycosylase OGG1. DNA Repair (Amst) 10(1):34–44

Richards JB, Valdes AM, Gardner JP, Paximadas D, Kimura M, Nessa A, Aviv A (2007) Higher serum vitamin D concentrations are associated with longer leukocyte telomere length in women. Am J Clin Nutr 86:1420–1425

Rivera M, Gutierrez B, Molina E, Torres-Gonzalez F, Bellon JA, Moreno-Kustner B, Cervilla JA (2009) High-activity variants of the uMAOA polymorphism increase the risk for depression in a large primary care sample. Am J Med Genet B Neuropsychiatr Genet 150B:395–402

Roux AV, Ranjit N, Jenny NS, Shea S, Cushman M, Fitzpatrick A, Seeman T (2009) Race/ethnicity and telomere length in the multi-ethnic study of atherosclerosis. Aging Cell 8:251–257

Rufer N, Brummendorf TH, Kolvraa S, Bischoff C, Christensen K, Wadsworth L, Lansdorp PM (1999) Telomere fluorescence measurements in granulocytes and T lymphocyte subsets point to a high turnover of hematopoietic stem cells and memory T cells in early childhood. J Exp Med 190:157–167

Rufer N, Dragowska W, Thornbury G, Roosnek E, Lansdorp PM (1998) Telomere length dynamics in human lymphocyte subpopulations measured by flow cytometry. Nat Biotechnol 16:743–747

Savage SA, Alter BP (2008) The role of telomere biology in bone marrow failure and other disorders. Mech Ageing Dev 129:35–47

Schwartz JE, Friedman HS, Tucker JS, Tomlinson-Keasey C, Wingard DL, Criqui MH (1995) Sociodemographic and psychosocial factors in childhood as predictors of adult mortality. Am J Public Health 85:1237–1245

Shalev I, Moffitt TE, Sugden K, Williams B, Houts RM, Danese A, Mill J, Arenseault L, Caspi A (2012) Exposure to violence during childhood is associated with telomere erosion from 5 to 10 years of age: a longutinal study. Mol Psychiatry April 24 (Epub ahead of print)

Shay JW, Wright WE (2005) Senescence and immortalization: role of telomeres and telomerase. Carcinogenesis 26:867–874

Simon NM, Smoller JW, McNamara KL, Maser RS, Zalta AK, Pollack MH, Wong KK (2006) Telomere shortening and mood disorders: preliminary support for a chronic stress model of accelerated aging. Biol Psychiatry 60:432–435

Steptoe A, Hamer M, Butcher L, Lin J, Brydon L, et al (2011) Educational attainment but not measures of current socioeconomic circumstances are associated with leukocyte telomere length in healthy older men and women. Brain Behav Immun 25(7):1292–1298

Surtees PG, Wainwright NW, Pooley KA, Luben RN, Khaw KT, Easton DF, Dunning AM (2011) Life stress, emotional health, and mean telomere length in the European Prospective Investigation into Cancer (EPIC)-Norfolk population study. J Gerontol A Biol Scin Med Sci 66(11):1152–1162

Tentolouris N, Nzietchueng R, Cattan V, Poitevin G, Lacolley P, Papazafiropoulou A, Benetos A (2007) White blood cells telomere length is shorter in males with type 2 diabetes and microalbuminuria. Diabetes Care 30:2909–2915

Tyrka AR, Price LH, Kao HT, Porton B, Marsella SA, Carpenter LL (2010) Childhood maltreatment and telomere shortening: preliminary support for an effect of early stress on cellular aging. Biol Psychiatr 67:531–534

Valdes AM, Andrew T, Gardner JP, Kimura M, Oelsner E, Cherkas LF, Spector TD (2005) Obesity, cigarette smoking, and telomere length in women. Lancet 366:662–664

van Baarle D, Nanlohy NM, Otto S, Plunkett FJ, Fletcher JM, Akbar AN (2008) Progressive telomere shortening of Epstein-Barr virus-specific memory T cells during HIV infection: contributor to exhaustion? J Infect Dis 198;1353–1357

van de Berg PJ, Griffiths SJ, Yong SL, Macaulay R, Bemelman FJ, Jackson S, van Lier RA (2010) Cytomegalovirus infection reduces telomere length of the circulating T cell pool J Immunol 184;3417–3423

van der Hulst M (2003) Long workhours and health. Scand J Work Environ Health 29:171–188

Von Zglinicki T (2003) Replicative senescence and the art of counting. Exp Gerontol 38:1259–1264

Weng NP, Hathcock KS, Hodes RJ (1998) Regulation of telomere length and telomerase in T and B cells: a mechanism for maintaining replicative potential. Immunity 9:151–157

Williams DR, Jackson PB (2005) Social sources of racial disparities in health. Health Aff (Millwood) 24:325–334

Wolkowitz OM, Epel ES, Reus VI, Mellon SH (2010) Depression gets old fast: do stress and depression accelerate cell aging? Depress Anxiety 27:327–338

Wolkowitz OM, Mellon SH, Epel ES, Lin J, Dhabhar FS, Su Y, Blackburn EH (2011a) Leukocyte telomere length in major depression: correlations with chronicity, inflammation and oxidative stress—preliminary findings. PLoS One 6:e17837

Wolkowitz OM, Mellon SH, Epel ES, Lin J, Reus VI, Rosser R, Blackburn EH (2011b) Resting leukocyte telomerase activity is elevated in major depression and predicts treatment response. Mol Psychiatr Jan 18 (Epub ahead of print)

Woo J, Suen EW, Leung JC, Tang NL, Ebrahim S (2009) Older men with higher self-rated socioeconomic status have shorter telomeres. Age Ageing 38:553–558

Xu Q, Parks CG, Deroo LA, Cawthon RM, Sandler DP, Chen H (2009) Multivitamin use and telomere length in women. Am J Clin Nutr 89:1857–1863

Young E, Korszun A (2010) Sex, trauma, stress hormones and depression. Mol Psychiatr 15:23–28

Zhu H, Wang X, Gutin B, Davis CL, Keeton D, Thomas J, Dong Y (2010) Leukocyte telomere length in healthy Caucasian and African-American adolescents: relationships with race, sex, adiposity, adipokines, and physical activity. J Pediatr 158;215–20

Chapter 8
Stress-related Behavioural Responses, Immunity and Ageing in Animal Models

Carmen Vida and Mónica De la Fuente

8.1 Introduction

Health maintenance during the ageing process, which allows for increased longevity, is determined by the preservation of tissue homeostasis at all physiological levels. Healthy ageing depends both on the individual's genetic make-up and lifestyle factors (Kirkwood 2008). The age-related loss of homeostasis and capacity to react appropriately to stress, as a consequence of the deterioration of physiological systems, have been proposed to explain the increase of morbidity and mortality with ageing (De la Fuente 2008; Fig. 8.1). Although the effects of stressors on behaviour and immunity are very heterogeneous since they depend on type, frequency, duration, intensity, animal models, perception of subject and coping by the stressed animal (Costa-Pinto and Palermo-Nieto 2010), it has been suggested that immunosenescence is a significant consequence of chronic stress and the actions of stress hormones (Bauer 2008; Lord et al. 2009). Moreover, this age-related impairment of the immune system in turn appears to be involved in the increased oxidation and inflammation status that occurs in the ageing process, increasing its rate and representing a vicious cycle of decline (De la Fuente and Miquel 2009).

In the context of the neuro-endocrine–immune network, it is known that humans with psychological distress, anxiety or depression (Arranz et al. 2007, 2009a), and experimental animals (Viveros et al. 2007) with an inadequate response to stress situations show premature immunosenescence. Since immune function is a marker of health, alterations in immunity are accompanied by increased morbidity (Wikby et al. 2008). To study the effect of individual responses to stress in the ageing process and thus the involvement of immunosenescence in this process, models with a shorter longevity than humans are needed. Thus, we have proposed several murine models of premature immunosenescence. One of them, which is characterised by an altered stress-related behaviour response, shows accelerated immunosenescence, an oxidation–inflammation state and premature ageing in the nervous and endocrine

C. Vida (✉) · M. De la Fuente
Department of Physiology, Faculty of Biology, Complutense University of Madrid, 28040 Madrid, Spain
e-mail: mondelaf@bio.ucm.es

J. A. Bosch et al. (eds.), *Immunosenescence*,
DOI 10.1007/978-1-4614-4776-4_8, © Springer Science+Business Media New York 2013

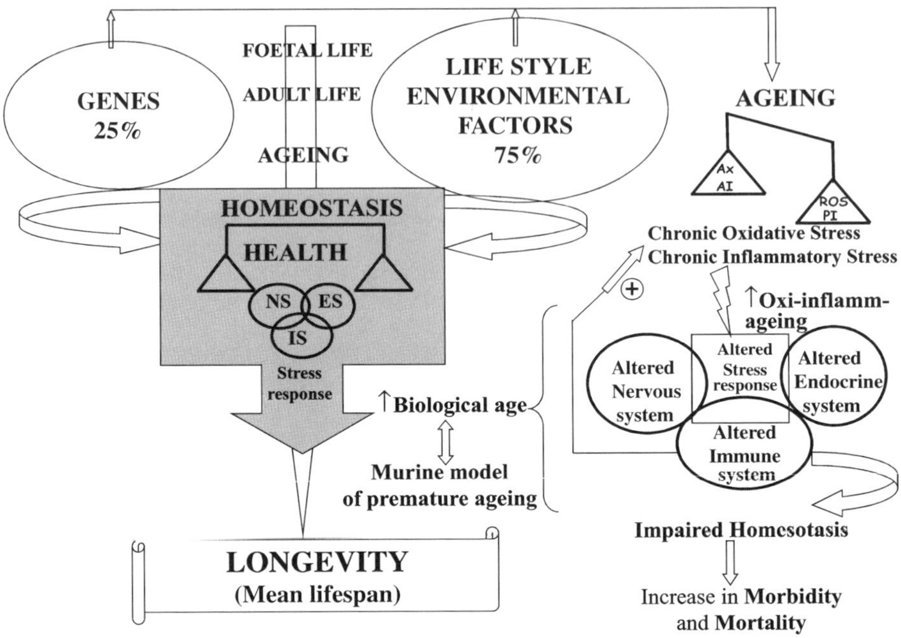

Fig. 8.1 A high longevity in each individual depends on health maintenance and homeostasis preservation, which is conditioned by genotype (approximately 25 % influence) and lifestyle and environmental factors (75 % influence). During the ageing process oxidative stress, affecting all cells and especially those of master regulatory systems, that is, the nervous, endocrine and immune system, and the communication among them, reduces the ability to maintain homeostasis. Moreover, the immune system could increase oxidative stress if it is not regulated thus accelerating the ageing process. This deterioration of homeostasis, which leads to the age-related increase in morbidity and mortality, is established at a different rate in each subject, and this rate is the result of individual epigenetic mechanisms acting on genes throughout the life of the subject. The use of animal models, specifically the prematurely ageing mouse (PAM) model, may be particularly useful to assess the relationships between a diminished ability to cope with stressful situations, ageing of homeostatic systems, accelerated biological ageing and a shorter longevity. *NS* Nervous System; *ES* Endocrine System; *IS* Immune System; *Ax* Antioxidant Compounds; *AI* Anti-Inflammatory Compounds; *ROS* Reactive Oxygen Species; *PI* Pro-Inflammatory Compounds

systems, as well as presenting a shorter life expectancy (De la Fuente 2010; Viveros et al. 2007). These theories of the ageing process and the model of prematurely ageing mice (PAM) are discussed in this chapter.

8.2 The Ageing Process and the Concepts of Longevity and Biological Age

The universal process of ageing may be defined as a progressive and general deterioration of the functions of the organism that leads to a lower ability to react to stress and preserve homeostasis. Although the accumulation of such adverse changes with the passing of time should not be considered a disease, it strongly increases the risk of disease, and finally results in death. The ageing process is finished at the end of the maximum lifespan or maximum longevity, which is the maximum time that a subject belonging to a particular species can live. In human beings this is about 122 years, the record for human longevity held by Mme Jean-Louise Calment, whereas in mouse and rat strains it is approximately 3 and 4 years, respectively. It is very important to distinguish the maximum lifespan from the mean longevity or mean lifespan, which can be defined as the mean time that the members of a population that have been born on the same date live. The maximum longevity is fixed in each species, but the mean lifespan of individual organisms, even when they are of the same genotype and are raised in a common environment protected from extrinsic hazards, shows marked variability. Although presently it is impossible to increase the maximum longevity of a species, the mean lifespan can be increased by environmental factors that allow the maintenance of good health and thus, approach the maximum lifespan in good condition. Presently, since the mean longevity is very high in developed countries, about 75–83 years, and we start the ageing process at about 18 years old, we spend most of our life ageing. Therefore, it is very important to know which lifestyle factors can increase or decrease this longevity and how they do achieve this. A higher mean longevity is achieved by preservation of good health and this depends approximately 25 % on genetic factors and 75 % on lifestyle and environmental factors (Fig. 8.1).

The ageing process is highly heterogeneous, and thus, there are different rates of physiological changes in the various systems of the organism and in the diverse members of a population of the same chronological age. This justifies the introduction of the concept of "biological ageing ", which determines the level of ageing experienced by each individual and therefore, his/her life expectancy. The biological age is related to mean longevity, subjects of a population with a higher rate of ageing show an older biological age and have a shorter lifespan. Since chronological age fails to provide an accurate indicator of the rate of ageing, it is necessary to select parameters useful as biomarkers of ageing to find out this rate and therefore, the probable longevity of each subject (Bae et al. 2008; Borkan and Norris 1980).

Although almost 400 single-cause theories have been proposed to explain the ageing process, we have recently published an integrated theory of ageing (De la Fuente and Miquel 2009). To answer the question of "how" ageing happens, a chronic oxidative stress condition (increase of oxidant compounds and decrease of antioxidant defences), which is linked to many age-related changes, has been proposed. In addition, emerging evidence shows the close link between oxidation and inflammation, and with ageing the pro-inflammatory compound levels are higher than those of the anti-inflammatory (termed inflammageing; Franceschi et al. 2007), leading to inflammatory stress. Therefore, an oxidative and an inflammatory state have been suggested as the cause of the loss of function that appears with senescence. Thus, after adult age the establishment of what has been called "oxi-inflamm-ageing" occurs (De la Fuente and Miquel 2009).

8.3 Neuro-endocrine–immune Communication in Ageing: The Immune System as a Marker of Biological Age and Predictor of Longevity

It is known that the three master regulatory systems, namely the nervous, endocrine and immune systems, are intimately linked and interdependent. Thus, there is a neuro-endocrine–immune axis that allows the preservation of homeostasis and health. The communication between these three systems has permitted the understanding of why situations of depression, emotional stress and anxiety are accompanied by a greater vulnerability to infections, cancers and autoimmune diseases (Ader et al. 2001; Besedovsky and Del Rey 2007; Irwin and Miller 2007). In fact, many neurotransmitters and hormones modulate immune cell functions and consequently behaviours, emotional states and stressful life experiences significantly affect the immune response (Ader et al. 2001). With ageing there is an adjustment in these physiological regulatory systems and in their communication, with the loss of homeostatic capacity and the resulting increase of morbidity and mortality that appears with the passage of time (De la Fuente 2008; Fabris 1990; Gemma 2010). For example, the nervous system changes with ageing showing a decrease of synaptic density (Hedden and Gabrieli 2004). The age-related decrease of neurogenesis, which clearly affects the hippocampus, explains the learning and cognitive impairment in aged subjects (Couillard-Depres et al. 2011), the hippocampus neurogenesis and plasticity being also altered by stress (Kim et al. 2006). In the endocrine system several changes accompany healthy ageing, these include, for example, the decrease of the growth hormone/insulin-like factor-1 axis (somatopause) and of the sexual hormones, namely estradiol (menopause), testosterone (andropause) and dehydroepiandrosterone (adrenopause) (Makrantonaki et al. 2010). Moreover, the age-related disturbances of the hypothalamic–pituitary–adrenal (HPA) axis are responsible for decreasing stress adaptability in old subjects, this being, at least in part, the cause of their health impairment (Aguilera 2011) and peripheral immunodepression (Woiciechowsky et al. 1999). Thus, the inadequate response to stress that occurs with ageing can be understood in the context of compromised neuro-endocrine–immune communication, this altered response being one of the conditions leading to an acceleration of ageing accompanied by the poor functioning of the immune system (Bauer 2008; Gouin et al. 2008). In addition, chronic stressful conditions modify immune functions and their interaction with the nervous system, causing detrimental effects on memory, neural plasticity and neurogenesis (Yirmiya and Goshen 2011).

With age, there is an increase in susceptibility to infections and increased risk of cancer, which indicates an inadequate immune response and influences age-related morbidity and mortality. The profound impact of ageing on immunity is widely accepted (see Chaps. 1 and 2), with age-related changes of the immune system that may include not only diminished but also enhanced functions such as pro-inflammatory cytokine production in unstimulated cells. Thus, the components of the immune system undergo striking age-associated re-structuring, termed immunosenescence (Weiskopf et al. 2009). Some of the key and most marked changes are a pronounced age-related decrease in T-cell functions (Haynes and Maue 2009), especially in the CD4 T-helper cell, which affects cell mediated and humoral immunity and causes an impaired B-cell function (Eaton et al. 2004; Frasca and Blomberg 2009). Phagocytic cells, including neutrophils and macrophages, show functions that decrease with ageing and the antitumoral activity of natural killer (NK) cells, in most of the work published, shows an age-related decrease (reviewed in Shaw et al. (2010)). The network of cytokines produced in response to immune challenge has also shown changes with ageing, notably a shift towards Th2 (effecting humoral antibody mediated immunity) responses and reduced anti-inflammatory cytokine production (Arranz et al. 2010a, 2010b).

Importantly, it has been demonstrated that the competence of the immune system is an excellent marker of health and several age-related changes in immune functions, have been established as markers of biological age and therefore as predictors of longevity and have been termed the immune risk profile or phenotype (Wikby et al. 2008). In order to identify the above parameters as markers of biological age, it has been necessary to confirm that the levels shown in particular subjects reveal their real health and senescent conditions. This has been achieved in the following two ways: (a) ascertaining that the individuals with those parameters indicative of a greater biological age, die before their counterparts. This can be confirmed only in longitudinal studies; (b) finding that the subjects reaching a very advanced age for their species, preserve these immune functions at levels similar to those of adults. For example, this can be tested in extremely long-lived subjects, such as centenarians or the extremely long-lived mice used in our own studies. Whilst biologically older individuals showing the immune competence levels characteristic of chronologically older subjects have been found to die prematurely, centenarians (Alonso-Fernandez et al. 2008) and long-lived mice (Arranz et al. 2010a, 2010b) exhibit a high degree of preservation of several immune functions, which may be related to their ability to reach a very advanced age in a healthy condition. All the above results confirm that the immune system is a good marker of biological age and a predictor of longevity (Wikby et al. 2008).

8.4 The Oxidation–Inflamamtion Theory of Immunosenescence and Ageing

According to our recently proposed theory of oxidation–inflammation in ageing (De la Fuente et al. 2005; De la Fuente and Miquel 2009), the age-related changes in the organism are linked to a chronic oxidative and inflammatory stress, which leads to the damage of cell components, including proteins, lipids and DNA. This affects all cells and especially those of the regulatory systems, including the immune system, which partially explains their impaired function. Moreover, the immune system, due to its capacity of producing oxidant and inflammatory compounds in order to eliminate foreign agents, could, if it is not well regulated, increase the general oxidative and inflammatory stress, and thus increase the rate of ageing.

Thus, immunosenescence could be involved in oxi-inflamm-ageing and affect the functions of other regulatory systems, resulting in age-related homeostatic decline and the consequent increase in morbidity and mortality. In this context, a relationship has been found between the redox and inflammatory state of the immune cells, their functional capacity and the life span of a subject. Thus, when subject shows a high oxidative stress in its immune cells, these cells have an impaired function and that animal shows a decreased longevity (Viveros et al. 2007). In contrast, subjects who achieve greater longevity, such as human centenarians and extremely long-lived mice, show a preserved redox state and immune functions (Arranz et al. 2010a, 2010b; Alonso-Fernandez and De la Fuente 2011). In summary, aged individuals that maintain a good regulation of the leukocyte redox state and consequently, a good function of their immune cells, with levels similar to those of healthy adults, have the greatest chance of achieving very high longevity.

8.5 Murine Models of Premature Neuro-endocrine–immune Ageing

Support for the role of the immune system in oxi-inflamm-ageing, in the context of neuroimmunomodulation, may be obtained by the study of animal models in which, individuals showing premature immunosenescence and a high oxidative and inflammatory stress in their immune cells (and in other cells), as well as a premature alteration of the nervous system (shown principally by behavioural tests) show decreased longevity in relation to other members of the group of the same chronological age. This has been studied using several murine models investigated and developed during the last few years (De la Fuente and Gimenez-Llort 2010; De la Fuente 2010) and summarised below.

8.5.1 Menopausal Models

Menopausal women as well as ovariectomised rats and mice (a model for human menopause) constitute a model for assessing premature ageing, since they show premature immunosenescence, with decreased leukocyte chemotaxis, lymphocyte proliferation and NK cell activity (Baeza et al. 2010a, 2011; De la Fuente et al. 2004), and a higher oxidative stress condition with higher GSSG/GSH ratio (Baeza et al. 2010b, 2011) as well as a decrease in anti-inflammatory cytokines such as IL-10 (Baeza et al. 2011). Moreover, ovariectomy causes the premature ageing of several behavioural responses such as sensorimotor abilities (loss of muscular vigour, impaired equilibrium and traction capacities) and reduction of exploratory activity (Baeza et al. 2010a). In addition, ovariectomised female rats and mice show a redox state and function in leucocytes similar to those in males (Baeza et al. 2011; De la Fuente et al. 2004). In mammalian species, males have a higher oxidative state and a lower function in their immune cells than those of females and also have a lower mean life span than the latter (Baeza et al. 2011; De la Fuente et al. 2004; Guayerbas and De la Fuente 2003).

8.5.2 *Obesity Models*

Obese subjects show a higher incidence of infections and some types of cancer, suggesting an impaired immune function (Lamas et al. 2002a). In general, the few studies of immunity in obese compared to non-obese subjects of the same chronological age, show a worse immune function, which has been observed in both genetically and diet-induced obese rats (De Castro et al. 2009; Lamas et al. 2002a, 2002b). Moreover, obesity is associated with an inflammatory state (Ye 2011), and immune cells from obese rats show premature immunosenescence, with a decreased proliferation and NK activity with respect to non-obese animals of the same chronological age (De Castro et al. 2009; De la Fuente an De Castro 2012; Lamas et al. 2002a, 2002b) as well as an oxidative stress situation (De Castro et al. 2010).

8.5.3 *Alzheimer's Disease Model*

The age-related changes in the neuro-endocrine–immune network influence both the progress of ageing and its related diseases such as neurodegenerative disorders. Alzheimer's disease (AD) is the most common of these disorders, with the main pathological hallmarks being the aberrant protein aggregates, amyloid plaques, comprising the amyloid β peptide and neurofibrillary tangles that consist of hyperphosphorylated tau protein. Synaptic and cholinergic deficits, reactive gliosis, an inflammatory profile and an oxidative stress situation as well as psychological symptoms of dementia and the impairment of cognition and behaviour are other neurodegenerative changes. The triple-transgenic mice for AD (3 x Tg-AD) harbouring $PS1_{\mathrm{M146V}}$, APP_{Swe} and tau_{P301L} transgenes, represent a unique animal model, which mimics both amyloid and tau AD neuropathologies, besides presenilin overexpression, in an age-dependent manner and in disease-relevant brain regions (Oddo et al. 2003). In this model, the key role of the neuro-immuno-endocrine network in the etiopathogenesis of AD has been shown. Thus, 3 x TgAD mice suffer an age-related impairment at the level of behaviour (lower ability to cope with stressors such as novelty, increased emotionality and anxiety-like behaviours, neophobia and reduced exploratory capacity), endocrine (higher plasma corticosterone levels) and immune parameters (decreased proliferation of lymphocytes and reduced NK cell activity), with males being more affected than females and showing higher mortality rates (Gimenez-Llort et al. 2008, 2012).

8.6 Models of Poor Response to Stress, Anxiety and Depression

It is accepted that an inadequate response to stress is one of the conditions leading to an acceleration of ageing, accompanied by an impaired immune system and other physiological systems. Moreover, the changes in cellular trafficking as well as cell-mediated immunity observed in ageing are similarly found following stress or chronic glucocorticoid exposure (Bauer 2005). Thus, it has been shown that mice with chronic hyper-reactivity to stress and anxiety show a premature ageing of the immune and nervous systems, a higher oxidative stress and a shorter life span (Viveros et al. 2007). These animals show premature ageing, and this model will be explained in more detail later. It has also been observed recently that mice exposed to the stressful condition of isolation have behavioural responses that reveal a certain degree of depression and a more evident immunosenescence than control animals of the same age housed in groups (Arranz et al. 2009b). These animal models show a significant premature immunosenescence and oxidative stress similar to those in human subjects suffering chronic anxiety or depression (Arranz et al. 2007, 2009a).

8.7 A Model of Premature Ageing in Mice Based on an Altered Stress-related Behavioural Response

The lifespan of rodent strains appears to be inversely related to the intensity of their behavioural and neuro-endocrine responses to stressful stimuli in an exploration test (Dellu et al. 1994), and reduced longevity could be caused by an accelerated age-dependent neurodegeneration (Gilad and Gilad 2000). In this context, several studies from our laboratory have shown that inter-individual differences among members of outbred Swiss and inbred BALC/c mouse populations, both male and female, may be related to their behaviour in a simple T-maze test. Moreover, animals which exhibit immobility or "freezing behaviour" (high levels of anxiety) when placed in a new environment, for example the T-maze, fail the test and show a worse immune function than those mice that performed the test correctly (De la Fuente et al. 1998; Guayerbas et al. 2000; 2002a, 2002b, 2002c; Viveros et al. 2001). These animals with premature immunosenescence also have a shorter life span (Guayerbas et al. 2002a, 2002c; Guayerbas and De la Fuente 2003). Thus, a model of premature ageing in mice based on an altered stress-related behavioural response in an exploratory test and a premature immunosenescence was established (Viveros et al. 2007). The studies carried out to characterise this model are discussed below.

8.7.1 Simple T-maze Exploration Test

Mice of the same strain, sex and chronological age are tested individually in a simple T-maze test (Guayerbas et al. 2000). The performance of the spontaneous exploratory behaviour of each mouse (marked for individual monitoring) is evaluated measuring the time elapsed until the animal crosses the intersection of the three arms with both hind legs. The test has to be performed once a week, for 4 weeks, in order to sort out the non-prematurely ageing mice (NPAM), which complete the exploration of the "vertical" arm of the maze four times in 10 s or less, from the PAM, which required over 10 s. Those animals showing an intermediate response in the T-maze are removed from the study.

8.7.2 Behavioural Characterisation

Behavioural tests have been major components of batteries designed to assess biological ageing in animal and human populations. Moreover, the performance in certain behavioural tests is considered as a marker of neurological ageing (Dellu et al. 1994), which is related to individual longevity (Gilad and Gilad 2000). Different behavioural tests have been carried out with chronologically adult–mature Swiss and BALC/c PAM and NPAM (both sexes) (Guayerbas et al. 2000, 2002a, 2005b; Viveros et al. 2001). In addition, certain behaviour characteristics of PAM, have been also investigated in young mice (Pérez-Álvarez et al. 2005). Our battery of tests provides relevant information about the strength-coordination (tightrope test) and diverse aspects of the adaptive response to stress, emotionality and anxiety (the hole board, the open field and the plus-maze test). PAM have shown an impaired neuromuscular vigour and coordination, a decreased locomotor activity and adaptive response to stressful situations, as well as an increased emotional reactivity and anxiety when compared to NPAM of the same age (the results have been summarised in Table 8.1).

The tightrope test, a method for evaluation of neuromuscular coordination and vigour (Miquel and Blasco 1978), is positively correlated to lifespan in rodents (Ingram and Reynolds 1986). Cross-sectional studies in adult Swiss and BALC/c mice (both sexes), reveal that PAM spend more time in performing this task than NPAM, which shows a decreased neuromuscular coordination and vigour in PAM (Guayerbas et al. 2000; Pérez-Álvarez et al. 2005). Moreover, this neuromuscular capacity decreased with age in both groups of mice, but more markedly in PAM (Guayerbas et al. 2002a). In the hole board and in the elevated plus-maze test, Swiss PAM (both sexes) show an increased grooming frequency, a decreased internal-central ambulation and a lower percentage of entries and time in open-arms in comparison to NPAM, which show an increased emotionality and anxiety. Interestingly, when the animals are submitted to a higher stress situation in the open field (bright white light), PAM show lower motor activity than NPAM, being the PAM females the most affected for external and internal ambulation, and showing the highest grooming frequency. Regarding to the gender differences, female mice (NPAM and PAM) show higher levels of emotionality than males, being more marked in the case of PAM.

Ageing is associated with alterations in neuro-endocrine responses to stress, such as an altered function of the hypothalamus–pituitary–adrenal axis (Orentreich et al. 1984), which is crucial for the regulation of stress and anxiety-related responses.

Table 8.1 Behavioural characterization of adult prematurely and non-prematurely ageing male and female Swiss mice. PAM, prematurely ageing mice; NPAM, non-prematurely ageing mice. Symbols represent statistical differences between the two categories of mice (PAM vs. NPAM), in both (♂) males and (♀) females mice: decreased (↓) P < 0.05; (↓↓) P < 0.01; increased (↑) P < 0.05; or (=) not change. (–) behavioral test not analyzed

Adult Swiss mice	PAM vs. NPAM	
A. Sensorimotor abilities	♂	♀
Thigtrope test (60 s trial)		
Muscular vigour (% mice falling off, latency)	↓↓	↓↓
Motor coordination	↓	↓↓
Traction	↓	↓
B. Exploratory and anxiety-like behaviors	♂	♀
Holeboard test		
External ambulation	=	=
Internal ambulation	↓↓	↓
Total ambulation	↓↓	↓
Central ambulation	↓↓	↓
Grooming	↑	↑
Open field test		
External ambulation (squares entered)	=	↓
Internal ambulation (squares entered)	=	↓↓
Total ambulation	↓↓	↓↓
Central ambulation	=	↓↓
Grooming	↑	↑
Plus-maze test		
% time in open arms	↓	↓
% open-arm entries	↓	↓
Closed-arms entries	↑	↑
Porsolt test		
Immobility	–	↓↓

Interestingly, PAM exhibited an increased baseline corticosterone levels and a blunted stress response when compared to NPAM (Pérez-Álvarez et al. 2005).

All these findings demonstrate that the PAM resemble those behavioural characteristics found in chronologically aged mice and confirm that the T-maze test could provide a simple and fast approach for the determination of murine biological age (Viveros et al. 2001).

8.7.3 Characterization of Monoaminergic Systems

There is an evidence indicating that ageing is accompanied by some alterations in the neurotransmission systems. In aged rodents, many studies have shown reductions of the levels of neurotransmitters and of the activities of the enzymes involved in their synthesis, as well as an age-related behavioural impairment, which is the result of dysfunction in the neurotransmission, but not the loss of neurons of the Central nervous system (CNS) (Magnone et al. 2000). Thus, with the aim to provide a neurochemical characterization of PAM, a comparative study with NPAM regarding their noradrenergic, serotonergic and dopaminergic systems were carried out in discrete brain regions, which are relevant for the behavioural responses, such as hypothalamus, hippocampus, striatum, frontal cortex and midbrain. For this purpose, the levels of noradrenaline (NA), serotonin (5-HT), dopamine (DA) and their respective metabolites (3-methoxy-4-hydroxyphenyl glycol (MHPG), 5-hydroxyindol-3-acetic acid (5-HIAA) and 3,4-dihydroxyphenylacetic acid (DOPAC) and homovanillic acid (HVA), respectively) were analysed in the brains of the same animals (males and females) used in the previous behavioural study (Viveros et al. 2001).

The results (De la Fuente et al. 2003) showed that PAM of both sexes show decreased NA levels in hippocampus, frontal cortex and midbrain in comparison to NPAM. Although the data on the sensitivity of noradrenergic systems to the ageing process are very heterogeneous, a decrease in the NA content of at least some brain regions, mainly in hippocampus has been proposed. PAM of both sexes also show decreased levels of 5-HT in hippocampus, striatum, midbrain and hypothalamus in comparison to NPAM, as well as an increase in the turnover rate (5-HIAA/5-HT) in hypothalamus and in hippocampus. The influence of age on the levels 5-HT in different strains of rodents have shown different results, but several studies show decreased levels of 5-HT and an increased 5-HIAA/5-HT ratio in several brain regions (Lee et al. 2001). Since there is abundant evidence about the involvement of the serotonergic system in the modulation of anxiety, the altered emotional responses in PAM, which present higher levels of emotionality and anxiety, could be related with these altered serotonergic indices. With respect to the dopaminergic system, the majority of the studies have shown a decrease of DA levels and its metabolites (DOPAC and HVA), as well as an increase of the DA turnover rate (DOPAC/DA and HVA/DA) with ageing. These alterations are in accordance with the majority of the changes found in the dopaminergic system of PAM, which (in both sexes) show a marked reduction in the DA content in most brain regions analysed (hypothalamus, hippocampus, striatum and frontal cortex), as well as a decrease in the levels of HVA in the hypothalamus and striatum, whereas in the hippocampus decreased levels of DOPAC and HVA are only observed in female PAM. The turnover rates (DOPAC/DA and HVA/DA) of PAM (both sexes) are either increased (hippocampus and frontal cortex) or unchanged with respect to NPAM, depending on the brain region analysed (De la Fuente et al. 2003). An age-related decrease in the levels of DA and its metabolites in striatum, as well as a correlation between this fact and diminished motor function in aged rodents have been observed. Thus, it is likely that the altered dopaminergic indices in PAM, particularly in striatum, are related to the impaired motor function, such as neuromuscular vigour and coordination in the tightrope test, and the decreased locomotor activity in three standard behavioural tests (Viveros et al. 2001) observed in these animals when compared to NPAM.

In spite of the sex differences in the age-related changes in the noradrenergic, serotonergic and dopaminergic systems, which depend on the brain region analysed, in most cases the differences between PAM and NPAM involve both sexes, with the exception of the hypothalamus, a typically sexual dimorphic area, where some differences only affect the male mice. In conclusion, the neuro-chemical modifications found in the monoaminergic systems in brain regions of PAM, which involve both sexes, clearly resemble some of the alterations reported for ageing animals, and this brain neurochemistry characteristic of older animals seems to be related to the

Table 8.2 Comparative changes in the level of monoamines and their respective metabolites from noradrenergic, serotonergic and dopaminergic systems in different brain regions (hippocampus, hypothalamus, striatum, frontal cortex and midbrain) of non-prematurely and prematurely ageing adult Swiss mice, both males and females. PAM, prematurely ageing mice; NPAM, non-prematurely ageing mice; NA, noradrenaline; MHPG, 3-methoxy-4-hydroxyphenyl; 5-HT, serotonine; DA, dopamine; DOPAC, 3,4-dihydroxyphenylacetic acid; HVA, homovanillic acid; 5-HIAA, 5-hydroxyindol-3-acetic acid. Symbols represent statistical differences between the two categories of mice (PAM vs. NPAM), in both (♂) males and (♀) females mice: decreased (↓) $P < 0.05$; (↓↓) $P < 0.01$; (↓↓↓) $P < 0.001$; increased (↑) $P < 0.05$; (↑↑) $P < 0.01$; (↑↑↑) $P < 0.001$; (=) or no change

Adult Swiss mice	PAM vs. NPAM	
	♂	♀
Hippocampus		
NA	↓	↓
MHPG	=	=
5-HT	↓↓↓	↓↓↓
HIAA/5-HT	↑↑↑	↑↑↑
DA	↓↓↓	↓↓↓
DOPAC	=	↓
HVA	=	↓
HVA/DA	↑↑	↑↑
DOPAC/DA	↑	=
Hypothalamus		
NA	↓↓	↓↓
5-HT	↓	↓
DA	↓	↓
Striatum		
NA	=	=
5-HT	↓↓↓	↓↓↓
DA	↓	↓
Frontal cortex		
NA	↓	↓
5-HT	=	=
DA	↓↓↓	↓↓↓
Midbrain		
NA	↓	↓
5-HT	↓	↓
DA	=	=

behavioural features found in PAM (De la Fuente et al. 2003). All these results on the monoaminergic systems have been summarised in Table 8.2.

8.7.4 Characterization of Immune Functions

Ageing has been associated with immunological changes, which presently are considered as good markers of health, biological age and longevity (De la Fuente and

Miquel 2009; Wikby et al. 2008). Moreover, the immunological changes observed in healthy ageing may be closely related to psychological stress, which may lead to an earlier onset of ageing-related diseases and premature ageing. Indeed, there is an evidence suggesting a link between chronic psychological stress and impaired immune function (Bauer 2008). In this context, preliminary studies showed a relation between the T-maze performance and the functions of immune cells from old and adult Swiss mice, animals with slow performance showing a less competent immune system (De la Fuente et al. 1998; Correa et al. 1999). This immunosenescence of PAM was confirmed after performing several investigations on a wide range of functions in immune cells from peritoneum, thymus, spleen and axillary nodes from PAM and NPAM of different strains (Swiss and BALB/c), chronological ages (young, adult, mature and old), both sex as well as in cross-sectional and longitudinal studies (Table 8.3). In fact, the results show, in general, an increased macrophage and lymphocyte adherence in PAM compared with NPAM, which suggests an impaired capacity of these cells to move to the infectious focus, this function increasing with ageing. Moreover, the spontaneous mobility and chemotaxis capacity of macrophages and lymphocytes, the phagocytic capacity and bacteriocidal activity of macrophages, the lymphoproliferative response to mitogens (Con A and LPS), the IL-1β and IL-2 release as well as the NK activity, which are the functions that decrease with ageing, are lower in PAM in comparison to NPAM (Alvarado et al. 2005, 2006b; Alvarez et al. 2006; De la Fuente 2010; Guayerbas et al. 2002a, 2002b, 2002c; 2005a; Guayerbas and De la Fuente 2003; Puerto et al. 2002; Viveros et al. 2001). Moreover, young PAM are more susceptible than NPAM to the effect of an acute stress (which includes forced swim) on the mitogen-induced lymphoproliferative responses, decreasing this activity (Pérez-Álvarez et al. 2005). The results on the immune functions in PAM versus NPAM are shown in Table 8.3.

Considering the relevance of optimal immune functions for successful ageing, the present data justify the view that PAM have worst preserved immune functions compared to NPAM, and they show values more similar to those of older animals. Since the immune functions studied have been proposed as markers of health and biological age, the impairment in these functions observed in PAM could play a central role in the shorter life span of these animals. The above-mentioned facts demonstrate the premature immunosenescence of PAM and support the increasing evidence on the key role played by the immune system in premature ageing (De la Fuente 2008; De la Fuente and Miquel 2009).

8.7.5 *Inflammatory and Oxidative Stress*

Progressive dysregulation of immune responses associated with ageing may be a result of increased oxidative stress (De la Fuente et al. 2005). In view of the link between oxidative and inflammatory stress and the ageing process, several oxidant and pro-inflammatory compounds, as well as antioxidant defences, have been evaluated in the model of premature ageing in mice, in order to characterize redox state

Table 8.3 Changes in immune functions in (A) peritoneal leukocytes (macrophages, lymphocytes and natural killer cells) from chronologically young, adult and old Swiss female mice and in (B) axillary nodes, spleen and thymus leukocytes from adult and old female Swiss and BALC/c mice, both in prematurely and non-prematurely ageing mice. PAM, prematurely ageing mice; NPAM, non-prematurely ageing mice. Symbols represent statistical differences between the two categories of mice (PAM vs. NPAM), in young, adult and old Swiss and BALC/c mice: decreased (↓) $P < 0.05$; (↓↓) $P < 0.01$; (↓↓↓) $P < 0.001$; increased (↑) $P < 0.05$; (↑↑) $P < 0.01$; or (=) no hange. (*) decreased (↓) $P < 0.05$ in BALC/c mice; (**) decreased (↓↓) $P < 0.01$ in BALC/c mice. (−) not analyzed

A) Peritoneal leukocytes	PAM vs. NPAM		
Chronological age	Young	Adult	Old
Macrophage functions			
Adherence capacity	–	↑↑	↑↑
Chemotaxis	↓↓↓	↓↓↓	↓↓
Phagocytosis capacity	=/↓*	↓↓	=/↓↓*
Intracellular O_2 anion levels	↓	↓↓↓	↓
Intracellular ROS levels	↓↓	↓	↓
IL-1ß release	–	↓	↓
Lymphocyte functions			
Adherence capacity	–	↑	↑
Chemotaxis	↓↓↓	↓↓↓	↓↓
Proliferation to Con A (%)	↓↓	↓↓	↓
IL-2 release	↓↓	↓	=
Natural Killer (NK) functions			
NK activity (lysis %)	↓↓↓	↓	↓

B) Leukocytes	PAM vs. NPAM					
Chronological age	Axillary nodes		Spleen		Thymus	
	Adult	Old	Adult	Old	Adult	Old
Swiss mice						
Chemotaxis	↓↓	↓	=	↓	=	=
Basal proliferation (cpm)	–	=	–	↓↓	–	↓↓
Proliferation to Con A (%)	↓↓	↓	↓↓	↓↓↓	↓↓	↓↓↓
NK activity (lysis %)	↓↓	=	↓	=	=	↓
IL-2 release	↓↓↓	↓	–	–	–	–
BALC/c mice						
Chemotaxis	=	↓↓	↓↓	↓↓	↓↓	=
Basal proliferation (cpm)	=	=	–	=	–	=
Proliferation to Con A (%)	↓↓	↓↓	↓↓	↓↓	↓	↓
NK activity (lysis %)	↓	↓	↓	↓↓↓	↓↓	↓↓↓
IL-2 release	↓↓	↓	–	–	–	–

Table 8.4 Comparative changes in several oxidative stress parameters in peritoneal leukocytes from prematurely versus non-prematurely ageing female Swiss mice in young and adult mice. PAM, prematurely ageing mice; NPAM, non-prematurely ageing mice. Symbols represent statistical differences between the two categories of mice (PAM vs. NPAM): decreased (↓) $P < 0.05$; (↓↓) $P < 0.01$; (↓↓↓) $P < 0.001$; increased (↑) $P < 0.05$; (↑↑) $P < 0.01$; (↑↑↑) $P < 0.001$; or (=) no change. (*) increased (↑) $P < 0.05$ in old Swiss mice; (**) increased (↑↑↑) $P < 0.001$ in BALC/c adult mice; (***) increased (↑) $P < 0.001$ in old Swiss mice; (–) not analyzed.

Peritoneal leukocytes	PAM vs. NPAM	
	Young	Adult
Oxidant/Pro-inflammatory compounds		
Extracellular superoxide anion levels	↑↑	↑/↑* ↑/↑↑↑**
Nitric oxide (NO) levels	↑↑	↑
Oxidized glutathione (GSSH)	↑	↑↑
Oxidized/reduced glutathione (GSSH/GSH)	↑↑↑	↑↑↑
Prostaglandin E2 (PGE2)	↑	↑↑
Xanthine oxidase (XO) activity	–	↑↑
Tumor necrosis factor alpha (TNF-α) levels	↑	↑/↑***
Antiaoxidant Defenses		
Reduced glutathione (GSH) levels	↓↓	↓↓
Glutathion peroxidase (GPx) activity	=	↓↓
Glutathion reductase (GR) activity	↓↓↓	–
Superoxide Dismutase (SOD) activity	↓↓↓	↓
Catalase (CAT) activity	↓	↓
Oxidative Damage		
Malondialdehyde (MDA) levels	↑↑	↑
8oxo-7,8dihydro-2deoxiguanosine (8oxodG)	–	↑↑

of PAM in comparison to NPAM. These parameters of the oxidative and pro-inflammatory stress status have been studied in leukocytes (from peritoneum, axillary nodes, spleen and thymus), as well as in other tissues of Swiss and BALC/c PAM and NPAM of different chronological ages (Alvarado et al. 2006a, 2006b; Guayerbas et al. 2002b; Viveros et al. 2007). In general, PAM show higher levels of oxidant and pro-inflammatory compounds as well as decreased levels of antioxidant defences in comparison to NPAM (results summarised in Table 8.4).

Regarding pro-inflammatory cytokine release, such as tumor necrosis factor-α (TNF-α) and prostaglandin E2 (PGE2), in Swiss female PAM increase more than in NPAM. Oxidants such as extracellular superoxide anion, nitric oxide (NO), oxidized glutathione (GSSG) and the GSSG/GSH ratio (a marker of oxidative stress situation) in young and middle-aged peritoneal leukocytes of female Swiss PAM, show higher levels than in those of NPAM (Alvarado et al. 2006a, 2006b). In addition, these leukocytes from PAM also show, in comparison to NPAM, a decrease in reduced glutathione levels (GSH), a key non-enzymatic antioxidant, as well as in the enzymatic antioxidant defences, namely superoxide dismutase, catalase, glutathione peroxidase and reductase activities which are the main antioxidant enzymes that eliminate the excess of reactive oxygen species (ROS). Therefore, PAM show oxidative stress in their leukocytes, which induces oxidative damage of biomolecules

Table 8.5 Xanthine oxidase (XO) activity (mU XO/mg protein) in different tissues (cerebral cortex, liver, spleen and kidney) from chronologically mature prematurely (PAM) and non-prematurely (NPAM) ageing female BALC/c mice. Data are mean and standard deviation for eight animals. Statistical differences between the groups were analyzed by the *Student's t* test. * $P < 0.05$; ** $P < 0.01$ with respect to the corresponding values in NPAM

Xanthine Oxidase (XO) activity (mU XO/mg protein)		
Tissue	BALC/c mature mice	
	NPAM	PAM
Cerebral cortex	29.83 ± 8.24	34.20 ± 6.39*
Spleen	4.78 ± 1.67	7.25 ± 3,58
Liver	54.36 ± 9.60	65.69 ± 10.9**
Kidney	9.53 ± 1.59	11.74 ± 1,69*

such as lipids and DNA, a characteristic of ageing, and contributes to the inappropriate function of these cells (De la Fuente and Miquel 2009). In fact, the oxidative stress damage to lipids and nuclear DNA measured by malondialdehyde (MDA) and 8-oxo,7,8-dyhidro-2′-deoxyguanosine (8-oxodG) levels, respectively) in peritoneal leukocytes from adult and young Swiss PAM are higher than those found in cells from NPAM (Alvarado et al. 2006a, 2006b). Moreover, an imbalance between oxidant and antioxidants, leading to an oxidative stress situation, has been observed in several tissues such as brain, liver, heart and kidney in Swiss and BALB/c PAM with respect to those in NPAM (Viveros et al. 2007).

All these findings demonstrate that the PAM suffer a situation of oxidative and inflammatory stress in their leukocytes and tissues that are characteristic of mice of an older chronological age. Because oxidative stress may lead to loss of homeostasis in immune cells, all these findings could explain the immunosenescence exhibited by PAM. Moreover, as mentioned above, it has been demonstrated that the PAM have a shorter life span than NPAM (Guayerbas et al. 2002a, 2002c; Guayerbas and De la Fuente 2003). Therefore, all these facts confirm that PAM are biologically older, at the same chronological age, than NPAM (Viveros et al. 2007).

8.7.6 Xanthine Oxidase in PAM and NPAM

Xanthine oxidase (XO), an enzyme characterized by its generation of ROS, such as superoxide anion and hydrogen peroxide, which increase in tissues and leukocytes of older mice (Arranz et al. 2010a; Vida et al. 2009, 2011), shows higher activity in cerebral cortex, liver and kidney of mature BALC/c PAM than in NPAM (Table 8.5). This increase in XO activity was also found in peritoneal leukocytes of adult PAM (in Swiss and BALC/c mice) than of NPAM (Table 8.4), being more evident in phagocytes than in lymphocytes (unpublished data). This adds support to the suggestion that phagocytes are the immune cells most involved in the ageing rate of organisms (De la Fuente and Miquel 2009).

8.8 Conclusion

The use of animal models, especially the PAM model, may be particularly useful to assess the relationships between a diminished ability of coping with stressful situation, immunosenescence, a chronic oxidative stress, an accelerated biological age and a shorter longevity.

Acknowledgments The authors would like to thank Dr. Miquel for his critical revision of the article. This work was supported by grants of the MICINN (BFU2011-30336), Research Group of UCM (910379ENEROINN) and RETICEF (RD06/0013/0003) (ISCIII).

References

Ader R, Felten DL, Cohen N (2001) Psychoneuroimmunology. Academic Press, San Diego

Aguilera G (2011) HPA axis responsiveness to stress: implications for healthy aging. Exp Gerontol 46:90–95

Alonso-Fernandez P, De la Fuente M (2011) Role of the immune system in aging and longevity. Curr Aging Sci 4:78–100

Alonso-Fernandez P, Puerto M, Maté I, Ribera JM, De la Fuente M (2008) Neutrophils of centenarians show function levels similar to those of young adults. J Am Geriatr Soc 56:2244–2251

Alvarado C, Álvarez P, Jiménez L, De la Fuente M (2005) Improvement of leukocytes functions in young prematurely aging mice after 5-week ingestion of a diet supplemented with biscuits enriched in antioxidants. Antioxid Redox Signal 7:1203–1209

Alvarado C, Álvarez P, Jiménez L, De la Fuente M (2006a) Oxidative stress in leukocytes from young prematurely aging mice is reversed by supplementation with biscuits rich in antioxidants. Dev Comp Immunol 30:1168–1180

Alvarado C, Álvarez P, Puerto M, Gausserès N, Jiménez L, De la Fuente M (2006b) Dietary supplementation with antioxidants improves functions and decreases oxidative stress of leukocytes from prematurely aging mice. Nutrition 22:767–777

Alvarez P, Alvarado C, Puerto M, Schlumberger A, Jimenez L, De la Fuente M (2006) Improvement of leukocyte functions in prematurely aging mice after five weeks of diet supplementation with polyphenol-rich cereals. Nutrition 22:913–921

Arranz L, Caamao JH, Lord JM, De la Fuente M (2010a) Preserved immune functions and controlled leukocyte oxidative stress in naturally long-lived mice: possible role of nuclear factor kappa β. J Gerontol Ser A 65:941–950

Arranz L, De Vicente A, Muoz M, De la Fuente M (2009a) Impairment immune function in a homeless population with stress-related disorders. Neuroimmunomodulation 16:251–260

Arranz L, Giménez-Llort L, De Castro NM, Baeza I, De la Fuente M (2009b) Social isolation during old age worsens cognitive, behavioral and immune impairment. Revista Espaola de Geriatría y Gerontología 44:137–142

Arranz L, Guayerbas N, De la Fuente M (2007) Impairment of several immune functions in anxious women. J Psychosom Res 62:1–8

Arranz L, Lord JM, De la Fuente M (2010b) Preserved ex vivo inflammatory status and cytokine responses in naturally long-lived mice. AGE 32:451–466

Bae CY, Kang YG, Kim S, Cho C, Kang HC, Yu BY, Lee S, Cho KH, Lee DC, Lee K, Kim JS, Shin KK (2008) Development of models for predicting biological age (BA) with physical, biochemical, and hormonal parameters. Arch Gerontol Geriatr 47:253–265

Baeza I, De Castro NM, Arranz L, Fdez-Tresguerres J, De la Fuente M (2011) Ovariectomy causes in peritoneal leukocytes of age mice similar levels to males of immunosenescence and oxi-inflamm-ageing. Biogerontology. 12:227–238
Baeza I, De Castro NM, Gimenez-Llort L, & De la Fuente M (2010a) Ovariectomy, a model of menopause in rodents, causes a premature aging of the nervous and immune systems. J Neuroimmunol 219:90–99
Baeza I, Fedez-Tresguerres J, Ariznavarreta C, De la Fuente M (2010b) Effects of growth hormone, melatonin, oestrogens and phytoestrogens on the oxidized glutathione (GSSG)/reduced glutathione (GSH) ratio and lipid peroxidation in aged ovariectomized rats. Biogerontology 11:687–701
Bauer ME (2005) Stress, glucocorticoids and ageing of the immune system. Stress 8:69–83
Bauer ME (2008) Chronic stress and immunosenescence: a review. Neuroimmunomodulation 15:241–250
Besedovsky HO, Del Rey A (2007) Physiology of psychoneuroimmunology: a personal view. Brain Behav Immun 21:34–44
Borkan A, Norris AH (1980) Assessment of biological age using a profile of physical parameters. J Gerontol 35:177–184.
Correa R, Blanco B, Del Río M, Víctor VM, Guayerbas N, Medina S, De la Fuente M (1999) Effect of a diet supplemented with thioproline on murine macrophage function in a model of premature ageing. BioFactors 10:195–200
Costa-Pinto FA, Palermo-Nieto J (2010) Neuroimmune interactions in stress. Neuroimmunomodulation 17:196–199
Couillard-Depres S, Iglseder B, Aigner L (2011) Neurogenesis, cellular plasticity and cognition: the impact of stem cells in the adult and aging brain. Gerontology 57:559–564
De Castro NM, Baeza I, Arranz L, Vida C, Hernandez O, Ubeda N, De la Fuente M (2009) Adult obese Zucker rats show an impairment of lymphoproliferation and natural killer activity with respect to Wistar rats. Acta Physiologica 57:559–564
De Castro NM, Baeza I, Vida C, Ubeda N, Manso R, De la Fuente M. (2010) Oxidative stress in genetically obese rats. A possible model of premature aging. Proc Nutr Soc 69 (OCE3):E258. doi: 10.1017/S0029665100000479
De la Fuente M (2008) Role of neuroimmunomodulation in aging. Neuroimmunomodulation 15:213–223
De la Fuente M (2010) Murine models of premature ageing for the study of diet-induced immune changes: improvement of leukocyte functions in two strains of old prematurely ageing mice by dietary supplementation with sulphur-containing antioxidants. Proc Nutr Soc 69:651–659
De la Fuente M, Baeza I, Guayerbas N, Puerto M, Castillo C, Salazar V, Ariznavarreta C, F-Tresguerres JA (2004) Changes with aging in several leukocyte functions of male and female rats. Biogerontology 5:389–400
De la Fuente M, De Castro NM (2012) Obesity as a model of premature immunosenescence. Curr Immunol Rev 8:63–75
De la Fuente M, Gimenez-Llort L (2010) Models of aging of neuroimmunomodulation: strategies for its improvement. Neuroimmunomodulation 17:213–216
De la Fuente M, Hernanz A, Medina S, Guayerbas N, Fernández B, Viveros MP (2003) Characterization of monoaminergic systems in brain regions of prematurely ageing mice. Neurochem Int 43:165–172
De la Fuente M, Hernanz A, Vallejo MC (2005) The immune system in the oxidation stress conditions of aging and hipertensión: favorable effects of antioxidants and physical exercise. Antioxid Redox Signal 7:1356–1366
De la Fuente M, Miano M, Victor VM, Del Río M, Fernández MD, Díez A, Miquel J (1998) Relation between exploratory activity and immune function in aged mice: a preliminary study. Mech Ageing Dev 102:263–277
De la Fuente M, Miquel J (2009) An update of the oxidation-inflammation theory of aging: involvement of the immune system in oxi-inflamm-aging. Curr Pharm Des 15:3003–3026

Dellu F, Mayo W, Vallée M, Le Moal M, Simon H (1994) Reactivity to novelty during youth as a predictive factor of cognitive impairment in the elderly: a longitudinal study in rats. Brain Res 653:51–56

Eaton SM, Burns EM, Kusser K, Randall TD, Haynes L (2004) Age-related defects in CD4 T cell cognate helper function lead to reductions in humoral responses. J Exp Med 200:1613–1622

Fabris N. (1990) A neuroendocrine-immune theory of aging. Int J Neurosci 51:373–375

Franceschi C, Capri M, Monti D, Giunta S, Olivieri F, Sevini F, Panouraia MP, Invidia L, Celani L, Scurti M, Cevenini E, Castellani GC, Salvioli S (2007) Inflammaging and anti-inflammaging: a systemic perspective on aging and longevity emerged from studies in humans. Mech Ageing Dev 128:92–105

Frasca D, Blomberg BB (2009) Effects of aging on B cell function. Curr Opin Immunol 21:425–430

Gemma C (2010) Neuroimmunomodulation and aging. Aging Dis 1:169–172

Gilad GM, Gilad VH (2000) Strain, stress, neurodegeneration and longevity. Mech Ageing Dev 78:75–83

Gimenez-Llort L, Arranz L, Maté I, De la Fuente M (2008) Gender-specific neuroimmunoendocrine aging in a triple-transgenic 3xTg-AD mouse model for Alzheimer×s Disease and its relation with longevity. Neuroimmunomodulation 15:331–343

Gimenez-Llort L, Maté I, Manassra R, Vida C, De la Fuente M (2012) Peripheral immune system and neuroimmune communication impairment in a mouse model of Alzheimer's disease. Ann N Y Acad Sci 1262:74–84

Gouin JP, Hantsoo L, Kiecolt-Glaser JK (2008) Immune dysregulation and chronic stress among older adults: a review. Neuroimmunomodulation 15:254–262

Guayerbas N, Catalán M, Victor VM, Miquel J, De la Fuente M (2002a) Relation of behaviour and macrophage function to life span in a murine model of premature immunosenescence. Behav Brain Res 134:41–48

Guayerbas N, De la Fuente M (2003) An impairment of phagocytic function is linked to a shorter life span in two strains of prematurely aging mice. Dev Comp Immunol 27:339–350

Guayerbas N, Puerto M, Alvarado C, De la Fuente M (2005a) Effect of diet supplementation with N-acetylcysteine on leucocyte functions in prematurely aging mice. J Appl Biomed 3:199–205

Guayerbas N, Puerto M, Fernández MD, De la Fuente M (2002b) A diet supplemented with thiolic anti-oxidants improves leukocyte function in two strains of prematurely ageing mice. Clin Exp Pharmacol Physiol 29:1009–1014

Guayerbas N, Puerto M, Hernanz A, Miquel J, De la Fuente M (2005b) Thiolic antioxidant supplementation of the diet reverses age-related behavioural dysfunction in prematurely ageing mice. Pharmacol Biochem Behav 80:45–51

Guayerbas N, Puerto M, Victor VM, Miquel J, De la Fuente M (2002c) Leukocyte function and life span in a murine model of premature immunosenescence. Exp Gerontol 37:249–256

Guayerbas N, Sánchez AI, Gamallo A, Miquel J, De la Fuente M (2000) Mouse performance in an exploratory activity test as a longevity biomarker. In: Tur-Marí JA, Orellana JM (eds) Animal research welfare. A partnership. Laboratory Animals Ltd, London, pp 159–162

Haynes L, Maue AC (2009) Effects of aging on T cell function. Curr Opin Immunol 21:414–417

Hedden T, Gabrieli JDE (2004) Insights into the ageing mind: a view from cognitive neuroscience. Nat Rev Neurosci 5:87–96

Ingram DK, Reynolds MA (1986) Assessing the predictive validity of psychomotor tests as measures of biological age in mice. Exp Aging Res 12:155–162

Irwin MR, Miller AH (2007) Depressive disorders and immunity: 20 years of progress and discovery. Brain Behav Immun 21:374–383

Kim JJ, Song EY, Kosten TA (2006) Stress effects in the hippocampus: synaptic plasticity and memory. Stress 9:1–11

Kirkwood T.B.L. (2008) Gerontology: healthy old age. Nature, 455:739–740

Lamas O, Marti A, Martinez JA (2002a) Obesity and immunocompetence. Eur J Clin Nutr 56:S42–S45

Lamas O, Martinez JA, Marti A (2002b) T helper lymphopenia and decreased mitogenic response in cafeteria diet-induced obese rats. Nutr Res 22:496–507

Lee JJ, Chang CK, Liu LM, Chi TC, Yu HJ, Cheng JT (2001) Changes in endogenous monoamines in aged rats. Clin Exp Pharmacol Physiol 28:285–289

Lord JM, Phillips AC, Arlt W (2009) Synergistic effects of ageing and stress on neutrophil function. In: Fulop T, Franceschi C, Hirokawa K, Pawelec G (eds) Handbook of immunosenescence Springer, New York pp 475–498

Magnone MC, Rossolini G, Piantanelli L, Migani P (2000) Neurochemical parameters of the main neurotransmission systems in aging mice. Arch Gerontol Geriatr 30:269–279

Makrantonaki E, Schonknecht P, Hossini AM, Kaiser E, Katsouli MM, Adjaye J, Schroder J, Zouboulis CC (2010) Skin and brain age together: the role of hormones in the ageing process. Exp Gerontol 45:801–813

Miquel J, Blasco M (1978) A simple technique for evaluation of vitality loss in aging mice, by testing their muscular coordination and vigor. Exp Gerontol 13:389–396

Oddo S, Caccamo A, Shepherd JD, Murphy MP, Golde TE, Kayed R, Metherate R, Mattson MP, Akbari Y, LaFerla FM (2003) Triple-transgenic model of Alzheimer's disease with plaques and tangles: intracellular Aβ and synaptic dysfunction. Neurone 39:409–421

Orentreich N, Brind JL, Rizer RL, Vogelman JH (1984) Age changes and sex differences in serum dehydroepiandrosterone sulfate concentrations throughout adulthood. J Clin Endocrinol Metab 59:551–555

Pérez-Álvarez L, Baeza I, Arranz L, Marco EM, Borcel E, Guaza C, Viveros MP, De la Fuente M (2005) Behavioral, endocrine and immunological characteristics of a murine model of premature aging. Dev Comp Immunol 29:965–976

Puerto M, Guayerbas N, Víctor VH, De la Fuente M (2002) Effects of N-acetylcysteine on macrophage and lymphocyte functions in a mouse model of premature ageing. Pharmacol Biochem Behav 73(4):797–804

Shaw AC, Joshi S, Greenwood H, Panda A, Lord JM (2010) Aging of the innate immune system. Curr Opin Immunol 22:507–513

Vida C, De Castro NM, Corpas I, De la Fuente M, González E (2009) Changes in xanthine oxidase activity and lipid peroxidation levels in prematurely aging mice. Acta Physiol 195:100–101

Vida C, Rodriguez-Téres S, Heras V, Corpas I, De al Fuente M, Gonzalez E (2011) The age-related increase in xanthine oxidase expression and activity in several tissues from mice is not shown in long-lived animals. Biogerontolgy 12:551–564

Viveros MP, Fernández B, Guayerbas N, De la Fuente M (2001) Behavioral characterization of a mouse model of premature immunosenescence. J Neuroimmunol 114:80–88

Viveros MP, Arranz L, Hernanz A, Miquel J, De la Fuente M (2007) A model of premature aging in mice based on altered stress-related behavioral response and immunosenescence. Neuroimmunomodulation 14:157–162

Weiskopf D, Weinberger B, Grubeck-Loebenstein B (2009) The aging of the immune system. Transplant Int 22:1041–1050

Wikby A, Mansson IA, Johansson B, Strindhall J, Nilsson SE (2008) The immune risk profile is associated with age and gender: findings from three Swedish population studies of individuals 20–100 years of age. Biogerontology 9:299–308

Woiciechowsky C, Schoning B, Lanksch WR, Volk HD, Docke WD (1999) Mechanisms of brain-mediated systemic anti-inflammatory syndrome causing immunodepression. J Mol Med 77:769–780

Ye J (2011) Obesity, inflammation and the metabolic syndrome. In: Serrano M, Ordovas JM, Gutierrez JA (eds) Obesity Elsevier, Barcelona, pp 169–188

Yirmiya R, Goshen I (2011) Immune modulation of learning, memory, neural plasticity and neurogenesis, Brain Behav Immun 25:181–213

Chapter 9
Socio-economic Status and Immunosenescence

Allison E. Aiello and Jennifer Beam Dowd

9.1 Introduction

Across the world, wealth and health go hand in hand. Those with fewer educational and economic resources live shorter lives and experience the onset of chronic disease and loss of functioning at an earlier age on average than their more advantaged peers (Adler and Stewart 2010; Mackenbach 2010). Recent estimates show over a 6-year difference in life expectancy at age 25 for those below the US poverty line compared to those in the top 35 % of the income distribution (Braveman et al. 2010). In Europe, similar socio-economic differences are seen but sometimes of different magnitude across countries. Educational inequalities range from a relative mortality risk that is four times higher for men with the lowest compared to the highest educational attainment in the Czech Republic, Hungary, and Lithuania, to less than two times that of those with the most education in Sweden and Spain (Mackenbach et al. 2008). While the biological mechanisms underlying these socio-economic health inequalities or "disparities" are not well understood, they are often discussed in the context of accelerated aging (Crimmins et al. 2009). Among the biological systems that have been investigated, studies of cardiovascular, metabolic, and neuroendocrine markers are the most common (Dowd et al. 2009b; Seeman et al. 2008). Despite the importance of immunosenescence in aging, research on the contribution of the immune system to health disparities until recently has been quite limited.

In this chapter, we will give an overview of recent epidemiological work identifying associations between socio-economic factors and various markers related to immunosenescence, including pathogen burden and immune response to persistent infections such as cytomegalovirus (CMV). CMV infection, in particular, may play

A. E. Aiello (✉)
Department of Epidemiology, Center for Social Epidemiology & Population Health, University of Michigan 3659 SPH Tower, 109 Observatory Ann Arbor, MI 48109-2029, USA
e-mail: aielloa@umich.edu

J. B. Dowd
Hunter College, City University of New York (CUNY), New York, NY, USA

J. A. Bosch et al. (eds.), *Immunosenescence*,
DOI 10.1007/978-1-4614-4776-4_9, © Springer Science+Business Media New York 2013

a direct role in immune system aging, and has even been called the "driving force" behind age-associated alterations to the T-cell immune system (Koch et al. 2006). Infection with this virus increases CMV specific T-cell proliferation, reducing the availability of other T-cells carrying receptors specific for other antigens. This process is referred to as taking up the "immunological space" and limits the capacity of the immune system to mount an efficient immune response in elderly individuals (Trzonkowski et al. 2004).

Next, we will touch on the lifecourse framework that motivates much epidemiological and social science research in this area. Overall, socio-economic status (SES) is strongly associated with the prevalence of many acute and persistent infections, but whether this pattern is due to increased infectious exposures or increased susceptibility is not clear. Individuals with lower SES also show increased IgG antibody response to persistent herpesviruses including CMV, Herpes Simplex Virus-1 (HSV-1), and Epstein-Barr Virus (EBV), possibly due to differential exposure to stress. There is evidence that these socio-economic differences in seroprevalence and antibody response arise in childhood, suggesting that early social factors might contribute to later immunosenescence. We conclude with our perspective on current research gaps and opportunities in this area.

9.2 SES and Infectious Burden

Lifelong antigenic history of exposure to pathogens and other antigens may play an important role in immunosenescence and its impact on frailty and survival in old age (De Martinis et al. 2005). Individual antigenic history in turn, is a function of historical trends, geographical location, as well as social, demographic, and behavioral factors. Research on the relationship between social factors and the distribution of infections in developed countries is relatively scarce, but mounting evidence suggests that an individual's social environment dramatically influences their lifelong exposure to pathogens, even in developed countries (Cohen et al. 2006). A recent review of the social determinants of infectious disease in Europe found that in addition to well-established associations between social factors and infections such as HIV and tuberculosis, socio-economic differences in prevalence were found for infections as varied as *Helicobacter pylori*, human papilomavirus (HPV), listeria, and tick-borne encephalitis (Semenza 2010). Recent evidence suggests the socio-economic patterning of infection may apply to everyday illnesses as well. In the United States, Stone et al. (2010) reported a socio-economic gradient in the presence of self-reported cold or flu symptoms for the previous day among a representative sample of adults.

Additional evidence is mounting regarding socio-economic differences in the prevalence of persistent pathogens. In a sample of Whitehall civil servants from the United Kingdom, employment grade was found to be related to infection with CMV, HSV-1, and *Chlamydia pneumoniae* (Steptoe et al. 2007). In the United States, race/ethnic differences in individual infection status for hepatitis A, B, and C viruses, *Toxoplasma gondii*, *Helicobacter pylori*, HSV-2, and CMV have been reported

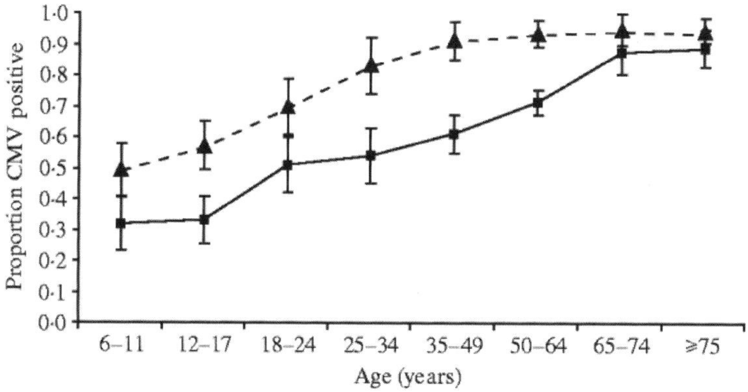

Fig. 9.1 Data from the Third National Health and Nutrition Examination Survey 1988–1994, showing seroprevalence of cytomegalovirus by age and income quartile. Results were adjusted for race/ethnicity, rural/urban residence, census region, sex, and household size. –■–, Top income quartile; –▲–, bottom income quartile. (Reprinted with permission from: Dowd et al. (2009a))

(Everhart et al. 2000; McQuillan et al. 2004; Staras et al. 2006). Differences in the seroprevalence of CMV by family income level were found in the United States as early as age 6, and were not explained by risk factors such as household crowding, parental smoking, or daycare attendance early in life (Dowd et al. 2009a, c) (Fig. 9.1). Beyond seroprevalence of individual infections, US adults of lower SES in the nationally representative National Health and Nutrition Examination Survey (NHANES) were also found to have a higher total burden of multiple chronic infections including CMV, HSV-1, *H. pylori*, and Hepatitis B (Dowd et al. 2009a; Zajacova et al. 2009). Similar results were found by Aiello et al. (2009) using data from the Multi-Ethnic Study of Atherosclerosis (MESA), where less education was associated with a greater probability of infection with multiple pathogens (CMV, HSV-1, *H. pylori*, and *C. pneumoniae).* A study among younger age populations in the United States showed significant differences in the seroprevalence of five different infections by SES emerging as early as age 6 (Dowd et al. 2009c). The mean persistent infection burden among US children whose parents had less than a high school education was significantly higher compared to children with parents that had more than a high school education. Taken together, these findings highlight the importance of understanding the connection between social factors, antigenic burden, and immunosenescence across the life course.

9.3 SES and Stress

Chronic stress has received widespread attention as a potential mechanism by which SES can “get under the skin” (Baum et al. 1999; Brunner 1997; Kristenson et al. 2004; Steptoe and Marmot 2002). Individuals with lower SES have both increased exposure to stressful events in their lives as well as fewer social and material resources with which to buffer stressful events that occur (Baum et al. 1999; Pearlin et al. 2005). Such increased stress exposure for lower SES individuals may emerge from physical environments characterized by crime, crowding, poor physical amenities, financial/occupational environments characterized by inadequate and/or unpredictable resources, nonexistent or little job security or personal autonomy and/or sociocultural environments characterized by discrimination, and impoverished social and psychological resources (e.g., fewer sociopolitical resources, individual perceptions of powerlessness, alienation, lack of self-esteem) (Seeman and Crimmins 2001). This stress, in turn, is thought to impact health outcomes via sustained activation of stress-related autonomic and neuroendocrine responses and impaired immunity. Cortisol has been the most commonly researched mechanism presumed to link low SES to poor health, but thus far, measures of SES have been inconsistently related to cortisol, providing little direct evidence that stress-related biological alterations explain the relationship between socio-economic position and health (Brandtstadter et al. 1991; Cohen et al. 1997; Dowd and Goldman 2006; Kristenson et al. 2004; Kunz-Ebrecht et al. 2004; Steptoe et al. 2003). Antibody response to latent infections such as CMV, which may be linked to both stress and immunosenescence, has been suggested as an alternative marker for investigating the interactions of the social environment, stress, and health (Dowd and Aiello 2009; Dowd et al. 2011).

9.4 Socio-economic Stressors and Latent Herpesvirus Antibodies

Adequate cell-mediated immunity is important for maintaining persistent infections such as CMV, EBV or HSV-1 in a latent state. Stress-related down regulation of cellular immunity can allow a herpesvirus to reactivate, releasing viral antigens into circulation (Glaser and Jones 1994). In turn, levels of antibodies to CMV, EBV, and HSV-1 provide an indirect measure of cell-mediated immune function. Findings from the psychoneuroimmunology literature support a strong and consistent relationship between stress and increased antibody response to herpesviruses (Herbert and Cohen 1993). Specifically, increases in herpesvirus antibody titers have been linked to academic stress in nursing, medical students and military cadets (Glaser et al. 1999; Sarid et al. 2002), caregiving for a family member with Alzheimer's disease (Glaser and Kiecolt-Glaser 1997), involvement in a poor quality marriage (Herbert and Cohen 1993), traumatic life events (McDade et al. 2007; Shirtcliff et al. 2009), as well as the psychological traits of loneliness, defensiveness, and anxiety (Esterling et al. 1993; Glaser et al. 1985). CMV antibodies, in particular, have been shown to increase in response to academic stress in several student populations (Glaser et al. 1985; Sarid et al. 2002), and the stress of spaceflight for astronauts (Mehta et al. 2000). Cross-sectionally, increased CMV antibody titers have been associated with depression and anxiety in older adults (Phillips et al. 2008). Increased severity of parental psychiatric symptoms has also been associated with an increase in the percentage of CD4 and CD8 cells associated with immune control over CMV in school-aged children (Caserta et al. 2008). Further, in women newly diagnosed or suspected of having breast cancer, lower EBV antibody titers were observed among those with higher education and social support, suggesting better cellular immunity while under stress in those with higher SES (Fagundes et al. 2012).

Animal studies support the idea that social stressors may contribute to reactivation of latent herpesviruses. Specifically, Padgett et al. (1998) found that mice whose hierarchy and social interactions were disrupted showed significant evidence of re-activation of latent HSV-1, while the use of a restraint stress did not activate the latent virus. Looking at social status in humans, Dowd et al. (2008) first reported associations between lower educational attainment and higher CMV and HSV-1 IgG antibody levels in an sample of elderly US Latinos, followed by similar associations for both education and income with CMV in a representative sample US adults aged 25 and older (Dowd and Aiello 2009). The results suggested that on average, each additional year of education was associated with a CMV antibody level equivalent to that of individuals almost 2 years younger (Dowd and Aiello 2009). These associations remained strong after adjustment for baseline health conditions, smoking, and BMI. Participants in the MESA study with lower levels of education were also more likely to have high levels of antibody to multiple pathogens including CMV and HSV-1 (Aiello et al. 2009). Stowe et al. (2010) confirmed associations between lower education and higher antibody titers to HSV-1 and EBV in a community sample of adults aged 25–90 in Texas. Differences in CMV antibody levels were recently found to emerge in childhood in the United States, with children below the poverty line having higher average CMV antibody levels compared to those above the poverty line (Dowd et al. 2012). Additional work has looked at the role of persistent infection in mediating the relationship between SES and cardiovascular disease in US adults, finding no mediating effect for HSV-1, but a significant mediating effect of CMV (Simanek et al. 2008). While these cross-sectional studies are only suggestive of stress pathways linking lower SES to elevated herpesvirus antibody titers, they represent an important new direction in the search for the biological links between the social environment and health.

9.5 Social Determinants and Immune Response Across the Life Course

Recent work in biodemography posited that reductions in life-time exposure to infection and inflammation was an important determinant of historical cohort declines in later-life morbidity and mortality. Crimmins and Finch argue that cohorts with lower infectious disease mortality in childhood can be characterized by a "cohort morbidity phenotype" that links their early life experience to later life cohort mortality patterns (Crimmins and Finch 2006; Finch and Crimmins 2004). McDade also highlights how early environments may model immune and inflammatory responses for the rest of the life course (McDade 2005b, 2000). More broadly, life course epidemiology has drawn attention to the potential long-term impacts of early life exposures for the development of chronic disease (Ben-Shlomo and Kuh 2002). Whether and what types of early pathogen exposures are important for immunological development is still an area of active research. The hygiene "hypothesis" posits that the absence of pathogenic challenges early in life can lead to poorly regulated inflammatory and cell-mediated immune responses in life (Strachan (1989). In support of this idea, McDade et al. (2005a) found that *greater* to exposure pathogens in infancy was associated with enhanced cell-mediated immune function in adolescence as measured by EBV antibody titers and reduced inflammation in young adulthood. Future longitudinal research on the consequences of both low and high pathogen exposure on the developing immune system and the later life consequences for immunosenescence is warranted. As the social distribution of the infectious exposures changes over time and place, such longitudinal studies could greatly increase understanding of how social differences in infectious exposures and immune function may translate into health disparities.

9.5.1 Socio-economic Differences in Infection: Exposure or Susceptibility?

Differences in the seroprevalence of common infections by social class should be largely explained by two factors: differences in exposure and/or differences in susceptibility given exposure. Current epidemiological evidence is lacking on the relative importance of these factors in the observed socio-economic differences in infections. Environmental factors associated with SES such as household crowding or increased use of public transportation could contribute directly to exposure risk. Other exposure opportunities that may vary by social factors include increased use of daycares for children at younger ages and larger family size, although these factors were not found to mediate the association between CMV seropositivity and SES in US children (Dowd et al. 2009a, c). Though breastfeeding has been implicated in exposure to CMV and other viruses among children (Stronati 2007), in the US the proportion of children ever breastfed is 23–26 % higher among women in the highest income group compared with the lowest income group, suggesting that CMV transmission via breast milk would decrease disparities in CMV exposure by SES (Fernandez-Twinn and Ozanne 2006).

Suppressed immune function as a result of stress, poor nutrition, smoking, or other environmental exposures could increase susceptibility to infections given equal levels of exposure. Evidence regarding socio-economic differentials in exposure to infectious agents is very difficult to directly assess for several reasons. First, the timing of infection is often difficult to determine without a longitudinal study design without frequent follow-up. Many of the infections examined in these studies are acquired early in life with little clinical manifestation among health individuals. Therefore, identifying the first period of exposure to these infections is difficult. Second, laboratory methods may not distinguish between recent and past exposure to infection. In many cases, IgG antibody levels based on Enzyme Linked Immunoassay

(ELISA) methods are utilized and do not provide any information on timing of the infection. IgM can provide better information on recent infection, but IgM is fleeting and often it is not possible to establish whether IgM increases represent new infection or re-exposure. For all of these reasons, it is difficult to identify the source of infection and how the individual was exposed.

Experimental studies have provided a basis of evidence for differences in susceptibility to infection among those exposed. Cohen and colleagues demonstrated that low social status was associated with an increased risk of respiratory infections in both humans and other primates using experimental exposure to infection (Cohen 2005, 1999; Cohen et al. 2004, 1997, 2008). The association between low subjective social status and susceptibility to experimentally administered rhinovirus and influenza virus was partially explained by poorer sleep duration (Cohen et al. 2008). In Scotland, low social class was also found to be associated with lower secretory immunoglobulin (sIgA), cited as a first line of defense against infection, in a large community sample, and this association was partially mediated by smoking status (Evans et al. 2000). In a sample of 25–49 year olds from the United States, those who were persistently unemployed had significantly lower levels of natural killer cell cytotoxicity (NKCC) compared to a matched employed sample (Cohen et al. 2007a). In addition, NKCC was significantly higher after the participants became employed, compared with their unemployed period, with substantial "recovery" of immune function compared with values from the steadily employed group. Similar to the research challenges in identifying infectious exposure timing and pathways, establishing differences in susceptibility in non-experimental studies is difficult. For example, differences in baseline immune function and genetic factors that may determine susceptibility are difficult to measure in large-scale population based studies. Nonetheless, the available experimental and non-experimental community-based studies suggest that lower social status may be associated with increased susceptibility to infection through a variety of pathways. Further research on baseline differences in immune function, genetic determinants, and the contribution of various pathways, including nutrition, to better understand variations in susceptibility are needed.

9.6 Gaps and Future Directions

Although many studies have clearly demonstrated socio-economic gradients in infection and immune function as described in this chapter, there are still several existing gaps in the literature. First, most of the available studies have only examined a small number of the possible infections that an individual may be exposed to and that may be associated with socio-economic factors. Future studies should strive to examine larger, more comprehensive sets of pathogens and possibly measures of the human microbiome as technologies become more accessible and affordable for application in large scale cohort studies. Next, to our knowledge no studies have directly tested the pathways such as psychosocial stressors and biological mediators that connect

SES and immune function in non-experimental settings. Studies that examine these pathways could provide important information on the underlying social and biological mechanisms linking SES and immunosenescence, and suggest potential targets for intervention at either the policy or clinical level. An additional challenge is that economic and social surveys rarely measure biological data (or measure it well), and vice versa. Inclusion of biomarkers in population-based social surveys including the Health and Retirement Study (HRS), the Survey of Health, Ageing, and Retirement in Europe (SHARE), and the National Longitidunal Study of Adolescent Health (Add Health) has grown in recent years, most including collection of inflammatory markers and some including addition measures such as antibodies to persistent herpesviruses or other infections (McDade et al. 2007; Williams and McDade 2009). Better measurement of biological data in social surveys should go hand in hand with better measurement of social and economic variables. The SES markers used in existing studies are quite diverse and include education, income, occupational grade or status, subjective social status, wealth, financial strain, material hardship, and duration of poverty, among others. Different components of SES have been shown to vary in their association with health risk for the same individuals, thus due care should be exercised when assessing associations of biological parameters with different measures of SES, especially across different country contexts (Duncan et al. 2002).

In this regard, population-based studies of immunological parameters are at a similar stage to earlier genetic studies when the focus was on individual genetic variation with little regard to social and environmental variables. There are still very few large-scale prospective cohorts that have collected information on environmental, psychological, and social exposures along with a comprehensive set of immunological parameters. In part, this is due to the financial and practical challenge of obtaining appropriate biological samples, such as peripheral mononuclear cells, needed to examine immunological parameters on large-scale basis. As a consequence, basic scientific information regarding population level variability in parameters of immunosenescence, including changes in the number and functionality of naïve T-cells and B cells, CD4:CD8 ratios, and memory T-cells is lacking. This type of descriptive epidemiology will be an important first to testing the influence of the social environment on immune parameters in prospective studies. One National Institute of Health (NIH) funded prospective cohort study of individuals 18 years and older, the Detroit Neighborhood Health Study, has been collecting peripheral blood mononuclear cells, testing for immune function, and assaying exposure to several infections linked to chronic disease. This study also has extensive measures on the social environment, stressors, and health indicators that can be examined in relationship to immune function and burden of infection. There has been a call for studies of immune function to be conducted at the population-level where immunological parameters are identified and assessed on wide-scale levels to create a profile of human immune function in the United States (Leslie 2010). It will be crucial to augment these types of undertakings with social and environmental factors that may shape population variability in immune profiles.

Moving beyond population-level description, studies examining alterations in cellular markers of immunosenescence and how these alterations relate to infection are needed. Prospective studies could provide data on the effect of socio-economic shocks (loss of jobs, housing, and foreclosure) on alterations in immune function over time both in terms of cellular markers of immunosenescence and immune response to persistent herpesvirus infections. Moreover, studies that combine genetic and epigenetic measures related to immune function to assess how these factors impact immunosenescence or whether these types of genetic determinants interact with the social environment to influence immune function are needed. Another social variable that may impact immune function is the neighborhood environment. Individuals living in disadvantaged neighborhoods may be more likely to suffer from poor quality housing, stress inducing and dangerous neighborhoods, litter, pollution, lack of healthy foods, and high levels of street noise and traffic volume. Studies that systematically assess the role that socio-economic stressors at both the neighborhood and individual-level play in shaping age-related alterations in immune function are needed to identify key features of the social context that may have public health significance for addressing differentials in immunological alterations at the population level. Where prospective and multi-level data are lacking, dynamic mathematical modeling may help to identify how immune function and pathogen burden vary within, between, and throughout generations of low socio-economic and race/ethnic groups.

9.7 Conclusions

In conclusion, a growing body of research has demonstrated socio-economic gradients in persistent infections, such as CMV, HSV-1, EBV, *C. pneumoniae, H. pylori* etc. Aiello et al. 2009; Dowd and Aiello 2009; Dowd et al. 2009a, c, 2008; Everhart et al. 2000; McQuillan et al. 2004; Simanek et al. 2008; Staras et al. 2006; Stowe et al. 2010; Zajacova et al. 2009. The results of these studies have demonstrated socio-economic gradients in the burden of infection that emerge in childhood and persist across the life course. This evidence strongly suggests that the timing of first exposure to infection and subsequently pathogen burden varies by SES, with potential life-long health implications for affected groups. This research also highlights the importance of understanding the connection between antigenic burden across the life-course and immunosenescence. Experimental studies and some non-experimental studies have demonstrated variability in susceptibility to infection by levels of social status in humans. Future studies that collect data on variability in immune function, genetic determinants, and the contribution of various pathways, including diet and physical activity, are needed to better understand what factors mediate socio-economic differentials in susceptibility to infection. Levels of antibodies to herpesviruses, such as CMV and HSV-1, have provided an indirect measure of cell-mediated immune function in large population-based studies of the impact of socio-economic position on immune function. These studies have consistently demonstrated a link between lower socio-economic position and increases in antibody levels, supporting the hypothesis that socio-economic stressors may influence immune function at the population level. Further population-based studies examining stress-related pathways and more traditional markers of immunosenescence, such as inverted CD4/CD8 ratios, terminal, memory, and effector cells, are needed. Although there are many gaps in the research yet to be addressed, the available literature has consistently supported links between socio-economic position and various markers of immune function and infection. Policy and preventive efforts should focus on reducing factors that influence exposures and susceptibility as well as the development of treatments and vaccines targeted at persistent infections to reduce socio-economic disparities in the burden of infection over the life course.

References

Adler NE, Stewart J (2010) Health disparities across the lifespan: meaning, methods, and mechanisms. Ann NY Acad Sci 1186 (The Biology of Disadvantage: Socioeconomic Status and Health):5–23

Aiello AE, Diez-Roux A, Noone A-M, Ranjit N, Cushman M, Tsai MY et al (2009) Socioeconomic and psychosocial gradients in cardiovascular pathogen burden and immune response: the multi-ethnic study of atherosclerosis. Brain Behav Immun 23(5):663–671

Baum A, Garofalo JP, Yali AM (1999) Socioeconomic status and chronic stress: does stress account for SES effects on health? Ann NY Acad Sci 896(1):131–144

Ben-Shlomo Y, Kuh D (2002) A life course approach to chronic disease epidemiology: conceptual models, empirical challenges and interdisciplinary perspectives. Int J Epidemiol 31(2):285–293

Brandtstadter J, Baltes-Gotz B, Kirschbaum C, Hellhammer D (1991) Developmental and personality correlates of adrenocortical activity as indexed by salivary cortisol: observations in the age range of 35–65 years. J Psychosom Res 35(2–3):173–185

Braveman PA, Cubbin C, Egerter S, Williams DR, Pamuk E (2010) Socioeconomic disparities in health in the United States: what the patterns tell us. Am J Public Health 100(S1):S186–196

Brunner E (1997) Socioeconomic determinants of health: stress and the biology of inequality. Br Med J 314(7092):1472–1476

Caserta MT, O'Connor TG, Wyman PA, Wang H, Moynihan J, Cross W et al (2008) The associations between psychosocial stress and the frequency of illness, and innate and adaptive immune function in children. Brain Behav Immun 22(6):933–940

Cohen S (1999) Social status and susceptibility to respiratory infections. Ann NY Acad Sci 896(1):246–253

Cohen S (2005) The Pittsburgh common cold studies: psychosocial predictors of susceptibility to respiratory infectious illness. Int J Behav Med 12(3):123–131

Cohen S, Alper CM, Doyle WJ, Adler N, Treanor JJ, Turner RB (2008) Objective and subjective socioeconomic status and susceptibility to the common cold. Health Psychol 27(2):268–274

Cohen S, Doyle WJ, Turner RB, Alper CM, Skoner DP (2004) Childhood socioeconomic status and host resistance to infectious illness in adulthood. Psychosom Med 66(4):553–558

Cohen F, Kemeny ME, Zegans LS, Johnson P, Kearney KA, Stites DP (2007a) Immune function declines with unemployment and recovers after stressor termination. Psychosom Med 69(3):225–234

Cohen S, Line S, Manuck SB, Rabin BS, Heise ER, Kaplan JR (1997) Chronic social stress, social status, and susceptibility to upper respiratory infections in nonhuman primates. Psychosom Med 59(3):213–221

Cohen S, Schwartz JE, Epel E, Kirschbaum C, Sidney S, Seeman T (2006) Socioeconomic status, race, and diurnal cortisol decline in the coronary artery risk development in young adults (CARDIA) study. Psychosom Med 68(1):41–50

Cohen JM, Wilson ML, Aiello AE (2007b) Analysis of social epidemiology research on infectious diseases: historical patterns and future opportunities. J Epidemiol Commun Health 61(12): 1021–1027

Crimmins EM, Finch CE (2006) Infection, inflammation, height, and longevity. PNAS 103(2): 498–503

Crimmins EM, Kim JK, Seeman TE (2009) Poverty and biological risk: the earlier aging of the poor. J Gerontol Series A: Biol Sci Med Sci 64A(2):286–292

De Martinis M, Franceschi C, Monti D, Ginaldi L (2005) Inflamm-ageing and lifelong antigenic load as major determinants of ageing rate and longevity. FEBS Lett 579(10):2035–2039

Dowd JB, Goldman N (2006) Do biomarkers of stress mediate the relation between socioeconomic status and health? J Epidemiol Commun Health 60(7):633–639

Dowd JB, Haan MN, Blythe L, Moore K, Aiello AE (2008) Socioeconomic gradients in immune response to latent infection. Am J Epidemiol 167(1):112–120

Dowd JB, Aiello AE (2009) Socioeconomic differentials in immune response. Epidemiol 20(6): 902–908

Dowd JB, Aiello AE, Alley DE (2009a) Socioeconomic disparities in the seroprevalence of cytomegalovirus infection in the US population: NHANES III. Epidemiol Infect 137(01):58–65

Dowd JB, Simanek AM, Aiello AE (2009b) Socio-economic status, cortisol and allostatic load: a review of the literature. Int J Epidemiol 38(5):1297–1309

Dowd JB, Zajacova A, Aiello A (2009c) Early origins of health disparities: burden of infection, health, and socioeconomic status in U.S. children. Soc Sci Med 68(4):699–707

Dowd JB, Aiello AE, Chyu L, Huang YY, McDade TW (2011) Cytomegalovirus antibodies in dried blood spots: a minimally invasive method for assessing stress, immune function, and aging. Immunty and Ageing 8(1):3

Dowd JB, Palermo TM, Aiello AE (2012) Family poverty is associated with cytomegalovirus antibody titers in U.S. Children. Health Psychol 31(1):5–10

Duncan GJ, Daly MC, McDonough P, Williams DR (2002) Optimal indicators of socioeconomic status for health research. Am J Public Health 92(7):1151–1157

Esterling BA, Antoni MH, Kumar M, Schneiderman N (1993) Defensiveness, trait anxiety, and Epstein-Barr viral capsid antigen antibody titers in healthy college students. Health Psychol 12(2):132–139

Evans P, Der G, Ford G, Hucklebridge F, Hunt K, Lambert S (2000) Social class, sex, and age differences in mucosal immunity in a large community sample. Brain Behav Immun 14(1):41–48

Everhart JE, Kruszon-Moran D, Perez-Perez GI, Tralka TS, McQuillan G (2000) Seroprevalence and ethnic differences in Helicobacter pylori infection among adults in the United States. J Infect Dis 181(4):1359–1363

Fagundes CP, Bennett JM, Alfano CM, Glaser R, Povoski SP, Lipari AM, Agnese DM, Yee LD, Carson WE III, Malarkey WB, Chen M, Kiecolt-Glaser JK (2012) Social support and socioeconomic status interact to predict Epstein-Barr virus latency in women awaiting diagnosis or newly diagnosed with breast cancer. Health Psychol 31(1):11–19

Fernandez-Twinn DS, Ozanne SE (2006) Mechanisms by which poor early growth programs type-2 diabetes, obesity and the metabolic syndrome. Physiol Behav 88(3):234–243

Finch CE, Crimmins EM (2004) Inflammatory exposure and historical changes in human life-spans. Science 305(5691):1736–1739

Glaser R, Kiecolt-Glaser JK, Speicher CE, Holliday JE (1985) Stress, loneliness, and changes in herpesvirus latency. J Behav Med 8(3):249–260

Glaser R, Jones JF (eds) (1994) Herpesvirus Infections. Marcel Dekker, New York

Glaser R, Kiecolt-Glaser JK (1997) Chronic stress modulates the virus-specific immune response to latent herpes simplex virus type 1. Ann Behav Med 19(2):78–82

Glaser R, Friedman SB, Smyth J, Ader R, Bijur P, Brunell P et al (1999) The differential impact of training stress and final examination stress on herpesvirus latency at the United States Military Academy at West Point. Brain Behav Immun 13(3):240–251

Herbert TB, Cohen S (1993) Stress and immunity in humans: a meta-analytic review. Psychosom Med 55(4):364–379
Koch S, Solana R, Rosa OD, Pawelec G (2006) Human cytomegalovirus infection and T cell immunosenescence: a mini review. Mech Age Dev 127(6):538–543
Kristenson M, Eriksen HR, Sluiter JK, Starke D, Ursin H (2004) Psychobiological mechanisms of socioeconomic differences in health. Soc Sci Med 58(8):1511–1522
Kunz-Ebrecht SR, Kirschbaum C, Steptoe A (2004) Work stress, socioeconomic status and neuroendocrine activation over the working day. Soc Sci Med 58(8):1523–1530
Leslie M (2010) Biomedical research. Immunology uncaged. [News]. Science 327(5973):1573
Mackenbach JP (2010) Has the English strategy to reduce health inequalities failed? Soc Sci Med 71(7):1249–1253
Mackenbach JP, Stirbu I, Roskam AJR, Schaap MM, Menvielle G, Leinsalu M et al (2008) Socioeconomic inequalities in health in 22 European countries. New Engl J Med 358(23):2468
McDade TW (2005a) The ecologies of human immune function. Ann Rev Anthropol 34(1):495–521
McDade TW (2005b) Life history, maintenance, and the early origins of immune function. Am J Hum Biol 17(1):81–94
McDade TW, Stallings JF, Angold A, Costello EJ, Burleson M, Cacioppo JT et al (2000) Epstein-Barr virus antibodies in whole blood spots: a minimally invasive method for assessing an aspect of cell-mediated immunity. Psychosom Med 62(4):560–568
McDade TW, Williams S, Snodgrass JJ (2007) What a drop can do: dried blood spots as a minimally invasive method for integrating biomarkers into population-based research. Demography 44(4):899–925
McDade T, Rutherford J, Adair L, Kuzawa C (2010) Early origins of inflammation: microbial exposures in infancy predict lower levels of C-reactive protein in adulthood. Proc Royal Soc Biol Sci 277(1684):1129–1137
McQuillan GM, Kruszon-Moran D, Kottiri BJ, Curtin LR, Lucas JW, Kington RS (2004) Racial and ethnic differences in the seroprevalence of 6 infectious diseases in the United States: data from NHANES III, 1988–1994. Am J Public Health 94(11):1952–1958
Mehta SK, Stowe RP, Feiveson AH, Tyring SK, Pierson DL (2000) Reactivation and shedding of cytomegalovirus in astronauts during spaceflight. J Infect Dis 182(6):1761–1764
Padgett DA, Sheridan JF, Dorne J, Berntson GG, Candelora J, Glaser R (1998) Social stress and the reactivation of latent herpes simplex virus type 1. Proc Natl Acad Sci USA 95(12):7231–7235
Pearlin LI, Schieman S, Fazio EM, Meersman SC (2005) Stress, health, and the life course: some conceptual perspectives. J Health Soc Behav 46(2):205–219
Phillips AC, Carroll D, Khan N, Moss P (2008) Cytomegalovirus is associated with depression and anxiety in older adults. Brain Behav Immun 22(1):52–55
Sarid O, Anson O, Yaari A, Margalith M (2002) Human cytomegalovirus salivary antibodies as related to stress. Clin Lab 48(5–6):297–305
Seeman T, Merkin SS, Crimmins E, Koretz B, Charette S, Karlamangla A (2008) Education, income and ethnic differences in cumulative biological risk profiles in a national sample of US adults: NHANES III (1988–1994). Soc Sci Med 66(1):72–87
Seeman TE, Crimmins E (2001) Social environment effects on health and aging: integrating epidemiologic and demographic approaches and perspectives. Ann NY Acad Sci 954(1):88–117
Semenza JC (2010) Strategies to intervene on social determinants of infectious diseases. Euro Surveill 15(27):32–39
Shirtcliff EA, Coe CL, Pollak SD (2009) Early childhood stress is associated with elevated antibody levels to herpes simplex virus type 1. Proc Natl Acad Sci USA 106(8):2963–2967
Simanek AM, Dowd JB, Aiello AE (2008) Persistent pathogens linking socioeconomic position and cardiovascular disease in the US. Int J Epidemiol 38(3):775–787
Staras SAS, Dollard SC, Radford KW, Dana Flanders W, Pass RF, Cannon MJ (2006) Seroprevalence of cytomegalovirus infection in the United States, 1988–1994. Clin Infect Dis 43(9):1143–1151
Steptoe A, Marmot M (2002) The role of psychobiological pathways in socio-economic inequalities in cardiovascular disease risk. Eur Heart J 23(1):13–25

Steptoe A, Kunz-Ebrecht S, Owen N, Feldman PJ, Willemsen G, Kirschbaum C et al (2003) Socioeconomic status and stress-related biological responses over the working day. Psychosom Med 65(3):461–470

Steptoe A, Shamaei-Tousi A, Gylfe A, Henderson B, Bergstrom S, Marmot M (2007) Socioeconomic status, pathogen burden, and cardiovascular disease risk. Heart 93:1567-1570. doi:10.1136/hrt.2006.113993

Stone AA, Krueger AB, Steptoe A, Harter JK (2010) The socioeconomic gradient in daily colds and influenza, headaches, and pain. Arch Intern Med 170(6):570–572

Stowe RP, Peek MK, Perez NA, Yetman DL, Cutchin MP, Goodwin JS (2010) Herpesvirus reactivation and socioeconomic position: a community-based study. J Epid Commun Hlth 64(8):666–671

Strachan DP (1989) Hay fever, hygiene, and household size. BMJ 299(6710):1259–1260

Stronati M (2007) Breastfeeding and cytomegalovirus infections. J Chemother 19 (Suppl 2):49–51

Trzonkowski P, Mysliwska J, Godlewska B, Szmit E, Lukaszuk K, Wieckiewicz J et al (2004) Immune consequences of the spontaneous pro-inflammatory status in depressed elderly patients. Brain Behav Immun 18(2):135–148

Williams SR, McDade TW (2009) The use of dried blood spot sampling in the national social life, health, and aging project. J Gerontol B Psychol Sci Soc Sci 64B(Suppl 1): i131–i136. doi:10.1093/geronb/gbn022

Zajacova A, Dowd JB, Aiello AE (2009) Socioeconomic and race/ethnic patterns in persistent infection burden among U.S. adults. J Gerontol A Biol Sci Med Sci 64A(2):272–279. doi:10.1093/gerona/gln012

Chapter 10
Exercise and Immunosenescence

Richard J. Simpson and Guillaume Spielmann

10.1 Introduction

Life expectancy in the developed world has increased exponentially over the last century, predisposing the elderly population to an unprecedented risk of infectious disease and malignancy. This apparent increase in disease susceptibility appears to be due to inexorable declines in the normal functioning of the immune system, with this age-related effect commonly being referred to as immunosenescence. The decline in functioning of the immune system is thought to be multifactorial (see Chaps. 1 and 2) and in part reflects a manifestation of repeated exposures to external pathogens and persistent viral infections throughout the lifespan, although certain lifestyle factors (i.e. smoking, inactivity, socio-economic status) have also been linked with impaired immunity and biomarkers of ageing. Immunosenescence describes a progressive deterioration of systemic immunity, however, ageing is associated with a more pronounced functional decline in adaptive (i.e. T- and B-lymphocytes and their products) compared to innate immunity. Many of these age-related changes within the T-cell compartment (i.e. inverted CD4/CD8 T-cell ratio, low mitogen-induced T-cell proliferation, memory cell inflation, low IL-2 synthesis) have been clustered together over the years to form an "immune risk profile" (IRP) that has proven to be a useful set of biomarkers to predict morbidity and mortality in the elderly (Pawelec et al. 2010). However, circulating levels of inflammatory mediators such as IL-6, IL-1ra and C-reactive protein (CRP) have also been found to be useful prognostic markers in very old people (Jylha et al. 2007).

It is generally accepted that regular physical exercise is associated with increased longevity and a lower risk of developing cardiovascular disease (CVD), diabetes, metabolic syndrome, hypertension, infectious illnesses and cancer (Barlow et al. 2006; Blair et al. 1996; Evenson et al. 2003; Kodama et al. 2009; Lynch et al. 1996). The British epidemiologist Jerry Morris first documented the relationship between

R. J. Simpson · G. Spielmann
Laboratory of Integrated Physiology, Department of Health and Human Performance,
University of Houston, Houston, TX 77204, USA
e-mail: rjsimpson@uh.edu

J. A. Bosch et al. (eds.), *Immunosenescence*,
DOI 10.1007/978-1-4614-4776-4_10, © Springer Science+Business Media New York 2013

physical activity and disease in the 1950s, reporting that CVD was more prevalent among drivers than conductors who worked on the London double-decker buses, attributing this to the greater levels of energy expended by the conductors who were required to climb stairs (Morris et al. 1953). This early work initiated a large number of investigations exploring the impact of physical activity on cardiovascular health and disease, but it was not until the 1990s that the impact of exercise on the ageing immune system began to receive a considerable amount of research attention (Kusaka et al. 1992; Mazzeo 1994; Nieman and Henson 1994; Nieman et al. 1993; Shinkai et al. 1995). As the aetiology of many age-related diseases in contemporary medicine have been attributed to a dysfunctional immune system (Pawelec 2006), exploring the impact of exercise on ageing immunity has considerable merit because of its potential to serve as a simplistic and inexpensive method to combat immunosenescence (Simpson and Guy 2010).

Despite the common perception that older individuals who exercise regularly have enhanced immunity and are less likely to incur an infectious illness than their sedentary counterparts, the mechanisms by which regular exercise exerts positive effects on the ageing immune system is yet to be determined. Moreover, whether or not exercise can prevent, or even reverse, the impaired immune responses associated with immunosenescence is a topic of great interest. In this chapter, we summarize the current literature that has explored the effects of regular exercise on the immune system as it pertains to immunosenescence and the associated IRP and discuss potential mechanisms by which habitual exercise might help improve immunity in older humans.

10.2 Exercise and the Ageing Immune System

Interest in the effects of exercise on the immune system stemmed from the early work of David Nieman, who showed that individuals engaged in regular exercise of a moderate intensity reported fewer symptoms associated with upper respiratory tract infections (Nieman 1994; Nieman et al. 1993, 1989). This work triggered a number of investigations aimed at identifying the influence of exercise in shaping the ageing immune system, some of which have used single bouts of acute exercise to compare transient immune responses between young and older adults (Bruunsgaard et al. 1999b; Ceddia et al. 1999; Mazzeo et al. 1998; Simpson et al. 2008). While this generates useful data on the responsiveness of the immune system to an acute stressor in the aged, they provide little information on the chronic effects of exercise on the ageing immune system. Studies aimed at exploring the effects of long-term exercise on immunity in the aged typically adopt a cross-sectional design to discriminate between active and inactive participants, or a longitudinal randomised controlled trial to document immune changes in response to an exercise-training intervention (Simpson and Guy 2010). The most frequent outcome measures among these studies include T-cell subset numbers and phenotype characteristics, mitogen-induced T-cell proliferation, serum antibody titres following vaccination and the cytotoxic activity of natural killer (NK) cells (Haaland et al. 2008). Longitudinal studies tend to incorporate either aerobic or resistance exercise training intervention (or a combination of both) ranging from 8 weeks to 12 months and have examined a variety of subject groups ranging from healthy middle-aged adults to frail elderly nursing home residents. More often than not, the inconsistencies in experimental design among the longitudinal studies complicate the juxtaposition of the findings, and the reasons for selecting certain outcome measures to determine the effects of exercise on ageing immunity are not always justified.

The IRP evolved from the findings of the Swedish octogenarian (OCTO) and nonagenarian (NONA) longitudinal studies on bio-behavioural ageing, which began in the late 1990s. A cluster of biomarkers were identified as indicators of mortality after a 2, 4 and 6-year follow up period (Wikby et al. 1998) and were collectively coined the "immune risk profile" (IRP). Although the IRP is still evolving as more data are published from the OCTO and NONA studies, the main features of the IRP include: an inverted CD4:CD8 ratio <1.0 due to the expansion of CD8+ T-cells with a late stage differentiation and/or a senescent phenotype (i.e. CD27−/CD28−; KLRG1+/CD57+); poor T-cell proliferation response to mitogens in vitro; lower numbers and proportions of naïve and early differentiated T-cells (i.e. CD45RA + / CCR7+; CD28 + /CD27+), latent cytomegalovirus (CMV) infection and, to a lesser extent, Epstein Barr virus (EBV) infection; increased oligoclonality of virus-specific CD8+ T-cells; and low IL-2 production and IL-2 receptor (IL-2R) expression (DelaRosa et al. 2006; Pawelec et al. 2009). The significance of the IRP at predicting morbidity and mortality in the aged would suggest that these are an extremely useful set of outcome measures to examine the impact of regular exercise training in the elderly, and it has been suggested that habitual physical exercise may influence the likelihood of an individual entering and reverting from the IRP category in old age (Fig. 10.1). No study to date has compared the impact of exercise training between elderly individuals assigned to an IRP and non-IRP category, and this is clearly an important avenue for future research. Some studies have, however, explored the impact of exercise training on certain IRP biomarkers, although these mostly involve sedentary but otherwise healthy participants. The most common IRP biomarkers to appear in these studies include CD4:CD8 T-cell ratios, frequencies of T-cell subtypes and phenotypes, in vitro T-cell responses to mitogens and IL-2 production and IL-2R expression, while others have also considered the effects of exercise on in vivo immune responses (i.e. vaccine efficacy) and classic biomarkers of ageing such as leukocyte telomere length. The findings from these studies are outlined in the subsequent sections of this chapter.

10.2.1 Exercise and the Frequency of CD4+ and CD8+ T-cells

In normal healthy adults, CD4+ T-cells are numerically superior to CD8+ T-cells. An inverted CD4:CD8 ratio below 1.0 is indicative of memory CD8+ T-cell inflation, presumably due to excessive homeostatic clonal expansions of apoptosis resistant CD8+ T-cells. An inverted CD4:CD8 ratio was one of the first biomarkers to be included in the IRP and, as this is indicative of a severely restricted T-cell repertoire,

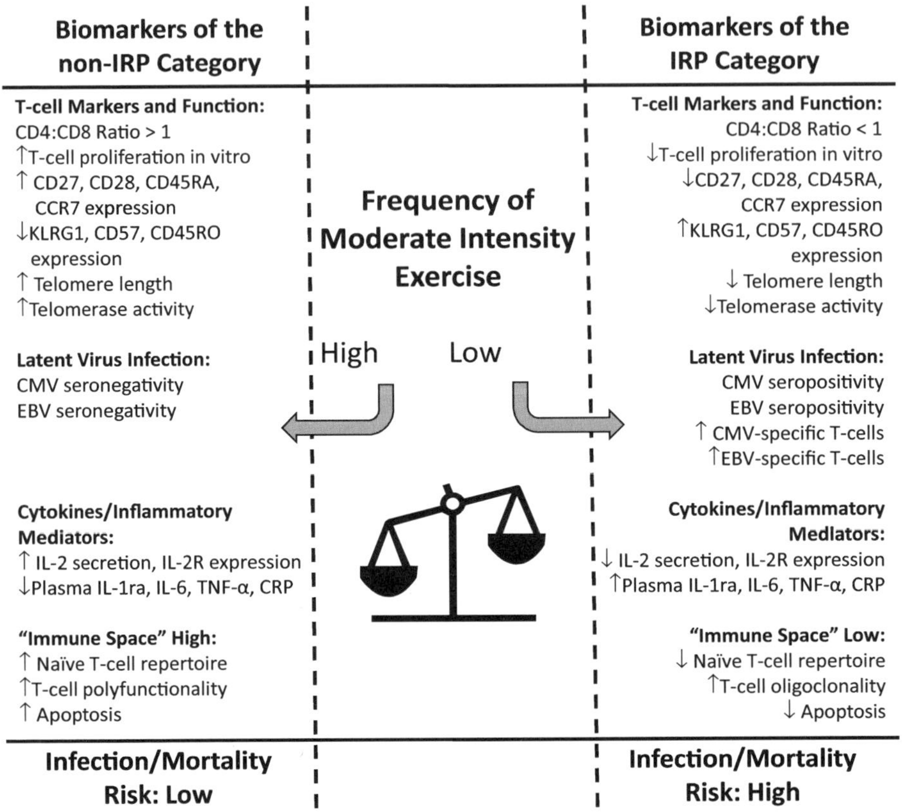

Fig. 10.1 Hypothetical model describing how the frequency of moderate intensity exercise can influence immune system biomarkers associated with the "immune risk profile" (IRP). Aging and latent viral infections are associated with an increase in the number of IRP biomarkers, which, in turn, is associated with increased infection risk and premature mortality. Individuals who are engaged in habitual physical exercise of a moderate intensity throughout their lifespan are more likely to remain in the non-IRP category in later life. Conversely, sedentary individuals are more likely enter the IRP category with aging. (Figure and legend adapted from Simpson R.J. 2011. Aging, persistent viral infections, and immunosenescence: Can exercise "make space"? Exerc Sport Sci Rev 39, 23–33)

the capabilities of the immune system to recognize and respond to newly evolving pathogens (i.e. influenza, rhinovirus, respiratory syncytial virus) are likely to become compromised when the CD4:CD8 ratio starts to fall. Despite this, it appears from the available literature that regular exercise has no effect on CD4:CD8 T-cell ratios in the resting blood of older adults. Cross-sectional data provided by Yan et al. (2001) showed no differences in the CD4:CD8 ratio due to physical activity status in men aged 20–73 years. Similarly, longitudinal exercise intervention studies using either strength or aerobic exercise training (or a mixture of both) from 6 weeks to 24 months have also reported no change (Campbell et al. 2008; Drela et al. 2004; Fahlman et al. 2000; Flynn et al. 1999; Okutsu et al. 2008; Woods et al. 1999). The lack of an exercise effect on CD4:CD8 T-cell ratios in the elderly is perhaps because none of these studies included subjects with CD4:CD8 ratios below 1.0 at baseline, leaving the possibility that exercise training may increase CD4:CD8 T-cell ratios only in IRP-assigned subjects. One study did report that elderly HIV infected patients improved their CD4:CD8 ratios from 0.63 to 0.81 after 9 months of resistance exercise training, although it is difficult to ascertain whether this was due to exercise or some other factor (i.e. an effective anti-retroviral therapy) as no control group was used in this study (Souza et al. 2008).

10.2.2 The Effects of Exercise on in vitro T-cell Function

The ability of T-cells to divide in vitro when stimulated with a mitogen is commonly used to assess T-cell functional responses. T-cell proliferation appears to be greater in physically active elderly compared to their sedentary counterparts, however, most longitudinal studies have failed to document changes in T-cell proliferation in response to aerobic or resistance exercise training interventions. Nieman et al. (1993) reported that the proliferative response of T-cells stimulated with PHA was greater in aerobically conditioned compared to sedentary older women aged 65–85 years, despite having the same total number of T-cells (Nieman et al. 1993). Similarly, Shinkai et al. (1995) found that PHA-induced T-cell proliferation was 44 % higher in older (mean 63 years) recreational male runners compared to sedentary older men (mean 66 years), despite no differences in absolute T-cell numbers. Although total T-cell numbers were similar between the trained and sedentary groups in these studies, it is very likely that the proportions of naïve (i.e. responsive) and senescent (i.e. unresponsive) T-cells will have been influenced by training status (Spielmann et al. 2011). As such, the potential altered proportions of responsive and unresponsive T-cells due to training may explain the differences in total T-cell proliferation observed between active and sedentary elderly in these studies (Nieman et al. 1993).

The sedentary women in the study by Nieman et al. (1993) also completed a 12-week exercise training intervention consisting of aerobic (30–40 min walking, 5 days per week) or flexibility exercise, but no intervention effect on T-cell proliferation was observed despite a 13 % increase in maximal oxygen uptake in the aerobic exercise group. Other longitudinal research studies have also failed to find an effect of exercise training on mitogen-induced T-cell proliferation (Campbell et al. 2008; Flynn et al. 1999; Kapasi et al. 2003). A 32-week exercise intervention combining both endurance and resistance type exercise did not improve in T-cell proliferation in frail elderly nursing home residents (Kapasi et al. 2003). Similarly, 10 weeks of resistance exercise training did not alter T-cell proliferation in previously sedentary elderly women (Flynn et al. 1999). Resistance training still appears to have no effect even when longer training interventions are used. Raso et al. (2007) found that moderate-intensity resistance training performed for 12 months did not change mitogen-induced T-cell proliferation in healthy older women. Aerobic exercise training also appears to have limited effects on in vitro T-cell function. Campbell et al.

(2008) reported that 12 months of aerobic exercise training failed to alter T-cell proliferative responses to mitogens in 115 overweight/obese postmenopausal women aged 50–75 years, despite high retention rates (6 % attrition over 12 months) and increases in VO2max of almost 14 % in the experimental group. There is one longitudinal study to report an effect of exercise on T-cell proliferation in the elderly. Woods et al. (1999) found that 6 months of supervised aerobic exercise (composed of brisk walking) conducted three times per week in the elderly (mean age 65 years) increased lymphocyte proliferation when stimulated with multiple doses of the mitogens Concanavalin A (Con A) and PMA in vitro, compared to a control group who performed 6 months of stretching/toning exercise.

Despite the clear differences in T-cell function that exist between active and inactive elderly, it is somewhat surprising that very few longitudinal studies report a positive effect of exercise training on T-cell function in previously sedentary elderly. The lack of an effect may be due to the relatively short time frame of the exercise intervention, or because most of the subjects used in these studies, although classified as sedentary, are otherwise healthy and probably have very few pre-existing characteristics associated with the IRP. For example, exercise-training studies performed in individuals carrying an illness mostly report an immune enhancement effect of exercise. Post-chemotherapy breast cancer survivors who participated in a 6-month supervised exercise training intervention consisting of both strength and aerobic activity had greater T-cell proiliferative responses to Con A and PH compared to breast cancer survivors who did not exercise (Hutnick et al. 2005). Again this may indicate that exercise training will be of greatest benefit to those with "weak" immune systems to begin with.

10.2.3 Exercise and the Frequency of Circulating T-cell Subsets

Increased proportions of effector–memory and senescent T-cells are considered to be a hallmark feature of an ageing immune system. Although T-cell clonal expansion and differentiation to form antigen-specific effector T-cells is a necessary process of adaptive immunity, the replicative potential of the T-cell is finite, and excess rounds of cell division can eventually cause the cell to become senescent (Spaulding et al. 1999). In this state, T-cells will no longer clonally expand upon further antigenic stimulation but will still retain effector cell functions (i.e. cytokine secretion, recognizing and killing virally infected cells) (Effros et al. 2003; Vescovini et al. 2007). Senescent T-cells express a number of signature cell-surface proteins, such as the killer cell lectin-like receptor G1 (KLRG1) and CD57 but lack expression of the co-stimulatory molecule CD28 (Brenchley et al. 2003; Ibegbu et al. 2005; Voehringer et al. 2002). Throughout the lifespan, naïve T-cells are gradually replaced with expanded clones of antigen-experienced effector and memory T-cells that exhibit a late stage differentiation (i.e. CD27−/CD28−) and/or a senescent (KLRG1+/CD57+) phenotype. The thymus gland atrophies rapidly up to the age of 35 years and only 10–15 % of functional thymic tissue remains at age 50 (Steinmann et al. 1985).

This results in a reduced output of naïve T-cells and a consequential lower ratio of naïve-to-memory T-cells in the periphery. This, in combination with the increased oligoclonality of the remaining T-cells (particularly for the CD8+ subset) due to homeostatic proliferation, reduces the repertoire of T-cells capable of responding to novel and seasonally evolving pathogens.

A potential mechanism by which exercise might exert protective or restorative effects on immunosenescence is to increase naïve T-cell output in the elderly. This would be achieved by either restoring thymic function (Aspinall and Mitchell 2008) or perhaps by stimulating extrathymic T-cell maturation in other tissues (i.e. the liver or intestines) (Torfadottir et al. 2006). This could generate an increased naïve-to-memory T-cell ratio and an expanded naïve T-cell compartment. No study to date has explored the impact of exercise on thymic or extrathymic T-cell development in the elderly; however, a few studies have determined the effects of regular exercise on the distribution of naïve and memory T-cells in the periphery. Using CD45RA and CD45RO as naïve and memory T-cell markers, respectively, Woods et al. (1999) reported that 6 months of supervised aerobic exercise (composed of brisk walking) conducted three times per week did not alter the proportions of naïve and memory cells (within either CD4+ or CD8+ T-cells) in elderly individuals aged 65 years. Unfortunately, however, the use of CD45RA and CD45RO are considered very crude markers of naïve and memory T-cells, due to the fact that CD45RA can be re-expressed by effector memory T-cells (Dunne et al. 2002). In a sample of 102 healthy adult men aged 18–61 years, Spielmann et al. (2011) found that maximal aerobic capacity (determined from a submaximal cycling test) was positively associated with naïve (KLRG1-/CD28+) and inversely associated with senescent (KLRG1+/CD28−) CD8+ T-cells in peripheral blood, even after adjusting for potential confounders such as age, percentage body fat and body mass index (BMI) (Spielmann et al. 2011). It was also reported that the well-accepted association between age and senescent T-cells no longer existed when age was adjusted for VO2max, indicating that aerobic fitness may have a stronger influence than age on the acquisition of a senescent T-cell phenotype. Ranking the subjects by their age-adjusted VO2max scores showed that those in the highest tertile had 17 % more naïve CD8+ T-cells and 57 % and 37 % less senescent CD4+ and CD8+ T-cells, respectively, compared to the lowest tertile (Fig. 10.2), indicating that regular physical exercise may help offset the natural age-related accumulation of senescent blood T-cells.

As the co-stimulatory molecule CD28 is progressively lost in response to activation and differentiation of the T-cell (Appay et al. 2002), a few longitudinal studies have examined the effects of an exercise training intervention on its surface expression. Kapasi et al. (2003) found no effect of 32-weeks endurance and resistance exercise in frail elderly nursing home residents (age 87 years) on CD28, CD45RA, CD45RO or HLA.DR (activated T-cell) expression within CD4+ or CD8+ T-cells. Similarly, Raso et al. (2007) found no differences in the absolute number of CD4+ or CD8+ T-cells expressing CD28 in sedentary older females (aged 60–77 years) who had completed 12-months of moderate intensity resistance training. In contrast,

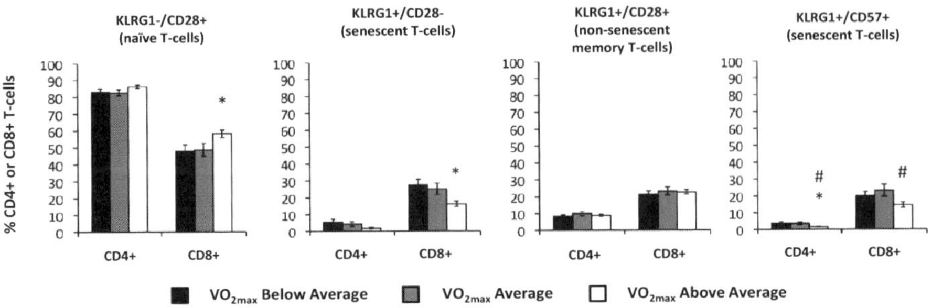

Fig. 10.2 The proportions of senescent (KLRG1+/CD57+; KLRG1+/CD28−), naïve (KLRG1−/CD28+), and non-senescent memory (KLRG1+/CD28+) cells within CD4+ and CD8+ T-cell subtypes in healthy adult men aged 18–61 years. Participants were ranked as below average, average and above average in accordance with their age-adjusted VO2max scores. The data presented is the mean ± SE. Statistically significant difference from below average indicated by *, and from average indicated by # ($p < 0.05$). (Adapted from Spielmann et al. (2011))

Shimizu et al. (2008) found increased numbers and percentages of CD4+ T-cells expressing CD28 after a 6-month supervised aerobic exercise program in elderly males and females aged 61–76 years. A limitation of this work; however, was the failure to document phenotype changes in CD8+ T-cells, which are known to undergo more pronounced age-related phenotypic changes than CD4+ T-cells. The discrepancies among these studies could be due to differences in the exercise protocols used and the age/condition of the subjects examined (Kapasi et al. 2003; Raso et al. 2007; Shimizu et al. 2008).

10.2.4 IL-2 Secretion and IL-2-receptor Expression with Exercise

T-cell activation following antigen recognition is dependent on the interaction between IL-2 and its receptor, which triggers cellular growth, differentiation and formation of effector-memory T-cells. The synthesis of IL-2 and expression of high affinity IL-2 receptors is impaired with age and associated with the IRP in the elderly (DelaRosa et al. 2006; Xu et al. 1993). It appears that IL-2 production is greater in older habitual exercisers in comparison to their sedentary counterparts. Compared to sedentary age-matched controls, Ogawa et al. (2003) found a greater number of CD8+ T-cells expressing IL-2 in elderly women who had been participating in an active walking programme for 4 years. A similar report was documented by Beshgetoor et al. (2004), who found that female masters athletes (>40 years) had twice as many CD8 + T-cells expressing intracellular IL-2 following mitogen stimulation compared to non-athletes of the same age, although exercise had no effect on the numbers of IL-2 expressing CD4+ T-cells (Beshgetoor et al. 2004; Ogawa et al. 2003). Shinkai et al. (1995) reported greater concentrations of IL-2 in the supernatants of cultured peripheral blood mononuclear cells (PBMCs) stimulated with LPS in elderly habitual runners compared to sedentary age-matched controls, which may have contributed to the greater T-cell proliferation that was also observed in the older runners (Shinkai et al. 1995). One longitudinal study reported that 24 months of aerobic exercise training in elderly women (age 62–86 years) resulted in an increased percentage of lymphocytes expressing intracellular IL-2, but not IL-4 or interferon-γ (IFN-γ) in response to in vitro stimulation with PMA/calcium ionophore compared to a group of women who had not yet undertaken the exercise program (Drela et al. 2004).

The effects of exercise training on the surface expression of the IL-2 receptor have received little research attention. Active elderly women (aged 60–98 years) were found to have a greater surface expression of the IL-2 alpha chain receptor CD25 when stimulated with CD3 monoclonal antibodies in vitro compared to inactive control subjects (Gueldner et al. 1997). However, using a longitudinal experimental design, Kapasi et al. (2003) found no change in CD25 expression on CD4+ or CD8+ T-cells following a 32-week exercise intervention in frail elderly. However, the ability to upregulate expression of the IL-2-R in response to activation is probably more indicative of T-cell function than basal expression levels of the receptor. Although the IL-2 beta chain receptor CD122 is believed to play a more important role in cell signalling than CD25 following the IL2/IL2-R interaction, no studies have assessed T-cell expression of CD122 in response to chronic exercise in older humans.

10.2.5 Immune Responses to Exercise in vivo

Despite the discernible differences in IRP-related biomarkers that exist between active and inactive older humans, the evidence currently available to suggest that exercise training can improve in vivo immune function in previously sedentary elderly is not compelling. Although it might appear from the in vitro work that exercise has limited immune restorative properties in previously sedentary elderly, attempts to improve in vivo immune responses using a longitudinal exercise intervention are usually more successful, with the majority of these studies using vaccines or recall antigens to induce an immune response. A 17-week exercise intervention in frail elderly was shown to prevent the decline in delayed-type hypersensitivity skin test response against seven recall antigens that were observed in non-exercising control subjects (Chin et al. 2000). Older adults (aged $<$64 years) immunized with a trivalent influenza vaccine before and after a 10-month aerobic exercise-training intervention had a greater mean fold increase in antibody titre to H1N1 and H3N2 strains of influenza A virus than the controls who did not exercise (Kohut et al. 2004). A recent study by Woods et al. (2009) found that older adults (average age 70 years) randomized to a 10-month cardiovascular exercise-training intervention had increased seroprotection 24 weeks after receiving the influenza vaccine, whereas those who completed flexibility training did not. Furthermore, an exercise training intervention of 25 weeks was found to enhance skin reaction to tuberculin-purified protein derivative and reduce serum IgG4 concentration in the elderly ($>$60 years) but not controls, indicating that moderate exercise facilitates T-helper 1 (Th1) immune responses in the elderly (Okutsu et al. 2008). Physical activity status also appears to improve immune responses to novel antigenic challenges in the elderly. Smith et al. (2004) cross-sectionally assessed delayed-type hypersensitivity skin response to the novel protein antigen keyhole-limpet haemocyanin (KLH) in active and inactive elderly subjects (60–79 years) who were immunized against KLH via intramuscular inoculation 21 days earlier. Elderly subjects identified as physically active had higher anti-KLH IgM, IgG and IgG1 serum antibody concentrations and enhanced delayed-type hypersensitivity responses compared to their inactive counterparts.

Although randomized exercise controlled trials that use vaccine efficacy or recall antigens as outcome measures appear to suggest enhanced immunity in the elderly, this may not necessarily be due to a long-term immune enhancement or restorative effect of exercise. Edwards et al. (2007, 2006) have shown that an acute stressor, either through mental stimulation or physical exercise (aerobic and eccentrically biased muscle-damaging bouts of exercise), given immediately prior to influenza intramuscular vaccination increases serum antibody responses compared to control subjects. This might indicate that certain in vivo immune responses (i.e. vaccine efficacy) to exercise in previously sedentary elderly are due to transient immune perturbations as opposed to a permanent restoration of the immune system caused by long-term exercise training.

10.2.6 Exercise and Telomeres

Telomeres are DNA nucleoprotein complexes that form the physical ends of linear eukaryotic chromosomes and erode progressively with each round of cell division. Excessive shortening of the telomere will trigger mechanisms for senescence causing the cell to undergo growth arrest. Leukocyte telomere length is a known indicator of morbidity and mortality in humans and a well-accepted marker of biological ageing, so there has been recent interest in the effects of physical exercise on telomere length (Cherkas et al. 2008; Ludlow et al. 2008; Shin et al. 2008). Cherkas et al. (2008) reported that leukocyte telomere length was positively correlated with self-reported scores of leisure time physical activity, even after adjusting for age, gender, BMI and smoking history in 2401 subjects. Differences in telomere length between active and inactive subjects were reported to be around 200 nt, which corresponds to 10 years of biological ageing (Cherkas et al. 2008). Ludlow et al. (2008) also reported that 50–70 year old study participants reporting to expend less than 990 kcal per week had shorter leukocyte telomeres than subjects expending between 991 and 2,340 kcal per week. Only one longitudinal study has been conducted to date, but found no effect of exercise on leukocyte telomere length in middle-aged obese women who completed 6 months of supervised aerobic exercise, despite increases in maximal oxygen uptake, leukocyte antioxidant activity and a reduced BMI (Shin et al. 2008). Although early work in this area indicates that exercise might delay human biological immune ageing (as evident by differences in telomere length), future studies should explore the effects of long-term exercise training on the telomere lengths of individual T-cell subsets and not just mixed populations of leukocytes.

10.3 Exercise and Low-grade Inflammation

Successful immune responses rely on the rapid development of localized and transient inflammatory responses to recruit immune cells, limit pathogenic dissemination and initiate wound repair and healing. However, ageing is associated with a more permanent form of low-grade chronic inflammation, which has been coined "inflammaging" (Franceschi et al. 2000a, b). This is characterized by increased concentrations of circulating inflammatory mediators and cytokines such as CRP, IL-1β, IL-6, IL-15 and tumor necrosis factor-α (TNF-α) (Bruunsgaard et al. 1999a; Ershler et al. 1993; Forsey et al. 2003). Although normally asymptomatic, inflammaging is thought to play a central role in the development of many chronic diseases in the elderly, such as CVD (Libby 2006), type 2 diabetes mellitus (Pedersen et al. 2003), Alzheimer's disease (Giunta et al. 2008), osteoporosis (Lencel and Magne 2011) and certain cancers (Allavena et al. 2008; Mantovani et al. 2008). Moreover, plasma levels of IL-1ra, IL-6 and CRP have been shown to predict mortality in nonagenerians after a 4-year follow up (Jylha et al. 2007).

Reducing inflammaging via exercise could be an efficient therapeutic approach to either prevent or delay the onset of those chronic diseases (i.e. CVD, diabetes) associated with low-grade inflammation, and thus reduce frailty and mortality in the elderly. Cross-sectional studies have shown an association between low-grade systemic inflammation and physical inactivity in healthy subjects (Colbert et al. 2004; Kullo et al. 2007; Pedersen and Bruunsgaard 2003; Taaffe et al. 2000). Regular exercise training has also been shown to reduce circulating levels of IL-6 in the elderly (Nicklas et al. 2008), and lower plasma CRP, TNF-α and IL-18 /IL-10 ratios in patients with type 2 diabetes (Kadoglou et al. 2007). In response to 6 months of moderate intensity aerobic exercise training, participants identified as being at risk for ischemic heart disease had a lower production of pro-inflammatory cytokines TNF-α and IFN-γ by mitogen-stimulated PBMCs, while exhibiting a higher secretion of anti-inflammatory cytokines IL-4, IL-10 and TGF-β1 compared to controls (Smith et al. 1999). Regular moderate activity during a 2-month period also reduced the pro-inflammatory profile of older adults by decreasing TNF-α, MCP-1 and nitric oxide synthase mRNA levels in PBMCs (Gano et al. 2011). Although the positive effects of exercise on circulating inflammatory markers may be confounded by weight loss, cross-sectional and longitudinal data indicate that exercise may exert anti-inflammatory effects independently from reductions in adiposity (Balducci et al. 2010; Fischer et al. 2007).

10.4 Modulating Immunosenescence with Exercise: Potential Mechanisms

While it is has been shown in epidemiological and experimental studies that regular moderate-intensity exercise is associated with a lower frequency of discrete IRP biomarkers, the precise mechanisms by which regular exercise exerts positive effects on the immune system are unknown. In addition to the potential changes within the immune compartment itself, immune enhancement may occur due to the modulatory effects that habitual exercise has on body fat distribution, lipid metabolism, plasma hormones and cytokines, cholesterol and peripheral circulation that are known to change with age. Indeed, increasing body mass and plasma leptin concentrations are inversely correlated with leukocyte telomere length (Valdes et al. 2005), while changes in the cholesterol content of plasma cell membranes can alter T-cell proliferative responses, major histocompatibility (MHC) class I and II expression and subsequent antigen presentation to T-lymphocytes (Shaikh and Edidin 2006). Exercise may also exert anti-inflammatory properties preventing chronic low-grade inflammation during the natural course of aging (Pedersen and Bruunsgaard 2003). While the maintenance of healthy blood circulation with regular exercise might help preserve T-cell trafficking and increase immune surveillance in the elderly; the preservation of growth hormone synthesis with exercise might help prevent impaired immune responses that are associated with ageing-induced growth hormone deficiency such as decreased thymocyte maturation and subsequent migration of naïve T-cells into the periphery (Dardenne et al. 2009).

One potential mechanism is for exercise to increase thymic mass in the elderly, possibly through increased IL-7 and/or growth hormone synthesis (Aspinall and Mitchell 2008; Dardenne et al. 2009). However, no studies to date have examined the effects of exercise on IL-7 secretion or thymic function (i.e. frequency of recent thymic emigrant T-cells in the periphery) in older adults, despite the recent observation that IL-7 is secreted by human skeletal muscle (Haugen et al. 2010) and may therefore be involved in the aetiology of exercise-induced immune enhancement. Indeed, the secretion of IL-7 from skeletal muscle may be stimulated with exercise, as Andersson et al. (2009) reported elevated plasma IL-7 levels in female soccer players after an acute bout of exercise.

Exercise may also moderate immunosenescence in an indirect manner due to its usefulness at reducing stress. Bosch et al. (2009) showed that an inverted CD4:CD8 ratio and CD8+ T-cells lacking expression of CD27 and CD28 was more profound in adults working under stressful conditions (Bosch et al. 2009). It was also reported that the frequency of CD8+ T-cells lacking surface expression of CD27 and CD28 was associated with higher overnight urinary cortisol secretion (Bosch et al. 2009). While cortisol could be having a direct effect on the acquisition of a senescent phenotype in these workers, periods of stress are known to trigger CMV reactivation (Sarid et al. 2002), which could lead to an inflation of the memory T-cell compartment and the premature acquisition of a senescent immune profile.

10.4.1 The Acute Exercise Hypothesis

It has been suggested that the long-term benefits of regular exercise may be a direct result of the repeated effects of acute exercise (Simpson 2011). A single bout of exercise elicits an almost immediate mobilization of NK cells and effector memory

CD8+ T-cells into the peripheral blood compartment, which causes around a two-fold increase in the blood lymphocyte count depending on the intensity and duration of exercise (Campbell et al. 2009; Simpson et al. 2010, 2007; Turner et al. 2010). During the early stages of exercise recovery (within 30–60 min), there is a rapid extravasation of the same lymphocyte subtypes, although the numbers of cells that leave the blood (presumably to recirculate to the peripheral tissues) is often greater than the number of cells that were mobilized causing a transient lymphocytopenia (Simpson et al. 2010, 2007; Turner et al. 2010). Given that the cell types preferentially mobilized by exercise have high effector functions (i.e. NK cells, effector–memory T-cells), exercise may simply enhance immunosurveillance by increasing the trafficking of cells among the peripheral tissues and the circulation.

Increased proportions of effector-memory and senescent T-cells are striking features of an ageing immune system and a hallmark of the IRP. Interestingly, these T-cell subtypes that are known to expand in the periphery as a consequence of ageing and latent CMV infection are also the most responsive cells to a single bout of exercise. For instance, the frequencies of CD8+ T-cells with a late stage differentiation/senescent phenotype (i.e. CD27−/CD28−; KLRG1+/CD57+) are expanded in the aged and the CMV-infected (Appay et al. 2008) and are mobilized in relatively greater numbers with exercise compared to less differentiated cell types (Simpson et al. 2010; Turner et al. 2010). Moreover, the mobilization of late stage differentiated T-cells are amplified in those infected with CMV (Turner et al. 2010), and aerobically fit people have a lower frequency of these cell types in resting blood compared to their untrained counterparts independently of age (Spielmann et al. 2011).

A theoretical framework (Fig. 10.3) has been published to illustrate the hypothesis that frequent bouts of acute exercise might help expand the "immune space" by causing senescent T-cells to undergo exercise-induced apoptosis (Simpson 2011). Through a looped feedback mechanism, which is likely to involve IL-7 and/or growth hormone synthesis, lower T-cell numbers due to apoptosis, may trigger the signals required to increase thymic output and the extrathymic maturation of naïve T-cells. The newly generated naïve T-cells then occupy the "vacated space", increasing the naïve T-cell repertoire of the host, which could potentially help restore the detrimental effects of ageing and persistent viral infections on T-cell immunity (Simpson 2011). Although this framework offers a potential pathway by which regular exercise might help negate some aspects of immunosenescence, future experimental studies are required before the acute-exercise hypothesis can be accepted as a possible pathway for exercise-induced immune enhancement. Although there is evidence from both human and animal studies that lymphocytes that egress the blood during the recovery phase of exercise are more prone to apoptosis (Simpson 2011), an important first step will be to determine if the senescent T-cells are more susceptible to apoptosis than the naïve T-cells. Moreover, whether or not repeated bouts of acute exercise can restore thymic output or stimulate extrathymic T-cell maturation, particularly in the elderly and those with persistent viral infections, also remains to be determined.

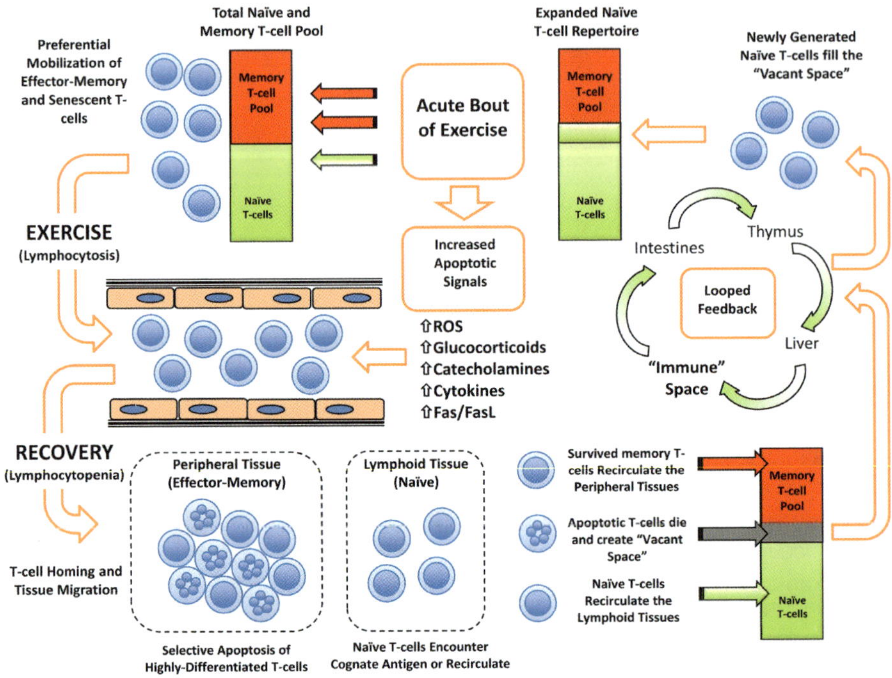

Fig. 10.3 Hypothetical model depicting the impact of repeated bouts of acute exercise on the "immune space". Acute exercise elicits the preferential mobilization of highly differentiated and viral-specific senescent T-cells from the peripheral tissues into the blood compartment (lymphocytosis) under the influence of catecholamines. Exercise, in turn, increases the production of reactive oxygen species (ROS), glucocorticoids and pro-inflammatory cytokines, thus exposing the senescent T-cells to a milieu of pro-apoptotic stimuli. Cell surface death receptors (Fas/FasL) are upregulated on senescent T-cells, which also incur mild oxidative DNA damage in the blood. These apoptosis susceptible cells, along with undamaged naïve and memory T-cells, egress the blood compartment during the recovery phase of exercise (lymphocytopenia) and migrate to specific tissues. A portion of these senescent T-cells subsequently undergo apoptosis in the peripheral tissues thus creating "vacant space". Consequently, lowered T-cell numbers drive the positive feedback loop increasing naïve T-cell output from the thymus or sites of extrathymic T-cell development (i.e. liver, intestines). These newly generated T-cells fill the "vacant space" and contribute to an expanded naïve T-cell repertoire. Repetitions of this process in response to habitual exercise reduce the frequency of senescent T-cells over time, lowering infection risk and increasing healthy longevity. (Reprinted from Simpson R.J. (2011). Aging, persistent viral infections, and immunosenescence: can exercise "make space"? Exerc Sport Sci Rev 39, S. 23–33)

10.5 Summary

Many immunological biomarkers that have been associated with the IRP are positively displayed in older individuals who are habitually active compared to those who are sedentary. These include greater T-cell responses to mitogens in vitro, longer leukocyte telomere lengths, greater naïve/memory T-cell ratios, enhanced in vivo immune responses to vaccines, and greater IL-2 production. The apparent positive effects of exercise come mostly from cross-sectional data, while the majority of the longitudinal studies that have been performed on previously sedentary elderly have failed to find an exercise effect on similar immune parameters, with the exception of in vivo immune responses to vaccines and recall antigens which appear to be enhanced following a period of exercise training. However, to state that exercise has limited immune restorative effects on immunity would be premature due to the large experimental inconsistencies that exist among the current literature and because most of these studies have involved sedentary but otherwise healthy elderly. The positive effects of exercise may be more palpable in those already deemed to have "weak" immune systems, such as those with chronic viral infections (i.e. HIV, hepatitis), cancer or even latent viruses such as CMV that are known to cause large expansions of the memory T-cell compartment. An exercise training intervention study in older people already assigned to the IRP category would be irradiating, as it would help to determine if exercise has effective immune restorative properties. While the potential rejuvenating properties of exercise on the immune system remain contentious, there is more conspicuous evidence from the cross-sectional studies to indicate that performing regular physical exercise during the natural course of ageing might help preserve immunity in a way that prevents an individual from falling into the IRP category in later life. However, more research work is required to determine if regular exercise can help restore, as well as prevent, the impaired immune responses associated with immunosenescence and the mechanisms underpinning these apparent adaptations to exercise.

References

Allavena P, Garlanda C, Borrello MG, Sica A, Mantovani A (2008) Pathways connecting inflammation and cancer. Curr Opin Genet Dev 18(1):3–10

Andersson H, Bohn SK, Raastad T, Paulsen G, Blomhoff R, Kadi F (2009) Differences in the inflammatory plasma cytokine response following two elite female soccer games separated by a 72-h recovery. Scand J Med Sci Sports 20(5):740–747

Appay V, Dunbar PR, Callan M, Klenerman P, Gillespie GM, Papagno L et al (2002) Memory CD8+ T cells vary in differentiation phenotype in different persistent virus infections. Nat Med 8(4):379–385

Appay V, van Lier RA, Sallusto F, Roederer M (2008) Phenotype and function of human T lymphocyte subsets: consensus and issues. Cytometry A 73(11):975–983

Aspinall R, Mitchell W (2008) Reversal of age-associated thymic atrophy: treatments, delivery, and side effects. Exp Gerontol 43(7):700–705

Balducci S, Zanuso S, Nicolucci A, Fernando F, Cavallo S, Cardelli P et al (2010) Anti-inflammatory effect of exercise training in subjects with type 2 diabetes and the metabolic syndrome is dependent on exercise modalities and independent of weight loss. Nutr Metab Cardiovasc Dis 20(8):608–617

Barlow CE, LaMonte MJ, Fitzgerald SJ, Kampert JB, Perrin JL, Blair SN (2006) Cardiorespiratory fitness is an independent predictor of hypertension incidence among initially normotensive healthy women. Am J Epidemiol 163(2):142–150

Beshgetoor D, Arrues S, McGuire K (2004) Effect of competitive training on T-cell mediated immune function in Master's female athletes. Int J Sports Med 25(7):553–558

Blair SN, Kampert JB, Kohl HW 3rd, Barlow CE, Macera CA, Paffenbarger RS Jr et al (1996) Influences of cardiorespiratory fitness and other precursors on cardiovascular disease and all-cause mortality in men and women. J Am Med Assoc 276(3):205–210

Bosch JA, Fischer JE, Fischer JC (2009) Psychologically adverse work conditions are associated with CD8+ T cell differentiation indicative of immunosenescence. Brain Behav Immun 23(4):527–534

Brenchley JM, Karandikar NJ, Betts MR, Ambrozak DR, Hill BJ, Crotty LE et al (2003) Expression of CD57 defines replicative senescence and antigen-induced apoptotic death of CD8+ T cells. Blood 101(7):2711–2720

Bruunsgaard H, Andersen-Ranberg K, Jeune B, Pedersen AN, Skinhoj P, Pedersen BK (1999a) A high plasma concentration of TNF-alpha is associated with dementia in centenarians. J Gerontol A Biol Sci Med Sci 54(7):M357–364

Bruunsgaard H, Jensen MS, Schjerling P, Halkjaer-Kristensen J, Ogawa K, Skinhoj P et al (1999b) Exercise induces recruitment of lymphocytes with an activated phenotype and short telomeres in young and elderly humans. Life Sci 65(24):2623–2633

Campbell JP, Riddell NE, Burns VE, Turner M, van Zanten JJ, Drayson MT et al (2009) Acute exercise mobilises CD8+ T lymphocytes exhibiting an effector-memory phenotype. Brain Behav Immun 23(6):767–775

Campbell PT, Wener MH, Sorensen B, Wood B, Chen-Levy Z, Potter JD et al (2008) Effect of exercise on in vitro immune function: a 12-month randomized, controlled trial among postmenopausal women. J Appl Physiol 104(6):1648–1655

Ceddia MA, Price EA, Kohlmeier CK, Evans JK, Lu Q, McAuley E et al (1999) Differential leukocytosis and lymphocyte mitogenic response to acute maximal exercise in the young and old. Med Sci Sports Exerc 31(6):829–836

Cherkas LF, Hunkin JL, Kato BS, Richards JB, Gardner JP, Surdulescu GL et al (2008) The association between physical activity in leisure time and leukocyte telomere length. Arch Intern Med 168(2):154–158

Chin APMJ, de Jong N, Pallast EG, Kloek GC, Schouten EG, Kok FJ (2000) Immunity in frail elderly: a randomized controlled trial of exercise and enriched foods. Med Sci Sports Exerc 32(12):2005–2011

Colbert LH, Visser M, Simonsick EM, Tracy RP, Newman AB, Kritchevsky SB et al (2004) Physical activity, exercise, and inflammatory markers in older adults: findings from the health, aging and body composition study. J Am Geriatr Soc 52(7):1098–1104

Dardenne M, Smaniotto S, de Mello-Coelho V, Villa-Verde DM, Savino W (2009) Growth hormone modulates migration of developing T cells. Ann N Y Acad Sci 1153:1–5

DelaRosa O, Pawelec G, Peralbo E, Wikby A, Mariani E, Mocchegiani E et al (2006) Immunological biomarkers of ageing in man: changes in both innate and adaptive immunity are associated with health and longevity. Biogerontology 7(5–6):471–481

Drela N, Kozdron E, Szczypiorski P (2004) Moderate exercise may attenuate some aspects of immunosenescence. BMC Geriatr 4:8

Dunne PJ, Faint JM, Gudgeon NH, Fletcher JM, Plunkett FJ, Soares MV et al (2002) Epstein-Barr virus-specific CD8(+) T cells that re-express CD45RA are apoptosis-resistant memory cells that retain replicative potential. Blood 100(3):933–940

Edwards KM, Burns VE, Allen LM, McPhee JS, Bosch JA, Carroll D et al (2007) Eccentric exercise as an adjuvant to influenza vaccination in humans. Brain Behav Immun 21(2):209–217

Edwards KM, Burns VE, Reynolds T, Carroll D, Drayson M, Ring C (2006) Acute stress exposure prior to influenza vaccination enhances antibody response in women. Brain Behav Immun 20(2):159–168

Effros RB, Dagarag M, Valenzuela HF (2003) In vitro senescence of immune cells. Exp Gerontol 38(11–12):1243–1249

Ershler WB, Sun WH, Binkley N, Gravenstein S, Volk MJ, Kamoske G et al (1993) Interleukin-6 and aging: blood levels and mononuclear cell production increase with advancing age and in vitro production is modifiable by dietary restriction. Lymphokine Cytokine Res 12(4):225–230

Evenson KR, Stevens J, Cai J, Thomas R, Thomas O (2003) The effect of cardiorespiratory fitness and obesity on cancer mortality in women and men. Med Sci Sports Exerc 35(2):270–277

Fahlman M, Boardley D, Flynn MG, Braun WA, Lambert CP, Bouillon LE (2000) Effects of endurance training on selected parameters of immune function in elderly women. Gerontology 46(2):97–104

Fischer CP, Berntsen A, Perstrup LB, Eskildsen P, Pedersen BK (2007) Plasma levels of interleukin-6 and C-reactive protein are associated with physical inactivity independent of obesity. Scand J Med Sci Sports 17(5):580–587

Flynn MG, Fahlman M, Braun WA, Lambert CP, Bouillon LE, Brolinson PG et al (1999) Effects of resistance training on selected indexes of immune function in elderly women. J Appl Physiol 86(6):1905–1913

Forsey RJ, Thompson JM, Ernerudh J, Hurst TL, Strindhall J, Johansson B et al (2003) Plasma cytokine profiles in elderly humans. Mech Ageing Dev 124(4):487–493

Franceschi C, Bonafe M, Valensin S (2000a) Human immunosenescence: the prevailing of innate immunity, the failing of clonotypic immunity, and the filling of immunological space. Vaccine 18(16):1717–1720

Franceschi C, Bonafe M, Valensin S, Olivieri F, De Luca M, Ottaviani E et al (2000b) Inflamm-aging. An evolutionary perspective on immunosenescence. Ann NY Acad Sci 908:244–254

Gano LB, Donato AJ, Pierce GL, Hamza M, Magerko KA, Roeca C et al (2011) Increased proinflammatory and oxidant gene expression in circulating mononuclear cells in older adults: amelioration by habitual exercise. Physiol Genomics 43: 895-902

Giunta B, Fernandez F, Nikolic WV, Obregon D, Rrapo E, Town T et al (2008) Inflammaging as a prodrome to Alzheimer's disease. J Neuroinflammation 5(51)

Gueldner SH, Poon LW, La Via M, Virella G, Michel Y, Bramlett MH et al (1997) Long-term exercise patterns and immune function in healthy older women. A report of preliminary findings. Mech Ageing Dev 93(1–3):215–222

Haaland DA, Sabljic TF, Baribeau DA, Mukovozov IM, Hart LE (2008) Is regular exercise a friend or foe of the aging immune system? A systematic review. Clin J Sport Med 18(6):539–548

Haugen F, Norheim F, Lian H, Wensaas AJ, Dueland S, Berg O et al (2010) IL-7 is expressed and secreted by human skeletal muscle cells. Am J Physiol Cell Physiol 298(4):C807–816

Hutnick NA, Williams NI, Kraemer WJ, Orsega-Smith E, Dixon RH, Bleznak AD et al (2005) Exercise and lymphocyte activation following chemotherapy for breast cancer. Med Sci Sports Exerc 37(11):1827–1835

Ibegbu CC, Xu YX, Harris W, Maggio D, Miller JD, Kourtis AP (2005) Expression of killer cell lectin-like receptor G1 on antigen-specific human CD8+ T lymphocytes during active, latent, and resolved infection and its relation with CD57. J Immunol 174(10):6088–6094

Jylha M, Paavilainen P, Lehtimaki T, Goebeler S, Karhunen PJ, Hervonen A et al (2007) Interleukin-1 receptor antagonist, interleukin-6, and C-reactive protein as predictors of mortality in nonagenarians: the vitality 90+ study. J Gerontol A Biol Sci Med Sci 62(9):1016–1021

Kadoglou NP, Iliadis F, Angelopoulou N, Perrea D, Ampatzidis G, Liapis CD et al (2007) The anti-inflammatory effects of exercise training in patients with type 2 diabetes mellitus. Eur J Cardiovasc Prev Rehabil 14(6):837–843

Kapasi ZF, Ouslander JG, Schnelle JF, Kutner M, Fahey JL (2003) Effects of an exercise intervention on immunologic parameters in frail elderly nursing home residents. J Gerontol A Biol Sci Med Sci 58(7):636–643

Kodama S, Saito K, Tanaka S, Maki M, Yachi Y, Asumi M et al (2009) Cardiorespiratory fitness as a quantitative predictor of all-cause mortality and cardiovascular events in healthy men and women: a meta-analysis. J Am Med Assoc 301(19):2024–2035

Kohut ML, Arntson BA, Lee W, Rozeboom K, Yoon KJ, Cunnick JE et al (2004) Moderate exercise improves antibody response to influenza immunization in older adults. Vaccine 22(17–18):2298–2306

Kullo IJ, Khaleghi M, Hensrud, DD (2007) Markers of inflammation are inversely associated with VO2 max in asymptomatic men. J Appl Physiol 102(4):1374–1379

Kusaka Y, Kondou H, Morimoto K (1992) Healthy lifestyles are associated with higher natural killer cell activity. Prev Med 21(5):602–615

Lencel P, Magne D (2011) Inflammaging: the driving force in osteoporosis? Med Hypotheses 76(3):317–321

Libby P (2006) Inflammation and cardiovascular disease mechanisms. Am J Clin Nutr 83(2): 456S–460S

Ludlow AT, Zimmerman JB, Witkowski S, Hearn JW, Hatfield BD, Roth SM (2008) Relationship between physical activity level, telomere length, and telomerase activity. Med Sci Sports Exerc 40(10):1764–1771

Lynch J, Helmrich SP, Lakka TA, Kaplan GA, Cohen RD, Salonen R et al (1996) Moderately intense physical activities and high levels of cardiorespiratory fitness reduce the risk of non-insulin-dependent diabetes mellitus in middle-aged men. Arch Intern Med 156(12):1307–1314

Mantovani A, Allavena P, Sica A, Balkwill F (2008) Cancer-related inflammation. Nature 454(7203):436–444

Mazzeo RS (1994) The influence of exercise and aging on immune function. Med Sci Sports Exerc 26(5):586–592

Mazzeo RS, Rajkumar C, Rolland J, Blaher B, Jennings G, Esler M (1998) Immune response to a single bout of exercise in young and elderly subjects. Mech Ageing Dev 100(2):121–132

Morris JN, Heady JA, Raffle PA, Roberts CG, Parks JW (1953) Coronary heart-disease and physical activity of work. Lancet 265(6795):1053–1057; contd.

Nicklas BJ, Hsu FC, Brinkley TJ, Church T, Goodpaster BH, Kritchevsky SB et al (2008) Exercise training and plasma C-reactive protein and interleukin-6 in elderly people. J Am Geriatr Soc 56(11):2045–2052

Nieman DC (1994) Exercise, infection, and immunity. Int J Sports Med 15(Suppl 3):S131–141

Nieman DC, Henson DA (1994) Role of endurance exercise in immune senescence. Med Sci Sports Exerc 26(2):172–181

Nieman DC, Henson DA, Gusewitch G, Warren BJ, Dotson RC, Butterworth DE et al (1993) Physical activity and immune function in elderly women. Med Sci Sports Exerc 25(7):823–831

Nieman DC, Johanssen LM, Lee JW (1989) Infectious episodes in runners before and after a roadrace. J Sports Med Phys Fitness 29(3):289–296

Ogawa K, Oka J, Yamakawa J, Higuchi M (2003) Habitual exercise did not affect the balance of type 1 and type 2 cytokines in elderly people. Mech Ageing Dev 124(8–9):951–956

Okutsu M, Yoshida Y, Zhang X, Tamagawa A, Ohkubo T, Tsuji I et al (2008) Exercise training enhances in vivo tuberculosis purified protein derivative response in the elderly. J Appl Physiol 104(6):1690–1696

Pawelec G (2006) Immunity and ageing in man. Exp Gerontol 41(12):1239–1242

Pawelec G, Derhovanessian E, Larbi A, Strindhall J, Wikby A (2009) Cytomegalovirus and human immunosenescence. Rev Med Virol 19(1):47–56

Pawelec G, Larbi A, Derhovanessian E (2010) Senescence of the human immune system. J Comp Pathol 142(Suppl 1):S39–44

Pedersen BK, Bruunsgaard H (2003) Possible beneficial role of exercise in modulating low-grade inflammation in the elderly. Scand J Med Sci Sports 13(1):56–62

Pedersen M, Bruunsgaard H, Weis N, Hendel HW, Andreassen BU, Eldrup E et al (2003) Circulating levels of TNF-alpha and IL-6-relation to truncal fat mass and muscle mass in healthy elderly individuals and in patients with type-2 diabetes. Mech Ageing Dev 124(4):495–502

Raso V, Benard G, AJ DASD, Natale VM (2007) Effect of resistance training on immunological parameters of healthy elderly women. Med Sci Sports Exerc 39(12):2152–2159

Sarid O, Anson O, Yaari A, Margalith M (2002) Human cytomegalovirus salivary antibodies as related to stress. Clin Lab 48(5–6):297–305

Shaikh SR, Edidin M (2006) Polyunsaturated fatty acids, membrane organization, T cells, and antigen presentation. Am J Clin Nutr 84(6):1277–1289

Shimizu K, Kimura F, Akimoto T, Akama T, Tanabe K, Nishijima T et al (2008) Effect of moderate exercise training on T-helper cell subpopulations in elderly people. Exerc Immunol Rev 14:24–37

Shin YA, Lee JH, Song W, Jun TW (2008) Exercise training improves the antioxidant enzyme activity with no changes of telomere length. Mech Ageing Dev 129(5):254–260

Shinkai S, Kohno H, Kimura K, Komura T, Asai H, Inai R et al (1995). Physical activity and immune senescence in men. Med Sci Sports Exerc 27(11):1516–1526

Simpson RJ (2011) Aging, persistent viral infections, and immunosenescence: can exercise "make space"? Exerc Sport Sci Rev 39(1):23–33

Simpson RJ, Cosgrove C, Chee MM, McFarlin BK, Bartlett DB, Spielmann G et al (2010) Senescent phenotypes and telomere lengths of peripheral blood T-cells mobilized by acute exercise in humans. Exerc Immunol Rev 16:36–51

Simpson RJ, Cosgrove C, Ingram LA, Florida-James GD, Whyte GP, Pircher H et al (2008) Senescent T-lymphocytes are mobilised into the peripheral blood compartment in young and older humans after exhaustive exercise. Brain Behav Immun 22(4):544–551

Simpson RJ, Florida-James GD, Cosgrove C, Whyte GP, Macrae S, Pircher H et al (2007) High-intensity exercise elicits the mobilization of senescent T lymphocytes into the peripheral blood compartment in human subjects. J Appl Physiol 103(1):396–401

Simpson RJ, Guy K (2010) Coupling aging immunity with a sedentary lifestyle: has the damage already been done?—a mini-review. Gerontology 56(5):449–458

Smith JK, Dykes R, Douglas JE, Krishnaswamy G, Berk S (1999) Long-term exercise and atherogenic activity of blood mononuclear cells in persons at risk of developing ischemic heart disease. J Am Med Assoc 281(18):1722–1727

Smith TP, Kennedy SL, Fleshner M (2004) Influence of age and physical activity on the primary in vivo antibody and T cell-mediated responses in men. J Appl Physiol 97(2):491–498

Souza PM, Jacob-Filho W, Santarem JM, Silva AR, Li HY, Burattini MN (2008) Progressive resistance training in elderly HIV-positive patients: does it work? Clinics 63(5):619–624

Spaulding C, Guo W, Effros RB (1999) Resistance to apoptosis in human CD8+ T cells that reach replicative senescence after multiple rounds of antigen-specific proliferation. Exp Gerontol 34(5):633–644

Spielmann G, McFarlin BK, O'Connor DP, Smith PJW, Pircher H, Simpson RJ (2011) Aerobic fitness is associated with lower proportions of senescent blood T-cells in man. Brain Behav Immun 25:1521–1529

Steinmann GG, Klaus B, Muller-Hermelink HK (1985) The involution of the ageing human thymic epithelium is independent of puberty. A morphometric study. Scand J Immunol 22(5):563–575

Taaffe DR, Harris TB, Ferrucci L, Rowe J, Seeman TE (2000) Cross-sectional and prospective relationships of interleukin-6 and C-reactive protein with physical performance in elderly persons: MacArthur studies of successful aging. J Gerontol A Biol Sci Med Sci 55(12):M709–715

Torfadottir H, Freysdottir J, Skaftadottir I, Haraldsson A, Sigfusson G, Ogmundsdottir HM (2006) Evidence for extrathymic T cell maturation after thymectomy in infancy. Clin Exp Immunol 145(3).407–412

Turner JE, Aldred S, Witard O, Drayson MT, Moss PM, Bosch JA (2010) Latent Cytomegalovirus infection amplifies CD8 T-lymphocyte mobilisation and egress in response to exercise. Brain Behav Immun Nov 24(8): 1362–1370

Valdes AM, Andrew T, Gardner JP, Kimura M, Oelsner E, Cherkas LF et al (2005) Obesity, cigarette smoking, and telomere length in women. Lancet 366(9486):662–664

Vescovini R, Biasini C, Fagnoni FF, Telera AR, Zanlari L, Pedrazzoni M et al (2007) Massive load of functional effector CD4+ and CD8+ T cells against cytomegalovirus in very old subjects. J Immunol 179(6):4283–4291

Voehringer D, Koschella M, Pircher H (2002) Lack of proliferative capacity of human effector and memory T cells expressing killer cell lectinlike receptor G1 (KLRG1) Blood 100(10):3698–3702

Wikby A, Maxson P, Olsson J, Johansson B, Ferguson FG (1998) Changes in CD8 and CD4 lymphocyte subsets, T cell proliferation responses and non-survival in the very old: the Swedish longitudinal OCTO-immune study. Mech Ageing Dev 102(2–3):187–198

Woods JA, Ceddia MA, Wolters BW, Evans JK, Lu Q, McAuley E (1999) Effects of 6 months of moderate aerobic exercise training on immune function in the elderly. Mech Ageing Dev 109(1):1–19

Woods JA, Keylock KT, Lowder T, Vieira VJ, Zelkovich W, Dumich S et al (2009) Cardiovascular exercise training extends influenza vaccine seroprotection in sedentary older adults: the immune function intervention trial. J Am Geriatr Soc 57(12):2183–2191
Xu X, Beckman I, Bradley J (1993) Age-related changes in the expression of IL-2 and high-affinity IL-2 binding sites. Int Arch Allergy Immunol 102(3):224–231
Yan H, Kuroiwa A, Tanaka H, Shindo M, Kiyonaga A, Nagayama A (2001) Effect of moderate exercise on immune senescence in men. Eur J Appl Physiol 86(2):105–111

Chapter 11
Obesity and Immunosenescence: Psychological, Behavioral, and Biochemical Pathways

Aric A. Prather, Kirstin Aschbacher, Robert H. Lustig and Elissa S. Epel

11.1 Introduction

> Thou see'st I have more flesh than another man, and therefore more frailty. (from Henry IV, William Shakespeare)

With improvements in medical care, individuals are living longer than any time in human history. However, with age comes increasing susceptibility to a cadre of age-related illnesses, including cardiovascular disease, diabetes, arthritis, and cancer. While chronological age remains the best predictor of disease onset, the rates at which these diseases befall aged individuals vary. Accordingly, a deeper understanding of factors that contribute to "biological aging" may illuminate novel therapeutic interventions with broad public benefit. In this regard, there is growing interest in the influence of obesity on the aging process.

Obesity and obesity-related disease prevalence have increased worldwide, especially in Western and other developed countries (Flegal et al. 1998, 2010; Yach et al. 2006). In the United States, the latest data show that by the end of 2010, 9.3 million more people will meet criteria for obesity (BMI $\geq 30\,kg/m^2$) than in 1990, 89 % of whom will be 50 years of age or older (Wang et al. 2007). The obese are more likely to develop several age-related diseases, including type 2 diabetes and cardiovascular disease, and suffer premature mortality than normal weight individuals (Flegal et al. 2007; Hubert et al. 1983; Stern and Haffner 1986), independent of chronological age. As such, there has been increasing interest in understanding how obesity promotes accelerated disease risk. In this regard, emerging evidence suggests that obesity may contribute to aging of the immune system (i.e., immunosenescence). Accordingly,

A. A. Prather (✉)
Robert Wood Johnson Foundation Health and Society Scholars Program,
Center for Health and Community, University of California, San Francisco, CA 94143, USA
e-mail: PratherA@chc.ucsf.edu

K. Aschbacher · E. S. Epel
Department of Psychiatry, University of California, San Francisco, CA 94143, USA

R. H. Lustig
Department of Pediatrics, University of California, San Francisco, CA 94143, USA

J. A. Bosch et al. (eds.), *Immunosenescence*,
DOI 10.1007/978-1-4614-4776-4_11, © Springer Science+Business Media New York 2013

this chapter reviews the current evidence linking obesity and four indicators of immunosenescence, including: (1) telomere attrition, (2) thymic involution, (3) diminished antibody response to vaccination, and (4) elevated inflammation. This will be followed by a selective review of key biochemical mediators proposed to link excess adiposity and biological aging. Finally, a brief discussion on the role of modifiable precursors of obesity such as psychological stress and behavioral factors, plausible intervention strategies, and future directions for research will be presented.

11.2 Markers of Biological Aging

11.2.1 Telomere Attrition

Recent evidence has begun to elucidate the consequences of telomere stability versus shortening, as well as the mechanisms of telomere regulation at the cellular level. In a landmark study conducted nearly 50 years ago, Hayflick (1965) and colleagues demonstrated that normal cells in vitro will not proliferate indefinitely. After extended periods of cell doubling, the capacity for cells to divide declines and eventually ceases, resulting in a phenomena known as "cellular senescence" (Campisi and d'Adda di Fagagna 2007). Telomeres are the primary molecular explanation for this halted cell division. Telomeres are DNA-protein complexes that cap chromosomal ends, conferring chromosomal stability (Blackburn 1991; Effros 2009; Effros et al. 2005; McElhaney and Effros 2009). With each cellular division, telomeric DNA terminal regions, that is, the ends of DNA strands, are not fully replicated. If not counteracted by cellular repair mechanisms (e.g., telomerase), progressive telomere attrition can lead to cell death (Effros 2009; Effros et al. 2005; McElhaney and Effros 2009). Telomerase is a critical enzyme produced by cells with the function of adding DNA sequence repeats to telomere regions, thereby preventing the constant telomere attrition incurred with each cell division. While our understanding of how short telomeres lead to cellular senescence remains incomplete, research suggests that critically short telomeres leads to activation of tumor suppressors, in particular p53, which results in the cell's inability to divide (Campisi and d'Adda di Fagagna 2007). Telomere attrition and related p53 signaling leads to diminished mitochondrial function, the cellular machinery that produces energy for the cell, and increased intracellular reactive oxygen species (ROS) (Sahin et al. 2011). ROS are produced as a normal byproduct of cell metabolism; however, at high levels, ROS can lead to genetic mutations and eventual cell death. It is important to note, however, that other biological mechanisms contribute to cellular aging independent of telomere shortening. For instance, dysfunction of a variety of tumor-suppressor genes can induce senescence in normal cells (Braig et al. 2005).

Telomere attrition not only causes cellular dysfunction and arrest; it has been associated with a variety of human diseases, including hypertension (Demissie et al. 2006), atherosclerosis (Samani et al. 2001), type 2 diabetes (Sampson et al. 2006; Zee et al. 2010), dementia (Martin-Ruiz et al. 2006) and various cancers (Risques et al. 2007; Willeit et al. 2010). Furthermore, telomere attrition predicts risk for cardiovascular and all-cause mortality (Cawthon et al. 2003; Epel et al. 2009; Fitzpatrick et al. 2011). The associations between telomeres and a range of psychosocial factors are covered in more depth elsewhere in this book; however, a review of what is currently known about obesity and telomere length is provided below.

11.2.2 Obesity and Telomere Length

Nearly 20 cross-sectional studies have investigated the association between adiposity (indexed primarily by BMI) and leukocyte telomere length and for the most part have found a significant inverse correlation (Al-Attas et al. 2010; Aviv et al. 2009; Cassidy et al. 2010; Fitzpatrick et al. 2007; Lee et al. 2011; Nordfjall et al. 2008; O'Donnell et al. 2008; Valdes et al. 2005; Zannolli et al. 2008). For instance, in a sample of over 2000 women from the Nurse's Health Study, higher BMI and greater waist circumference were associated with shorter leukocyte telomere length after adjusting for chronological age (Cassidy et al. 2010). However, not all studies support such an association (Bekaert et al. 2007; Benetos et al. 2001; Brouilette et al. 2007; Diaz et al. 2010; Mirabello et al. 2009; Nettleton et al. 2008; Risques et al. 2007; Zhu et al. 2010), possibly due, in part, to differences in sample composition, age, range of adiposity, and measures of adiposity. For instance, reliance on BMI as the sole measure of adiposity lacks specificity. BMI measures four compartments: bone, muscle, subcutaneous fat (which for the most part are not associated with metabolic disease), and visceral fat—the latter is specifically linked to medical comorbidities such as insulin resistance (IR), dyslipidemia, hypertension, glucose intolerance, and fatty liver. Biochemically, visceral fat mass is associated with systemic elevations in oxidative stress, and markers of inflammation (e.g., interleukin-6) (Ferroni et al. 2004). We further describe these mechanisms later in this review.

Visceral fat can be discerned from subcutaneous fat using anthropometrics such as waist circumference and waist to hip ratio, along with newer imaging technologies, such as magnetic resonance imaging (MRI), positron emission tomography (PET), dual energy x-ray absorptiometry (DEXA), and ultrasonography. A recent study compared different parameters of fat distribution to telomere length (Lee et al. 2011). The authors report that BMI, waist circumference, hip circumference, total body fat, and amount of visceral body fat were inversely related to leukocyte telomere length, while measures of subcutaneous fat were unrelated. In this study, being obese (BMI $\geq$ 30) was more strongly related to shorter telomere length in younger participants ($<$ 30 years old) than in older participants ($>$60 years old). It may be that early onset of obesity represents a greater threat than the obesity that commonly accrues with age. Although the causal direction has not been established, it is likely that the biochemical environment of visceral obesity—high IR, oxidative stress, and inflammation – promotes premature aging of immune cells, possibly by suppressing telomerase activity.

Several longitudinal studies have explored whether adiposity predicts telomere attrition over time (Aviv et al. 2009; Epel et al. 2009; Gardner et al. 2005; Kim et al. 2009). In one of the first studies to track telomere length longitudinally, Gardner et al. (2005) found that increases in BMI predicted decreases in telomere length over more than a decade. Epel et al. (2009) found that higher baseline BMI predicted greater decrease in telomere length over a 2.5 year period among women, but not men. In one study of women, duration of obesity (using retrospective reports of weight gain over the past 10–20 years) and history of frequent weight cycling were both associated with shorter telomere length (Kim et al. 2009). Thus, longitudinal studies replicate the cross-sectional findings that obesity is linked with telomere shortening. Longitudinal studies on the temporal dynamics of adipose accumulation versus telomere attrition are necessary to infer causation and mechanism.

11.2.3 Thymic Involution

One of the most prominent immune changes that occurs with aging is the shrinking of thymic cortex and medulla. Since T-cell maturation occurs in the thymus, involution of the thymus is marked by reductions in naïve T-cell output: A decline in the number of naïve T-cell results in a relative reliance on existing memory T-cells (McElhaney and Effros 2009). As naïve T-cells are crucial for the body's ability to respond to unfamiliar pathogens, their loss results in diminished cell-mediated immunity and a related increase in susceptibility to infectious agents.

11.2.4 Obesity and Thymic Involution

Animal models of obesity, such as leptin-deficient mice ($Lep^{ob/ob}$), display clear signs of thymic involution (Howard et al. 1999). Diet-induced obese mice show similar loss as well as fewer naïve and memory T-cells as compared to lean animals (Yang et al. 2009). Beyond diminished T-cell immunity, emerging evidence also indicates that thymic adipocytes may directly influence thymic fitness (Dixit 2010). Indeed, laboratory studies shows that thymic adipocyte cells engineered to constitutively express peroxisome proliferation-activated receptor gamma (PPAR-γ), a regulator of fatty acid storage and adipocyte development, promotes further thymic involution and restricted T-cell receptor diversity (Youm et al. 2010). This model system provides intriguing evidence that PPAR-γ may serve as an important biochemical mechanism linking diminished T-cell immunity and increasing adiposity that occurs with age.

In contrast to the knowledge from animal models, comparatively little is known about the impact of obesity on thymic aging in humans. In this regard, Yang and colleagues found the number of T-cell receptor excision circles (TREC) in peripheral blood (an index of recently generated T-cells from the thymus) to be significantly lower among obese humans as compared to normal weight participants (Yang et al.

2009), even after controlling for glycemic status. While more research in this area is needed, obesity-related decline in thymic function may represent an important indicator of immunosenescence, as diminished T-cell diversity may increase susceptibility to infectious agents (e.g., influenza), a serious health risk among older adults (Thompson et al. 2003).

11.2.5 Antibody Production Following Vaccination

Vaccination efficacy declines precipitously with age, with alarming implications for vulnerability of older adults to influenza or other infectious agents. Age-related declines in adaptive immunity have been proposed to be due to diminished T-cell functioning and its subsequent impact on essential B-cell activities (as reviewed in Haynes and Swain (2006), Miller (1996), Weinberger et al. (2008)). B-cells produce antigen-specific antibodies as part of an orchestrated immune response initiated by exposure to vaccine antigens (e.g., flu, hepatitis B). In addition to the production of antigen-specific antibodies, a successful vaccination response is marked by accumulation of memory B- and T-cells that remain in the blood stream, providing protection against additional exposures. Like with T-cells, however, there is a decline in number of naïve B-cells with age. Consequently, an older immune system is marked by reduced diversity in potential antibody responses (Allman and Miller 2005) and attenuated protection against infectious illness. Related, cross-sectional evidence shows telomere attrition in B-cells with age (reviewed in Weng (2008)), although the extent to which telomere erosion contributes to vaccination efficacy remains to be determined (High et al. 2010).

11.2.6 Obesity and Antibody Production

Much of what we have learned about the hazardous effects of obesity on the adaptive immune system is accrued through animal models. For example, diet-induced obese mice show increased morbidity and mortality following influenza infection compared to lean mice (Kaminski and Randall 2010; Karlsson et al. 2010a). Similar models have shown that obese mice display impaired dendritic cell and T-cell responsiveness (Verwaerde et al. 2006), fewer cytotoxic memory T-cells (Karlsson et al. 2010b), and reduced mitogen-induced IL-2 production (Mito et al. 2000) (a cytokine crucial to memory T-cell immunity).

In humans, several cross-sectional studies have documented an inverse relationship between adiposity and antibody response (Averhoff et al. 1998; Bock et al. 1996; Eliakim et al. 2006; Roome et al. 1993; Weber et al. 1986; Wood et al. 1993). For instance, after adjusting for age, obese patients were 2.1 times less likely to mount a clinically protective antibody response to the hepatitis B vaccination series than non-obese patients (Averhoff et al. 1998). Similar observations have been made with respect to influenza, hepatitis A, and tetanus vaccines (Eliakim et al. 2006; Potter et al. 1999; Van der Wielen et al. 2006; Weber et al. 1986); however, not all studies have been supportive (Phillips et al. 2006). That said, antibody responses are also lower among patients with obesity-related illness, such as diabetes (Alavian and Tabatabaei 2010; Kao et al. 2010). The mechanism underlying excess adiposity and impairment of antibody production or action remains unclear. To the extent that adiposity directly influences humoral immunity, it may be the case that adipocytes or adipocyte-derived soluble mediators negatively affect various aspects of antibody production, from antigen processing to antibody maintenance. As noted above, no study has yet to link telomere length and vaccination response (High et al. 2010). However, the link between obesity and vaccination response may also be due to mechanical issues related to vaccine administration, including an insufficient dose relative to body size, suboptimal absorption, or inappropriate injection site (i.e., areas with more or less adipose tissue) (Eliakim et al. 2006; Weber et al. 1986).

11.2.7 Inflammaging

Inflammation is a fundamental, highly conserved immune process regulated by inflammatory cytokines (e.g., interleukin (IL)-6, tumor necrosis factor (TNF)-alpha), and coordinates the destruction of invading pathogens and facilitate wound healing. While necessary for survival, inflammation is also a key contributor to several age-related diseases, such as cardiovascular disease (Ross 1999). Inflammatory activity increase with age, a topic discussed in more detail elsewhere in this book. Indeed, the term "inflamm-aging," (Franceschi et al. 2000) underscores the centrality of low-grade chronic inflammation to the aging process. This relationship is supported by a large literature showing that systemic levels of pro-inflammatory cytokines increase appreciably with age, partially due to the increase in terminally differentiated, senescent immune cells, which exhibit altered patterns of gene expression leading to a pro-inflammatory phenotype (Campisi 2001; Rodier et al. 2009). Moreover, emerging evidence suggests that inflammatory activity may contribute to telomere shortening, in part, through a down-regulation in telomerase activity (Akiyama et al. 2004), resulting in an increase in the number of senescent immune cells (Parish et al. 2009). Thus, inflammation appears to be both a contributor to and consequence of telomere attrition.

11.2.8 Obesity and Inflammaging

Adipose tissue is increasingly recognized as an endocrine organ that makes important contributions to the levels of inflammatory markers in the blood stream. Visceral fat is particularly metabolically active, producing several pro-inflammatory cytokines (Fontana et al. 2007; Hotamisligil et al. 1995). Indeed, it is estimated that approximately 30 % of the circulating IL-6 in humans originates from adipose tissue (Mohamed-Ali et al. 1997), some of which is released directly by adipocytes, while the rest is secreted by macrophages that have migrated into adipose tissue (Neels and Olefsky 2006; Weisberg et al. 2003; Xu et al. 2003).

While cross-sectional evidence clearly supports positive associations between BMI and other measures of obesity with markers of inflammation (Graziani et al. 2011; Ouchi et al. 2011), there are few prospective studies that have examined this relationship. Gentile et al. (2010) found that participants who gained the most weight over a 10-year study also displayed the greatest increase in inflammatory markers. Additionally, longitudinal weight loss interventions demonstrate that successful weight loss in older obese individuals' results in significant decreases in levels of inflammatory markers such as IL-6 and TNF-α (Miller et al. 2008; Nicklas et al. 2004). While it seems plausible that inflammation may serve as a link between obesity and accelerated biological aging, specifically telomere attrition (Balistreri et al. 2010), direct evidence in support of this pathway is still lacking.

11.3 Potential Biochemical Mechanisms Linking Obesity and Biological Aging

How does excess adiposity accelerate cellular and organismal aging? Growing evidence points to several complementary and overlapping mechanisms (Fig. 11.1). Excess energy storage in abdominal adipocytes promotes the release of a constellation of biochemical mediators (e.g., adipokines), which may contribute to telomere attrition, diminished adaptive immunity, and elevated levels of inflammation. Moreover, obesity is associated with increased likelihood of metabolic dysregulation (e.g., insulin resistance) and increased oxidative stress. While the directionality of these pathways still requires more direct evidence, it is plausible that adipocytes contribute to aging of the immune system, at least in part, through these biochemical pathways. A selective review of the biochemical mediators that link obesity and immunosenescence are discussed below.

11.3.1 Adipokines

11.3.1.1 Leptin

Leptin is an adipose-derived soluble factor released in the bloodstream shortly after eating that aids in regulating appetite, metabolism, and immune function. Interestingly, leptin deficient mice (i.e., $Lep^{ob/ob}$) exhibit several metabolic alterations, including severe obesity, hyperglycemia, IR, and elevated levels of pro-inflammatory cytokines (Fantuzzi 2005; Friedman 2002). These mice show thymic involution marked by T-cell apoptosis (Howard et al. 1999), which contributes to impaired

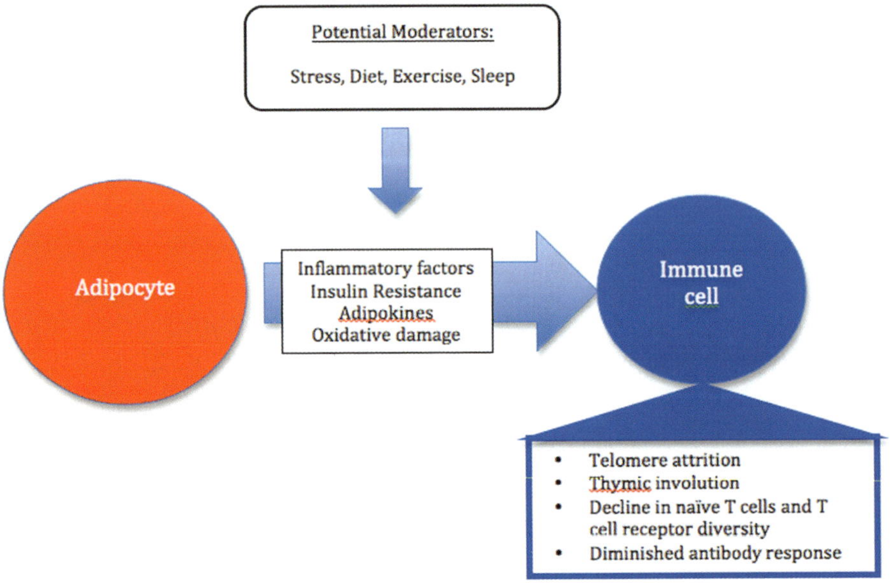

Fig. 11.1 Enlarged visceral adipocytes release high levels of cytokines, cortisol, and adipokines such as leptin and low levels of adiponectin. Activated macrophages release inflammatory molecules and contribute to insulin resistance (IR) of the adipocyte, which creates systemic IR. Together, this adverse biochemical milieu contributes to several aspects of immunosenescence. However, modifiable psychological and behavioral factors (e.g., stress, exercise) are proposed to moderate the associations between adipocytes and markers of immunosenescence

maintenance of memory T-cells (Karlsson et al. 2010b). Finally, transplantation of white adipose tissue from wild type to Lep$^{ob/ob}$ mice, thereby replenishing leptin levels, normalizes metabolic abnormalities (i.e., glycemia, insulin), improves thymic cellularity, and reduces systemic levels of inflammation (Klebanov et al. 2005; Sennello et al. 2006). These findings, albeit from an artificial animal model of obesity, suggest that leptin may be an important adipocyte-dependent mediator with far reaching influence on markers of immunosenescence; however, research exploring the influence of leptin in biological aging in humans has been limited.

Interestingly, in contrast to animal models of obesity where loss of leptin produces an obese phenotype, obese humans tend to display leptin at high levels, which may be indicative of leptin resistance (i.e., an inability for leptin to serve as a satiety biochemical signal) (Enriori et al. 2006). Several studies support an inverse association between circulating leptin levels and leukocyte telomere length (Aviv et al. 2006; Diaz et al. 2010; Valdes et al. 2005); however, not all findings withstand adjustment for demographic characteristics (e.g., age, gender) (Aviv et al. 2006; Diaz et al. 2010). Interestingly, leptin upregulates telomerase activity in human breast cancer cells in vitro (Ren et al. 2010), further highlighting the complex dynamics between leptin and markers of cellular aging. It may be that in vitro studies of leptin are not representative of obesity, as they do not capture the leptin resistance present in vivo.

Till date, leptin levels have not been examined in the context of thymus involution or vaccination response in humans. Future studies investigating associations of leptin levels with measures of immunosenescence over time in both obese and non-obese humans are needed.

11.3.1.2 Adiponectin

Adiponectin is a protein hormone secreted in response to metabolic demands by non-obese adipose tissue (smaller adipose cells). Adiponectin improves several metabolic processes, including glucose regulation, fatty acid catabolism, and insulin sensitivity, and serves as an anti-inflammatory hormone. Moreover, levels of adiponectin decrease with increasing obesity and are even lower in patients with obesity-related illnesses (Lara-Castro et al. 2007). Low levels of adiponectin are related to higher levels of oxidative stress and, when administered, it improves insulin sensitivity in mouse models (Yamauchi et al. 2001). Indeed, experimental evidence demonstrates that adiponectin downregulates TNF-α production and its underlying intracellular pathways related to pro-inflammatory cytokine production, such as pathways regulated by the transcription factor nuclear factor kappa-B (NF-κB) (reviewed in Ouchi and Walsh (2007)).

Few studies have explored the role of adiponectin in biological aging. In one study, low adiponectin was associated with shorter leukocyte telomere length among a sample of diabetic and non-diabetic middle-aged adults (Al-Attas et al. 2010); however, no association was observed in another study, using a larger healthy and more racially diverse sample (Diaz et al. 2010). It remains unclear as to whether adiponectin directly affects immune processes that underlie vaccination response or the aging thymus; however, adiponectin may have the capacity to slow telomere attrition through its ability to both dampen inflammatory signaling and IR.

11.3.2 Insulin Resistance

Obesity, especially visceral adiposity, is correlated with IR, a condition in which insulin no longer adequately lowers blood sugar, the precursor state to diabetes. Visceral adipocytes release high levels of soluble factors, such as adipokines (e.g., leptin), inflammatory cytokines, glycerol, and free fatty acids, all of which are associated with development of IR (Kahn et al. 2006). Though limited, growing evidence implicates IR in accelerated immune cell aging. For instance, a longitudinal study in humans showed that IR accounted for the predictive relationship between BMI and telomere attrition (Gardner et al. 2005). Similarly, in another study, IR accounted for the association between hypertension and shorter leukocyte telomere length (Demissie et al. 2006). In a separate mouse model, improvements in insulin sensitivity led to increased heart cell telomerase activity (Makino et al. 2009).

11.3.3 Oxidative Stress

The "Free Radical Theory of Aging" first proposed in the 1950s (Harman 1956), holds that accumulated free radical damage is a major factor in the determination of longevity. Oxidative stress occurs when there is an imbalance between the production of ROS and cellular mechanisms that protect against excess ROS, such as antioxidants (Turrens 2003). Under normal conditions, the majority of ROS is a byproduct of mitochondrial respiration, a process by which the mitochondria generate the chemical energy needed to fuel cellular activity (Ott et al. 2007). While necessary to cell functioning, this process can produce excess ROS under certain conditions, such as hyperglycemia (Marfella et al. 2001). Hyperglycemia, a correlate of obesity, also increases the production of pro-inflammatory cytokines (Esposito et al. 2002), and incidence of chronic metabolic disease, such as diabetes.

An important consequence of excess ROS generation is that oxidative damage to DNA and RNA can accumulate, impairing genetic code, disrupting normal cell function and promoting cell death. Telomeres are particularly vulnerable to genotoxic stress, including from excess ROS, which can promote the erosion of telomeres (von Zglinicki 2002). Moreover, oxidative stress is reliably elevated among obese individuals suggesting that oxidative stress is a plausible mechanism linking obesity and telomere shortening. Indeed, cross-sectional studies in humans find higher systemic oxidative stress in obese as compared to non-obese subjects (Keaney et al. 2003). Several studies have found that, compared to BMI or subcutaneous fat mass, visceral adiposity bears a stronger cross-sectional and prospective association with markers of lipid peroxidation, another marker of oxidative stress (Gletsu-Miller et al. 2009; Pou et al. 2007). Moreover, weight loss is associated with concomitant improvements in levels of oxidative stress (i.e., decreased ROS and increased antioxidant capacity, as reviewed in Vincent and Taylor 2006). Unfortunately, no studies have yet explored whether oxidative stress mediates the known association between obesity and telomere shortening.

11.3.4 Summary of Biochemical Mediators

We have reviewed how excessive adiposity, especially visceral adiposity, promotes IR, oxidative stress, and inflammation, all of which are closely linked, and contribute to the erosion of telomeres and promote other aspects of immune senescence. Through a series of bidirectional associations these mediators promote downward spiral into chronic disease, as witnessed by withering of lean mass, bone, and immune function observed with extended age. However, is all lost? Despite the daunting task of addressing the diverse metabolic consequences of obesity, the remainder of this chapter is devoted to the review of several modifiable lifestyle factors with the goal of highlighting potential targets for mitigating some of the deleterious effects of obesity on biological aging.

11.4 Psychological and Behavioral Mediators: The Promise of Interventions

The "age old" goal of slowing biological aging has been elusive. To the extent that excess adiposity accelerates the aging process, behavioral interventions targeted against some of the primary contributors of obesity, namely stress-eating, excessive caloric consumption, and limited physical activity, may slow the rate of immune system decline and age-related disease.

11.4.1 Psychological Stress and Eating Behavior

There are many direct and indirect pathways through which states of stress may be affecting obesity, immune senescence, and their relationship to each other. Stress promotes overeating, typically of processed food, also called "comfort food." This food tends to be energy dense, high fructose, and high fat food, which promote ROS formation (Adam and Epel 2007; Lim et al. 2010; Wardle et al. 2010). Moreover, chronic stress arousal increases cortisol and Neuropeptide Y production, both of which promote central adiposity (Epel et al. 2000; Kuo et al. 2007) and alterations in metabolism. These pathways represent specific mechanisms through which stress may modulate the link between obesity and biological aging (Epel 2009; Segerstrom and Miller 2004). Stress-eating is a potent behavior that can affect biochemical milieu. Stress-induced overeating can lead to increased aerobic metabolism and overproduction of ROS, thus promoting cellular damage and/or senescence. For instance, in rats, glucose infusion led to a decrease in antioxidants, increased oxidative stress, and a mild systemic inflammatory response (Ling et al. 2007). In humans, glucose infusion prior to mental challenge results in delayed cardiovascular recovery from the laboratory stressor (Synowski et al. 2011). Thus, while speculative at present, the combination of higher glucose and high stress arousal may promote greater biochemical stress. Moreover, stress-eating promotes poor nutrition, and the interaction between stress and certain nutrients, such as refined sugars and starches, can promote acute inflammatory states (Kiecolt-Glaser 2010).

Stress brought on by weight concerns may also accelerate biological aging. Women who reported high levels of dietary restraint, defined as chronic preoccupation with weight and attempts at restricting food intake, had shorter telomeres than women with low dietary restraint (Kiefer et al. 2008). A preliminary intervention for overweight women focused on stress reduction and mindful eating found a non-significant trend for increased telomerase in the treatment vs. wait list group. However, across groups, improved metabolic health correlated with immune cell telomerase activity (Daubenmier et al, under review). Specifically, reductions in chronic stress, dietary restraint, intake of dietary fat, and cortisol, were all related to increases in telomerase activity. While the mechanisms are not yet elucidated at a cellular level, this suggests interventions aimed at diet and stress eating might improve immune system longevity.

11.4.2 Caloric Restriction

Caloric restriction is effective in extending health and lifespan in several species (Barger et al. 2003). Indeed, a recent study found rhesus monkeys randomized to caloric restriction (i.e., 30 % restriction) showed a 50 % reduction in all-cause mortality compared to non-restricted rhesus monkeys (Colman et al. 2009). Caloric restriction appears to reduce metabolic rate, oxidative stress, and adiposity, while increasing mitochondrial biogenesis and insulin sensitivity (Civitarese et al. 2007; Redman et al. 2008). Similarly, caloric restriction has been associated with sustained T-cell receptor repertoire diversity and slowed thymic aging (Messaoudi et al. 2006; Yang et al. 2009). While caloric restriction shows promise, it remains unclear whether (1) humans have the capacity to maintain long-term caloric restriction and (2) such a lifestyle change can appreciably influences human biological aging. Furthermore, it remains unclear through what mechanisms does caloric restriction promote its effects. In this regard, members of our lab are exploring several markers of biological aging, including telomere length and telomerase activity, in a unique sample of adults who have chosen to adhere to caloric restriction for several years to decades. In addition, a randomized controlled trial of 25 % caloric restriction among 150 non-obese healthy men and women is currently ongoing (Rochon et al. 2011). Together, these studies will help elucidate the biological and psychological correlates of prolonged caloric restriction in humans and aid in clarifying whether such an intervention is fruitful in slowing biological aging.

11.4.3 Physical Activity

Moderate exercise is associated with improved glucose tolerance and insulin sensitivity (Borghouts and Keizer 2000; Kelley and Goodpaster 1999). It also promotes a better oxidative stress phenotype, marked by increased antioxidant activity, reduced low-density lipoprotein oxidation (Elosua et al. 2003), and lower levels of systemic inflammation (Abramson and Vaccarino 2002; Ford 2002), all of which may ultimately slow biological aging.

In human studies, older adults who are deemed physically fit based on their maximal O_2 uptake display a more robust antibody response to influenza vaccines than sedentary older individuals (Keylock et al. 2007). Moreover, several exercise interventions have been effective in improving vaccination responses and antibody maintenance (Kohut et al. 2004; Woods et al. 2009). Similar associations have been observed regarding telomere length. For instance, older adults practicing habitual endurance exercise display longer leukocyte telomeres than sedentary peers and fail to differ from younger exercise-trained adults (LaRocca et al. 2010). Moreover, self-reported vigorous physical activity, averaged over 3 days of daily diary assessment, was shown to buffer the deleterious effects of chronic stress on telomere length (Puterman et al. 2010). Finally, in a pilot study, individuals participating in a 3-month comprehensive lifestyle intervention, which included moderate aerobic exercise (walking 30 min/day, 6 days/week), showed a significant increase in telomerase activity, providing a plausible pathway through which exercise may slow cellular aging (Ornish et al. 2008). This intervention included a number of other components (e.g., low fat diet, stress management, social support) and had no control condition; thus while suggestive, future research exploring the specific longitudinal effects of physical activity on biological aging is greatly needed.

11.4.4 Sleep

Disordered sleep is proposed as both a contributor and a consequence of obesity. BMI has been found to increase over time among short sleepers (Gangwisch et al. 2005; Lopez-Garcia et al. 2008; Patel et al. 2006); however, other prospective samples have failed to replicate this association (Lauderdale et al. 2009; Stranges et al. 2008). At the biochemical level, acute sleep loss is associated with elevations in markers of systemic inflammation (Meier-Ewert et al. 2004; Shearer et al. 2001), dysregulation of adipokines (Pejovic et al. 2010; Spiegel et al. 2004a, b), and a decline in glucose tolerance (Shearer et al. 2001; Spiegel et al. 1999). Conversely, poor sleep is common among obese individuals, in part because high BMI is a strong predictor of obstructive sleep apnea (OSA) (Patil et al. 2007).

Poor sleep is associated with alterations in immune functioning, including markers of immunosenescence. For instance, partial and total sleep deprivation produces robust declines in antibody response to vaccination (Lange et al. 2003; Spiegel et al. 2002). A similar association was recently observed outside of the laboratory setting (Prather et al. 2011). While obesity remains a clear risk factor for a reduced vaccination response, no study has explored whether sleep moderates this relationship. With respect to the influence of sleep on telomere length, the data are limited. In one study, patients with OSA displayed shorter leukocyte telomeres compared to OSA free individuals, independent of cardiometabolic risk factors, OSA severity, daytime sleepiness, age, and BMI (Barcelo et al. 2010). In addition, preliminary findings from our group suggest that in a healthy sample of older women, poorer subjective sleep quality is associated with shorter telomere length, independent of chronological age and BMI (Prather et al., in press). Further research investigating the influences of sleep on immunosenescence both in OSA and normal sleepers is warranted.

11.5 Conclusions

In this review, we have presented illustrative examples of how obesity is related to four markers of immune system aging: telomere attrition, weak response to vaccination, thymic involution, and elevated inflammation. As shown in Fig. 11.1, it appears that these relationships are largely due to the pro-aging biochemical milieu created by adipocytes. However, there are likely many bidirectional relationships between the regulatory systems of energy balance and immune system function yet to be elucidated. Several intervention studies provide inspiration and preliminary empirical evidence that improvements in stress, salubrious health behaviors, and weight loss, may improve the robustness of our immune system and longevity.

References

Abramson JL, Vaccarino V (2002) Relationship between physical activity and inflammation among apparently healthy middle-aged and older US adults. Arch Intern Med 162(11):1286–1292

Adam TC, Epel ES (2007) Stress, eating and the reward system. Physiol Behav 91(4):449–458

Akiyama M, Yamada O, Hideshima T, Yanagisawa T, Yokoi K, Fujisawa K et al (2004) TNFalpha induces rapid activation and nuclear translocation of telomerase in human lymphocytes. Biochem Biophys Res Commun 316(2):528–532

Al-Attas OS, Al-Daghri NM, Alokail MS, Alfadda A, Bamakhramah A, Sabico S et al (2010) Adiposity and insulin resistance correlate with telomere length in middle-aged Arabs: the influence of circulating adiponectin. Eur J Endocrinol 163(4):601–607

Alavian SM, Tabatabaei SV (2010) The effect of diabetes mellitus on immunological response to hepatitis B virus vaccine in individuals with chronic kidney disease: a meta-analysis of current literature. Vaccine 28(22):3773–3777

Allman D, Miller JP (2005) B cell development and receptor diversity during aging. Curr Opin Immunol 17(5):463–467

Averhoff F, Mahoney F, Coleman P, Schatz G, Hurwitz E, Margolis H (1998) Immunogenicity of hepatitis B Vaccines. Implications for persons at occupational risk of hepatitis B virus infection. Am J Prev Med 15(1):1–8

Aviv A, Chen W, Gardner JP, Kimura M, Brimacombe M, Cao X et al (2009) Leukocyte telomere dynamics: longitudinal findings among young adults in the Bogalusa Heart Study. Am J Epidemiol 169(3):323–329

Aviv A, Valdes A, Gardner JP, Swaminathan R, Kimura M, Spector TD (2006) Menopause modifies the association of leukocyte telomere length with insulin resistance and inflammation. J Clin Endocrinol Metab 91(2):635–640

Balistreri CR, Caruso C, Candore G (2010) The role of adipose tissue and adipokines in obesity-related inflammatory diseases. Mediators of inflammation. doi:10.1155/2010/802078

Barcelo A, Pierola J, Lopez-Escribano H, de la Pena M, Soriano JB, Alonso-Fernandez A et al (2010) Telomere shortening in sleep apnea syndrome. Respir Med 104(8):1225–1229

Barger JL, Walford RL, Weindruch R (2003) The retardation of aging by caloric restriction: its significance in the transgenic era. Exp Gerontol 38(11–12):1343–1351

Bekaert S, De Meyer T, Rietzschel ER, De Buyzere ML, De Bacquer D, Langlois M et al (2007) Telomere length and cardiovascular risk factors in a middle-aged population free of overt cardiovascular disease. Aging Cell 6(5):639–647

Benetos A, Okuda K, Lajemi M, Kimura M, Thomas F, Skurnick J et al (2001) Telomere length as an indicator of biological aging: the gender effect and relation with pulse pressure and pulse wave velocity. Hypertension 37(2 Part 2):381–385

Blackburn EH (1991) Structure and function of telomeres. Nature 350(6319):569–573

Bock HL, Kruppenbacher J, Sanger R, Hobel W, Clemens R, Jilg W (1996) Immunogenicity of a recombinant hepatitis B vaccine in adults. Arch Intern Med 156(19):2226–2231

Borghouts LB, Keizer HA (2000) Exercise and insulin sensitivity: a review. Int J Sports Med 21(1):1–12

Braig M, Lee S, Loddenkemper C, Rudolph C, Peters AH, Schlegelberger B et al (2005) Oncogene-induced senescence as an initial barrier in lymphoma development. Nature 436(7051):660–665

Brouilette SW, Moore JS, McMahon AD, Thompson JR, Ford I, Shepherd J et al (2007) Telomere length, risk of coronary heart disease, and statin treatment in the West of Scotland Primary Prevention Study: a nested case-control study. Lancet 369(9556):107–114

Campisi J (2001) Cellular senescence as a tumor-suppressor mechanism. Trends Cell Biol 11(11):S27–31

Campisi J, d'Adda di Fagagna F (2007) Cellular senescence: when bad things happen to good cells. Nat Rev Mol Cell Biol 8(9):729–740

Cassidy A, De Vivo I, Liu Y, Han J, Prescott J, Hunter DJ et al (2010) Associations between diet, lifestyle factors, and telomere length in women. Am J Clin Nutr 91(5):1273–1280

Cawthon RM, Smith KR, O'Brien E, Sivatchenko A, Kerber RA (2003) Association between telomere length in blood and mortality in people aged 60 years or older. Lancet 361(9355):393–395

Civitarese AE, Smith SR, Ravussin E (2007) Diet, energy metabolism and mitochondrial biogenesis. Curr Opin Clin Nutr Metabol Care 10(6):679–687

Colman RJ, Anderson RM, Johnson SC, Kastman EK, Kosmatka KJ, Beasley TM et al (2009) Caloric restriction delays disease onset and mortality in rhesus monkeys. Science 325(5937):201–204

Daubenmier J, Lin J, Blackburn E, Hecht FM, Kristeller J, Maninger N et al (under review) Changes in stress and metabolic factors predict changes in telomerase activity among women enrolled in a randomized trial of a 4-month mindfulness intervention

Demissie S, Levy D, Benjamin EJ, Cupples LA, Gardner JP, Herbert A et al (2006) Insulin resistance, oxidative stress, hypertension, and leukocyte telomere length in men from the Framingham Heart Study. Aging Cell 5(4):325–330

Diaz VA, Mainous AG, Player MS, Everett CJ (2010) Telomere length and adiposity in a racially diverse sample. Int J Obes (Lond) 34(2):261–265

Dixit VD (2010) Thymic fatness and approaches to enhance thymopoietic fitness in aging. Curr Opin Immunol 22(4):521–528

Effros RB (2009) Kleemeier Award Lecture 2008– the canary in the coal mine: telomeres and human healthspan. J Gerontol A Biol Sci Med Sci 64(5):511–515

Effros RB, Dagarag M, Spaulding C, Man J (2005) The role of CD8+ T-cell replicative senescence in human aging. Immunol Rev 205:147–157

Eliakim A, Schwindt C, Zaldivar F, Casali P, Cooper DM (2006) Reduced tetanus antibody titers in overweight children. Autoimmunity 39(2):137–141

Elosua R, Molina L, Fito M, Arquer A, Sanchez-Quesada JL, Covas MI et al (2003) Response of oxidative stress biomarkers to a 16-week aerobic physical activity program, and to acute physical activity, in healthy young men and women. Atherosclerosis 167(2):327–334

Enriori PJ, Evans AE, Sinnayah P, Cowley MA (2006) Leptin resistance and obesity. Obesity (Silver Spring) 14(Suppl 5):254S-258S

Epel ES (2009) Psychological and metabolic stress: a recipe for accelerated cellular aging? Hormones (Athens) 8(1):7–22

Epel ES, McEwen B, Seeman T, Matthews K, Castellazzo G, Brownell KD et al (2000) Stress and body shape: stress-induced cortisol secretion is consistently greater among women with central fat. Psychosom Med 62(5):623–632

Epel ES, Merkin SS, Cawthon R, Blackburn EH, Adler NE, Pletcher MJ et al (2009) The rate of leukocyte telomere shortening predicts mortality from cardiovascular disease in elderly men. Aging (Albany NY) 1(1):81–88

Esposito K, Nappo F, Marfella R, Giugliano G, Giugliano F, Ciotola M et al (2002) Inflammatory cytokine concentrations are acutely increased by hyperglycemia in humans: role of oxidative stress. Circulation 106(16):2067–2072

Fantuzzi G (2005) Adipose tissue, adipokines, and inflammation. J Allergy Clin Immunol 115(5):911–919; quiz 920

Ferroni P, Basili S, Falco A, Davi G (2004) Inflammation, insulin resistance, and obesity. Curr Atheroscler Rep 6(6):424–431

Fitzpatrick AL, Kronmal RA, Gardner JP, Psaty BM, Jenny NS, Tracy RP et al (2007) Leukocyte telomere length and cardiovascular disease in the cardiovascular health study. Am J Epidemiol 165(1):14–21
Fitzpatrick AL, Kronmal RA, Kimura M, Gardner JP, Psaty BM, Jenny NS et al (2011) Leukocyte telomere length and mortality in the cardiovascular health study. J Gerontol A Biol Sci Med Sci 66(4):421–429
Flegal KM, Carroll MD, Kuczmarski RJ, Johnson CL (1998) Overweight and obesity in the United States: prevalence and trends, 1960–1994. InterJ Obes Metc Disorder 22(1):39–47
Flegal KM, Carroll MD, Ogden CL, Curtin LR (2010) Prevalence and trends in obesity among US adults, 1999–2008. J Am Med Assoc 303(3):235–241
Flegal KM, Graubard BI, Williamson DF, Gail MH (2007) Cause-specific excess deaths associated with underweight, overweight, and obesity. J Am Med Assoc 298(17):2028–2037
Fontana L, Eagon JC, Trujillo ME, Scherer PE, Klein S (2007) Visceral fat adipokine secretion is associated with systemic inflammation in obese humans. Diabetes 56(4):1010–1013
Ford ES (2002) Does exercise reduce inflammation? Physical activity and C-reactive protein among U.S. adults. Epidemiology 13(5):561–568
Franceschi C, Bonafe M, Valensin S, Olivieri F, De Luca M, Ottaviani E et al (2000) Inflamm-aging. An evolutionary perspective on immunosenescence. Ann NY Acad Sci 908:244–254
Friedman JM (2002) The function of leptin in nutrition, weight, and physiology. Nutr Rev 60(10 Pt 2):S1–14; discussion S68–84:85–17
Gangwisch JE, Malaspina D, Boden-Albala B, Heymsfield SB (2005) Inadequate sleep as a risk factor for obesity: analyses of the NHANES I. Sleep 28(10):1289–1296
Gardner JP, Li S, Srinivasan SR, Chen W, Kimura M, Lu X et al (2005) Rise in insulin resistance is associated with escalated telomere attrition. Circulation 111(17):2171–2177
Gentile M, Panico S, Rubba F, Mattiello A, Chiodini P, Jossa F et al (2010) Obesity, overweight, and weight gain over adult life are main determinants of elevated hs-CRP in a cohort of Mediterranean women. Eur J Clin Nutr 64(8):873–878
Gletsu-Miller N, Hansen JM, Jones DP, Go YM, Torres WE, Ziegler TR et al (2009) Loss of total and visceral adipose tissue mass predicts decreases in oxidative stress after weight-loss surgery. Obesity (Silver Spring) 17(3):439–446
Graziani F, Cialdella P, Liuzzo G, Basile E, Brugaletta S, Pedicino D et al (2011) Cardiovascular risk in obesity: different activation of inflammation and immune system between obese and morbidly obese subjects. Eur J Internal Med 22(4):418–423
Harman D (1956) Aging: a theory based on free radical and radiation chemistry. J Gerontol 11(3):298–300
Hayflick L (1965) The limited in vitro lifetime of human diploid cell strains. Exp Cell Res 37:614–636
Haynes L, Swain SL (2006) Why aging T cells fail: implications for vaccination. Immunity 24(6):663–666
High KP, D'Aquila RT, Fuldner RA, Gerding DN, Halter JB, Haynes L et al (2010) Workshop on immunizations in older adults: identifying future research agendas. J Am Geriat Soc 58(4):765–776
Hotamisligil GS, Arner P, Caro JF, Atkinson RL, Spiegelman BM (1995) Increased adipose tissue expression of tumor necrosis factor-alpha in human obesity and insulin resistance. J Clin Invest 95(5):2409–2415
Howard JK, Lord GM, Matarese G, Vendetti S, Ghatei MA, Ritter MA et al (1999) Leptin protects mice from starvation-induced lymphoid atrophy and increases thymic cellularity in ob/ob mice. J Clin Invest 104(8):1051–1059
Hubert HB, Feinleib M, McNamara PM, Castelli WP (1983) Obesity as an independent risk factor for cardiovascular disease: a 26-year follow-up of participants in the Framingham Heart Study. Circulation 67(5):968–977
Kahn SE, Hull RL, Utzschneider KM (2006) Mechanisms linking obesity to insulin resistance and type 2 diabetes. Nature 444(7121):840–846

Kaminski DA, Randall TD (2010) Adaptive immunity and adipose tissue biology. Trends Immunol 31(10):384–390

Kao TM, Hsieh SM, Kung HC, Lee YC, Huang KC, Huang LM et al (2010) Immune response of single dose vaccination against 2009 pandemic influenza A (H1N1) in the Taiwanese elderly. Vaccine 28(38):6159–6163

Karlsson EA, Sheridan PA, Beck MA (2010a) Diet-induced obesity impairs the T cell memory response to influenza virus infection. J Immunol 184(6):3127–3133

Karlsson EA, Sheridan PA, Beck MA (2010b) Diet-induced obesity in mice reduces the maintenance of influenza-specific CD8+ memory T cells. J Nutr 140(9):1691–1697

Keaney JF Jr, Larson MG, Vasan RS, Wilson PW, Lipinska I, Corey D et al (2003) Obesity and systemic oxidative stress: clinical correlates of oxidative stress in the Framingham Study. Arterioscler Thromb Vasc Biol 23(3):434–439

Kelley DE, Goodpaster BH (1999) Effects of physical activity on insulin action and glucose tolerance in obesity. Med Sci Sports Exerc 31(Suppl 11):S619–623

Keylock KT, Lowder T, Leifheit KA, Cook M, Mariani RA, Ross K et al (2007) Higher antibody, but not cell-mediated, responses to vaccination in high physically fit elderly. J Appl Physiol 102(3):1090–1098

Kiecolt-Glaser JK (2010) Stress, food, and inflammation: psychoneuroimmunology and nutrition at the cutting edge. Psychosom Med 72(4):365–369

Kiefer A, Lin J, Blackburn E, Epel E (2008) Dietary restraint and telomere length in pre- and postmenopausal women. Psychosom Med 70(8):845–849

Kim S, Parks CG, DeRoo LA, Chen H, Taylor JA, Cawthon RM et al (2009) Obesity and weight gain in adulthood and telomere length. Cancer Epidemiol Biomarkers Prev 18(3):816–820

Klebanov S, Astle CM, DeSimone O, Ablamunits V, Harrison DE (2005) Adipose tissue transplantation protects ob/ob mice from obesity, normalizes insulin sensitivity and restores fertility. J Endocrinol 186(1):203–211

Kohut ML, Arntson BA, Lee W, Rozeboom K, Yoon KJ, Cunnick JE et al (2004) Moderate exercise improves antibody response to influenza immunization in older adults. Vaccine 22(17–18):2298–2306

Kuo LE, Kitlinska JB, Tilan JU, Li L, Baker SB, Johnson MD et al (2007) Neuropeptide Y acts directly in the periphery on fat tissue and mediates stress-induced obesity and metabolic syndrome. Nat Med 13(7):803–811

Lange T, Perras B, Fehm HL, Born J (2003) Sleep enhances the human antibody response to hepatitis A vaccination. Psychosom Med 65(5):831–835

Lara-Castro C, Fu Y, Chung BH, Garvey WT (2007) Adiponectin and the metabolic syndrome: mechanisms mediating risk for metabolic and cardiovascular disease. Curr Opin Lipidol 18(3):263–270

LaRocca TJ, Seals DR, Pierce GL (2010) Leukocyte telomere length is preserved with aging in endurance exercise-trained adults and related to maximal aerobic capacity. Mech Ageing Dev 131(2):165–167

Lauderdale DS, Knutson KL, Rathouz PJ, Yan LL, Hulley SB, Liu K (2009) Cross-sectional and longitudinal associations between objectively measured sleep duration and body mass index: the CARDIA Sleep Study. Am J Epidemiol 170(7):805–813

Lee M, Martin H, Firpo MA, Demerath EW (2011) Inverse association between adiposity and telomere length: Fels Longitudinal Study. Am J Hum Biol 23(1):100–106

Lim JS, Mietus-Snyder M, Valente A, Schwarz JM, Lustig RH (2010) The role of fructose in the pathogenesis of NAFLD and the metabolic syndrome. Nature reviews. Gastroenterol Hepatol 7(5):251–264

Ling PR, Smith RJ, Bistrian BR (2007) Acute effects of hyperglycemia and hyperinsulinemia on hepatic oxidative stress and the systemic inflammatory response in rats. Crit Care Med 35(2):555–560

Lopez-Garcia E, Faubel R, Leon-Munoz L, Zuluaga MC, Banegas JR, Rodriguez-Artalejo F (2008) Sleep duration, general and abdominal obesity, and weight change among the older adult population of Spain. Am J Clin Nutr 87(2):310–316

Makino N, Maeda T, Oyama J, Higuchi Y, Mimori K (2009) Improving insulin sensitivity via activation of PPAR-gamma increases telomerase activity in the heart of OLETF rats. Am J Physiol Heart Circ Physiol 297(6):H2188–2195
Marfella R, Quagliaro L, Nappo F, Ceriello A, Giugliano D (2001) Acute hyperglycemia induces an oxidative stress in healthy subjects. J Clin Invest 108(4):635–636
Martin-Ruiz C, Dickinson HO, Keys B, Rowan E, Kenny RA, Von Zglinicki T (2006) Telomere length predicts poststroke mortality, dementia, and cognitive decline. Ann Neurol 60(2):174–180
McElhaney JE, Effros RB (2009) Immunosenescence: what does it mean to health outcomes in older adults? Curr Opin Immunol 21(4):418–424
Meier-Ewert HK, Ridker PM, Rifai N, Regan MM, Price NJ, Dinges DF et al (2004) Effect of sleep loss on C-reactive protein, an inflammatory marker of cardiovascular risk. J Am Coll Cardiol 43(4):678–683
Messaoudi I, Warner J, Fischer M, Park B, Hill B, Mattison J et al (2006) Delay of T cell senescence by caloric restriction in aged long-lived nonhuman primates. Proc Natl Acad Sci USA 103(51):19448–19453
Miller RA (1996) The aging immune system: primer and prospectus. Science 273(5271):70–74
Miller GD, Nicklas BJ, Loeser RF (2008) Inflammatory biomarkers and physical function in older, obese adults with knee pain and self-reported osteoarthritis after intensive weight-loss therapy. J Am Geriatr Soc 56(4):644–651
Mirabello L, Huang WY, Wong JY, Chatterjee N, Reding D, Crawford ED et al (2009) The association between leukocyte telomere length and cigarette smoking, dietary and physical variables, and risk of prostate cancer. Aging Cell 8(4):405–413
Mito N, Hosoda T, Kato C, Sato K (2000) Change of cytokine balance in diet-induced obese mice. Metabolism 49(10):1295–1300
Mohamed-Ali V, Goodrick S, Rawesh A, Katz DR, Miles JM, Yudkin JS et al (1997) Subcutaneous adipose tissue releases interleukin-6, but not tumor necrosis factor-alpha, in vivo. J Clin Endocrinol Metab 82(12):4196–4200
Neels JG, Olefsky JM (2006) Inflamed fat: what starts the fire? J Clin Invest 116(1):33–35
Nettleton JA, Diez-Roux A, Jenny NS, Fitzpatrick AL, Jacobs DR Jr (2008) Dietary patterns, food groups, and telomere length in the Multi-Ethnic Study of Atherosclerosis (MESA). Am J Clin Nutr 88(5):1405–1412
Nicklas BJ, Ambrosius W, Messier SP, Miller GD, Penninx BW, Loeser RF et al (2004) Diet-induced weight loss, exercise, and chronic inflammation in older, obese adults: a randomized controlled clinical trial. The Am J Clin Nutr 79(4):544–551
Nordfjall K, Eliasson M, Stegmayr B, Melander O, Nilsson P, Roos G (2008) Telomere length is associated with obesity parameters but with a gender difference. Obesity (Silver Spring) 16(12):2682–2689
O'Donnell CJ, Demissie S, Kimura M, Levy D, Gardner JP, White C et al (2008) Leukocyte telomere length and carotid artery intimal medial thickness: the Framingham Heart Study. Arterioscler Thromb Vasc Biol 28(6):1165–1171
Ornish D, Lin J, Daubenmier J, Weidner G, Epel E, Kemp C et al (2008) Increased telomerase activity and comprehensive lifestyle changes: a pilot study. Lancet Oncol 9(11):1048–1057
Ott M, Gogvadze V, Orrenius S, Zhivotovsky B (2007) Mitochondria, oxidative stress and cell death. Apoptosis: Int J Prog Cell Death 12(5):913–922
Ouchi N, Walsh K (2007) Adiponectin as an anti-inflammatory factor. Clin Chim Acta 380(1–2):24–30
Ouchi N, Parker JL, Lugus JJ, Walsh K (2011) Adipokines in inflammation and metabolic disease. Nature reviews. Immunology 11(2):85–97
Parish ST, Wu JE, Effros RB (2009) Modulation of T lymphocyte replicative senescence via TNF-{alpha} inhibition: role of caspase-3. J Immunol 182(7):4237–4243
Patel SR, Malhotra A, White DP, Gottlieb DJ, Hu FB (2006) Association between reduced sleep and weight gain in women. Am J Epidemiol 164(10):947–954

Patil SP, Schneider H, Schwartz AR, Smith PL (2007) Adult obstructive sleep apnea: pathophysiology and diagnosis. Chest 132(1):325–337

Pejovic S, Vgontzas AN, Basta M, Tsaoussoglou M, Zoumakis E, Vgontzas A et al (2010) Leptin and hunger levels in young healthy adults after one night of sleep loss. J Sleep Res 19(4):552–558

Phillips AC, Carroll D, Burns VE, Ring C, Macleod J, Drayson M (2006) Bereavement and marriage are associated with antibody response to influenza vaccination in the elderly. Brain Behav Immun 20(3):279–289

Potter JM, O'Donnel B, Carman WF, Roberts MA, Stott DJ (1999) Serological response to influenza vaccination and nutritional and functional status of patients in geriatric medical long-term care. Age Ageing 28(2):141–145

Pou KM, Massaro JM, Hoffmann U, Vasan RS, Maurovich-Horvat P, Larson MG et al (2007) Visceral and subcutaneous adipose tissue volumes are cross-sectionally related to markers of inflammation and oxidative stress: the Framingham Heart Study. Circulation 116(11):1234–1241

Prather AA, Hall M, Fury JM, Ross D, Muldoon MF, Cohen S et al (2011) Sleep duration and antibody response to hepatitis B vaccination. Paper presented at the annual meeting of the American Psychosomatic Society

Prather AA, Puterman E, Lin J, O'Donovan A, Krauss J, Tomiyama AJ et al (in press) Shorter leukocyte telomere length in midlife women with poor sleep quality. J Aging Res

Puterman E, Lin J, Blackburn E, O'Donovan A, Adler N, Epel E (2010) The power of exercise: buffering the effect of chronic stress on telomere length. PLoS One 5(5):e10837

Redman LM, Martin CK, Williamson DA, Ravussin E (2008) Effect of caloric restriction in non-obese humans on physiological, psychological and behavioral outcomes. Physiol Behav 94(5):643–648

Ren H, Zhao T, Wang X, Gao C, Wang J, Yu M et al (2010) Leptin upregulates telomerase activity and transcription of human telomerase reverse transcriptase in MCF-7 breast cancer cells. Biochem Biophys Res Commun 394(1):59–63

Risques RA, Vaughan TL, Li X, Odze RD, Blount PL, Ayub K et al (2007) Leukocyte telomere length predicts cancer risk in Barrett's esophagus. Cancer Epidemiol Biomarkers Prev 16(12):2649–2655

Rochon J, Bales CW, Ravussin E, Redman LM, Holloszy JO, Racette SB et al (2011) Design and conduct of the CALERIE study: comprehensive assessment of the long-term effects of reducing intake of energy. J Gerontol A Biol Sci Med Sci 66A(1):97–108

Rodier F, Coppe JP, Patil CK, Hoeijmakers WA, Munoz DP, Raza SR et al (2009) Persistent DNA damage signalling triggers senescence-associated inflammatory cytokine secretion. Nat Cell Biol 11(8):973–979

Roome AJ, Walsh SJ, Cartter ML, Hadler JL (1993) Hepatitis B vaccine responsiveness in Connecticut public safety personnel. JAMA 270(24):2931–2934

Ross, R (1999) Atherosclerosis– an inflammatory disease. New Eng J Med 340(2):115–126

Sahin E, Colla S, Liesa M, Moslehi J, Muller FL, Guo M et al (2011) Telomere dysfunction induces metabolic and mitochondrial compromise. Nature 470(7334):359–365

Samani NJ, Boultby R, Butler R, Thompson JR, Goodall AH (2001) Telomere shortening in atherosclerosis. Lancet 358(9280):472–473

Sampson MJ, Winterbone MS, Hughes JC, Dozio N, Hughes DA (2006) Monocyte telomere shortening and oxidative DNA damage in type 2 diabetes. Diabetes Care 29(2):283–289

Segerstrom SC, Miller GE (2004) Psychological stress and the human immune system: a meta-analytic study of 30 years of inquiry. Psychol Bull 130(4):601–630

Sennello JA, Fayad R, Pini M, Gove ME, Fantuzzi G (2006) Transplantation of wild-type white adipose tissue normalizes metabolic, immune and inflammatory alterations in leptin-deficient ob/ob mice. Cytokine 36(5–6):261–266

Shearer WT, Reuben JM, Mullington JM, Price NJ, Lee BN, Smith EO et al (2001) Soluble TNF-alpha receptor 1 and IL-6 plasma levels in humans subjected to the sleep deprivation model of spaceflight. Journal Allerg Clin Immunol 107(1):165–170

Spiegel K, Leproult R, L'Hermite-Baleriaux M, Copinschi G, Penev PD, Van Cauter E (2004a) Leptin levels are dependent on sleep duration: relationships with sympathovagal balance, carbohydrate regulation, cortisol, and thyrotropin. J Clin Endocrinol Metabol 89(11):5762–5771
Spiegel K, Leproult R, Van Cauter E (1999) Impact of sleep debt on metabolic and endocrine function. Lancet 354(9188):1435–1439
Spiegel K, Sheridan JF, Van Cauter E (2002) Effect of sleep deprivation on response to immunization. JAMA 288(12):1471–1472
Spiegel K, Tasali E, Penev P, Van Cauter E (2004b) Brief communication: sleep curtailment in healthy young men is associated with decreased leptin levels, elevated ghrelin levels, and increased hunger and appetite. Ann Intern Med 141(11):846–850
Stern MP, Haffner SM (1986) Body fat distribution and hyperinsulinemia as risk factors for diabetes and cardiovascular disease. Arteriosclerosis 6(2):123–130
Stranges S, Cappuccio FP, Kandala NB, Miller MA, Taggart FM, Kumari M et al (2008) Cross-sectional versus prospective associations of sleep duration with changes in relative weight and body fat distribution: the Whitehall II Study. Am J Epidemiol 167(3):321–329
Synowski SJ, Kop WJ, Warwick ZS, Waldstein SS (2011) Effects of acute glucose administration on cardiovascular reactivity to mental challenge. Paper presented at the American Psychosomatic Society
Thompson WW, Shay DK, Weintraub E, Brammer L, Cox N, Anderson LJ et al (2003) Mortality associated with influenza and respiratory syncytial virus in the United States. JAMA 289(2):179–186
Turrens JF (2003) Mitochondrial formation of reactive oxygen species. J Physiol 552(Pt 2):335–344
Valdes AM, Andrew T, Gardner JP, Kimura M, Oelsner E, Cherkas LF et al (2005) Obesity, cigarette smoking, and telomere length in women. Lancet 366(9486):662–664
Van Der Wielen M, Van Damme P, Chlibek R, Smetana J, von Sonnenburg F (2006) Hepatitis A/B vaccination of adults over 40 years old: comparison of three vaccine regimens and effect of influencing factors. Vaccine 24(26):5509–5515
Verwaerde C, Delanoye A, Macia L, Tailleux A, Wolowczuk I (2006) Influence of high-fat feeding on both naïve and antigen-experienced T-cell immune response in DO10.11 mice. Scand J Immunol 64(5):457–466
Vincent HK, Taylor AG (2006) Biomarkers and potential mechanisms of obesity-induced oxidant stress in humans. Int J Obes (Lond) 30(3):400–418
von Zglinicki T (2002) Oxidative stress shortens telomeres. Trends Biochem Sci 27(7):339–344
Wang YC, Colditz GA, Kuntz KM (2007) Forecasting the obesity epidemic in the aging U.S. population. Obesity (Silver Spring) 15(11):2855–2865
Wardle J, Chida Y, Gibson EL, Whitaker KL, Steptoe A (2010) Stress and adiposity: a meta-analysis of longitudinal studies. Obesity (Silver Spring)
Weber DJ, Rutala WA, Samsa GP, Bradshaw SE, Lemon SM (1986) Impaired immunogenicity of hepatitis B vaccine in obese persons. N Engl J Med 314(21):1393
Weinberger B, Herndler-Brandstetter D, Schwanninger A, Weiskopf D, Grubeck-Loebenstein B (2008) Biology of immune responses to vaccines in elderly persons. Clin Infect Dis 46(7):1078–1084
Weisberg SP, McCann D, Desai M, Rosenbaum M, Leibel RL, Ferrante AW Jr (2003) Obesity is associated with macrophage accumulation in adipose tissue. J Clin Invest 112(12):1796–1808
Weng NP (2008) Telomere and adaptive immunity. Mechs Ageing Dev 129(1–2):60–66
Willeit P, Willeit J, Mayr A, Weger S, Oberhollenzer F, Brandstatter A et al (2010) Telomere length and risk of incident cancer and cancer mortality. JAMA 304(1):69–75
Wood RC, MacDonald KL, White KE, Hedberg CW, Hanson M, Osterholm MT (1993) Risk factors for lack of detectable antibody following hepatitis B vaccination of Minnesota health care workers. JAMA 270(24):2935–2939
Woods JA, Keylock KT, Lowder T, Vieira VJ, Zelkovich W, Dumich S et al (2009) Cardiovascular exercise training extends influenza vaccine seroprotection in sedentary older adults: the immune function intervention trial. J Am Geriatr Soc 57(12):2183–2191

Xu H, Barnes GT, Yang Q, Tan G, Yang D, Chou CJ et al (2003) Chronic inflammation in fat plays a crucial role in the development of obesity-related insulin resistance. J Clin Invest 112(12):1821–1830

Yach D, Stuckler D, Brownell KD (2006) Epidemiologic and economic consequences of the global epidemics of obesity and diabetes. Nat Med 12(1):62–66

Yamauchi T, Kamon J, Waki H, Terauchi Y, Kubota N, Hara K et al (2001) The fat-derived hormone adiponectin reverses insulin resistance associated with both lipoatrophy and obesity. Nat Med 7(8):941–946

Yang H, Youm YH, Dixit VD (2009) Inhibition of thymic adipogenesis by caloric restriction is coupled with reduction in age-related thymic involution. J Immunol 183(5):3040–3052

Yang H, Youm YH, Vandanmagsar B, Rood J, Kumar KG, Butler AA et al (2009) Obesity accelerates thymic aging. Blood 114(18):3803–3812

Youm YH, Yang H, Amin R, Smith SR, Leff T, Dixit VD (2010) Thiazolidinedione treatment and constitutive-PPARgamma activation induces ectopic adipogenesis and promotes age-related thymic involution. Aging Cell 9(4):478–489

Zannolli R, Mohn A, Buoni S, Pietrobelli A, Messina M, Chiarelli F et al (2008) Telomere length and obesity. Acta Paediatr 97(7):952–954

Zee RY, Castonguay AJ, Barton NS, Germer S, Martin M (2010) Mean leukocyte telomere length shortening and type 2 diabetes mellitus: a case-control study. Transl Res 155(4):166–169

Zhu H, Wang X, Gutin B, Davis CL, Keeton D, Thomas J et al (2010) Leukocyte telomere length in healthy caucasian and African-American adolescents: relationships with race, sex, adiposity, adipokines, and physical activity. J Pediatr

Chapter 12
Sleep and Immunity in Older Age

Peter Hampson, Alessandra Rossi, Teresa Arora, Janet M. Lord and Shahrad Taheri

12.1 Sleep and Health

The general decline in sleep duration during the past 25 years has made suboptimal sleep quality a more common phenomenon (Jean-Louis et al. 2000). Reasons for reduced sleep include increased working hours, shift work and more time spent watching the television and exploring the internet (Bonnet and Arand 1995). Disruptions in sleep parameters have been related to a variety of health outcomes. Animal studies have shown that extreme levels of sleep deprivation are highly detrimental. Rats subjected to total sleep deprivation for up to 21 days displayed global deterioration ultimately resulting in death (Everson and Wehr 1993). Epidemiological studies in humans have indicated that both short and long sleep duration is associated with increased morbidity and mortality. Most notably, a study of 1.1 million participants in the American Cancer Prevention Society demonstrated that survival was optimal in people who slept 7 h per night. Sleep durations of 8.5 h or more, or less than 4.5 h were significantly associated with increased mortality as well as higher body mass index (BMI) scores (Kripke et al. 2002).

Variations in sleep timing, duration or quality have been shown to be associated with a number of other adverse health outcomes including the development of insulin resistance (Spiegel et al. 2005), type II diabetes mellitus (Beihl et al. 2009), cardiovascular disease (Knutson et al. 2009), cognitive decline (Van Dongen et al. 2003), and reduced immune function (Cohen et al. 2009). The incidence of these conditions also increases with age, impacting upon health and quality of life in old age. As sleep quality and duration also decline with age (Neubauer 1999), abnormalities in sleep may therefore be both a marker for, and a contributor to, age-related disease and decline in immune function. The direction of causality in such cross-sectional studies is difficult to establish other than through animal model studies of chronic

P. Hampson (✉) · A. Rossi · J. M Lord
School of Immunity and Infection, University of Birmingham,
Birmingham B15 2TT, UK
e-mail: p.hampson@bham.ac.uk

T. Arora · S. Taheri
School of Clinical and Experimental Medicine, University of Birmingham,
Birmingham B15 2TT, UK

J. A. Bosch et al. (eds.), *Immunosenescence*,
DOI 10.1007/978-1-4614-4776-4_12,

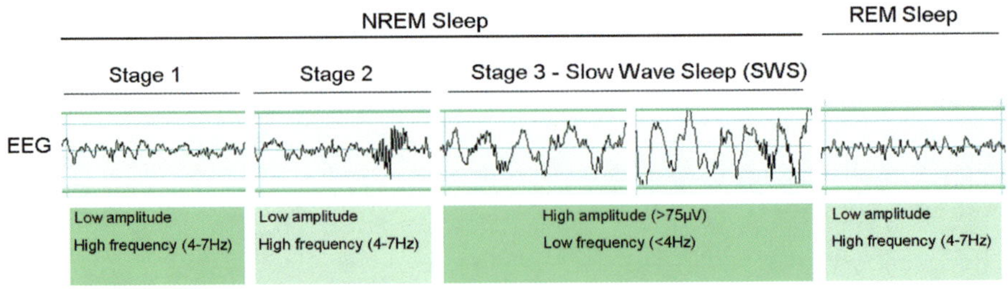

Fig. 12.1 An example of the different stages of sleep as measured by electroencephalogram (EEG). Different stages of sleep can be characterized by the frequency and amplitude of the EEG waves

sleep disruption, which are currently lacking. This chapter thus explores specifically the evidence for the possible contribution of sleep disruption to immunosenescence.

12.2 Sleep Biology

Sleep is an integral part of human biology. Typically, individuals spend one third of their adult life sleeping. The importance of sleep to survival, specifically the ability to combat life threatening infections, has been demonstrated in a rodent model of sleep deprivation (Everson and Wehr 1993). Human population studies have similarly observed an association between sleep duration and mortality (Kripke et al. 2002). The need for sleep cannot be resisted, no matter how long the attempt is to remain awake, demonstrating a strong biological drive. Superficially, sleep is not complex to understand—we sleep when we are tired and wake feeling rejuvenated. On closer examination, however, sleep is multifaceted and a state that is linked to development and a variety of physiological functions.

12.2.1 Sleep Stages

The gold standard for assessing sleep is by polysomnography. The electroencephalogram obtained through polysomnography has allowed differentiation of sleep into two distinct types, rapid eye movement (REM) sleep and non-rapid eye movement (NREM) sleep (Aserinsky and Kleitman 1953), and four defined sleep stages (Fig. 12.1). NREM can be further separated into three progressively deeper stage stages (N1, N2 and N3). Whereas stage N1 represents the transition between wakefulness and sleep, stage N2 represents a deeper sleep during which we become disengaged from our surroundings and body temperature begins to fall. Stage N3, also classed as slow wave sleep (SWS), is characterized by high amplitude ($> 75\mu V$), low frequency (<4 Hz) EEG waves. SWS is thought to be the stage most strongly related to sleep need and is accompanied by a gradual decrease in body temperature,

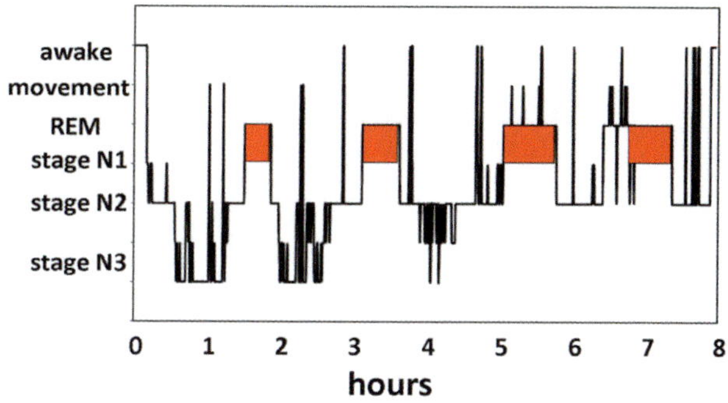

Fig. 12.2 A hypnogram of a normal healthy adult showing sleep stages and cycles over one night

respiratory rate, blood pressure and heart rate (Somers et al. 1993). The final stage of sleep, REM sleep, sees the re-emergence of low amplitude high frequency EEG activity as well as bursts of rapid eye movements, which give this stage of sleep its name. REM sleep is thought to be involved in memory processing and dreaming is more likely to occur in REM sleep (Mandai et al. 1989). All stages and types of sleep alternate cyclically and a healthy individual will obtain four to five cycles in a night with each cycle lasting approximately 90 min (Beersma 1998). During successive sleep cycles, episodes of REM sleep become progressively longer; in contrast, longest periods of SWS occur at the beginning of the night and become shorter with each cycle of sleep, as shown in Fig. 12.2.

12.2.2 Regulation of Sleep

Two processes regulate sleep quality and duration, called Process S (sleep/wake dependent component) and Process C (circadian component). Process S is appetitive and related to sleep debt, while Process C determines the timing of sleep and wakefulness (Borbely 1982). These processes are regulated by both neuronal and hormonal factors (Fig. 12.3). In relation to neuronal regulation, circadian rhythms are entrained to a 24-h period and are directed by the suprachiasmatic nucleus (SCN) in the brain which serves as the central circadian pacemaker (Reppert and Weaver 2001). The SCN synchronizes various internal biological rhythms through both internal and external cues and thus regulates sleep timing and duration. External cues (zeitbergers) include light, which influences the SCN through the retino-hypothalamic tract (Golombek and Rosenstein 2010). Internal cues include hormone secretion and body temperature.

Key hormonal regulators of sleep include pro-inflammatory cytokines (tumor necrosis factor alpha (TNFalpha) and interleukin-1beta (IL-1B),) and classical endocrine hormones such as cortisol, growth hormone (GH), prolactin and melatonin. Pro-inflammatory cytokines, GH and prolactin are believed to be somnogenic (Obal et al. 1995), while anti-inflammatory cytokines (IL4, IL10, IL13) and cortisol inhibit spontaneous sleep (Krueger and Majde 2003). Interestingly, ageing is associated with

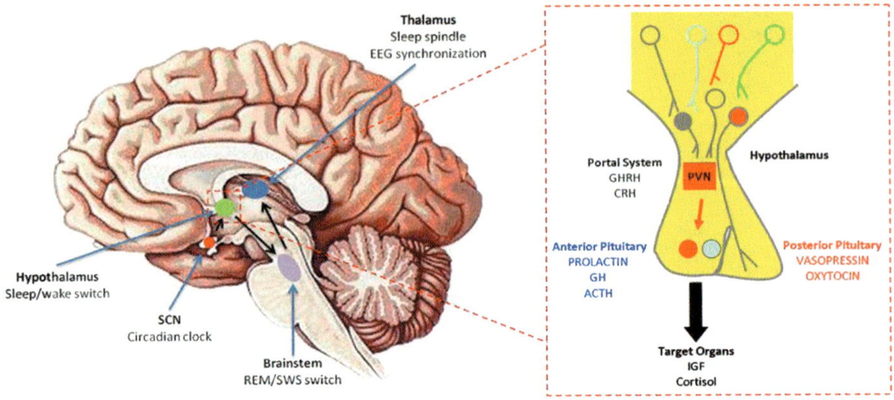

Fig. 12.3 The areas of the brain that control sleep (*left*) and hormonal regulation of sleep by the hypothalamus (*right*). *SCN* Suprachiasmatic Nucleus, *REM* Rapid Eye Movement, *SWS* Slow Wave Sleep, *PVN* Paraventricular Nucleus, *GHRH* Growth Hormone Releasing Hormone, *CRH* Corticotrophin Releasing Hormone, *GH* Growth Hormone, *ACTH* Adrenocorticotrophic Hormone, *IGF* Insulin-Like Growth Factor

an increase in the level of pro-inflammatory cytokines in the circulation (denoted as inflammageing) (Franceschi et al. 2007), including TNF-α and IL-1β, a significant decline in GH and melatonin along with a slightly increased level of cortisol (Arlt and Hewison 2004).

Each of these factors may change with age and thus impact upon sleep regulation in old age. A review of the data demonstrates that secretion of endogenous melatonin decreases with age as a result of the degeneration of the pineal gland (Ruzsas and Mess 2000). This decrease was found to be associated with decreased sleep efficiency and to a phase advance of the sleep–wake cycle (Touitou 2001). Intravenous administration of growth hormone releasing hormone (GHRH) in older adults leads to small improvements, such as a slight increase in stage 1 of NREM and a diminished number of awakenings (Guldner et al. 1997; Murck et al. 1997).

12.3 Sleep in the Older Adult

Sleep complaints are common in older individuals. Indeed, data from a large epidemiological study demonstrated that more than 80 % of older individuals (>65 years) reported at least one sleep problem on a regular basis (Foley et al. 1995). The same study followed up participants after 3 years and found that the 15 % who reported no baseline sleep problems had developed disturbed sleep at follow up, suggestive of an annual increase in incidence of 5 % in older adults. Although evidence shows that sleep duration decreases with age (see Fig. 12.4), older adults are prone to experiencing a variety of sleep alterations including:

- Inability to stay awake in the evening and early awakening (advanced circadian phase).
- Difficulty initiating and maintaining sleep (insomnia).

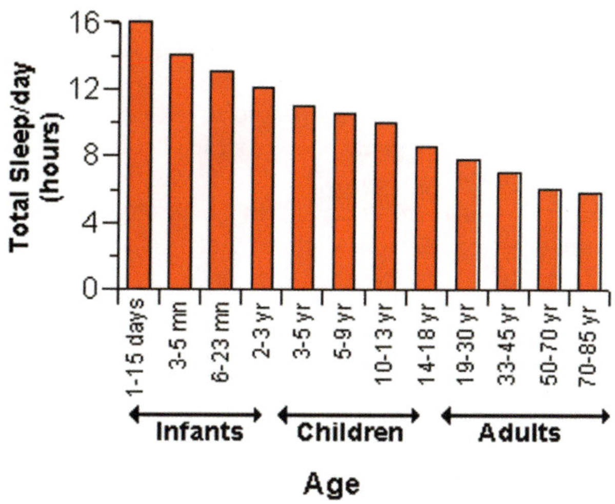

Fig. 12.4 Sleep duration across the lifespan in human subjects

- Sleep disturbances (poor sleep quality).
- Undiagnosed sleep disorders which individuals may attribute to the ageing process.
- Excessive daytime sleepiness, often leading to daytime napping.
- Feeling un-rested after a night's sleep.

The reasons underlying sleep disturbances in the elderly are complex due to their multifactorial nature. Some can be related to physiological changes that occur with ageing such as the need to urinate more frequently at night; others include chronic pain caused by chronic health conditions such as arthritis, depression, Alzheimer's disease; drug use, including prescribed, over the counter and sometimes, recreational drugs and alcohol can all affect sleep and a more sedentary lifestyle, use of stimulants such as caffeine and smoking and poor nutrition are also factors in sleep alterations with age (Vaz Fragoso and Gill 2007). However, despite all of these age-related variables, studies in healthy older adults suggest the presence of basic age-related physiological disruptions which alter the homeostatic and circadian mechanisms regulating sleep.

12.3.1 Changes in Sleep Architecture and Circadian Rhythms

Accumulating evidence from sleep laboratory studies, using polysomnography (an objective measure of sleep), have shown that sleep architecture is altered with chronological age. Van Cauter et al. (2000) successfully demonstrated that SWS declined rapidly with age, from midlife. Further, a meta-analysis of alterations in various sleep parameters across the lifespan also demonstrated that REM sleep declined with age and that reduced SWS and REM sleep appeared to be replaced by extended lighter states of sleep (stages 1 and 2) (Ohayon et al. 2004). It should be noted that because of changes in EEG that occur with ageing, it is possible to misclassify sleep stages. For example, an inability to generate large action potentials may result in reduced scoring of stage N3 because of the 75 μV criterion for scoring slow waves.

Not only is the architecture of sleep altered in the older adult but the sleep–wake cycle of an older adult also functions differently to younger counterparts. This is due to a disruption in the circadian rhythm, usually advanced in time, which is prevalent in older individuals (Van Someren et al. 2002). Alteration in the circadian system thus results in premature bed times and early morning rising in older individuals (also known as a 'lark-like' phenotype) (Vitiello 1997). Circadian rhythms are maintained in a 24-h cycle through the activity of suprachiasmatic nucleus, as described above. In old age, individuals experience a weakening influence of the SCN, subsequently showing less synchronized sleep–wake patterns, in part, due to decreased responsiveness to zietgebers (Hofman and Swaab 2006). The influence of the circadian clock may also decline with age progression, which can lead to more frequent nighttime awakenings and subsequent daytime sleepiness/fatigue. Interestingly, during ageing, there also seem to be changes in the circadian rhythmicity of hormones that are involved in the regulation of sleep. As discussed earlier, the reciprocal interaction of GHRH and corticotrophin releasing hormone (CRH) plays a key role in the regulation of sleep. While GHRH is released during the first half of the night promoting NREM sleep, CRH predominates during the second half of the night leading to an increase in REM sleep in the morning hours (Steiger et al. 1987). However, during ageing, there appears to be a reduction in GHRH release during the first half of the night, leading to changes in the CRH:GHRH ratio in favour of CRH (Steiger et al. 1987). This is thought to be involved in some of the changes in sleep patterns that are seen during ageing, including a reduction in SWS duration.

12.4 Sleep and Immunity

There is increasing evidence that sleep deprivation has a negative impact on immune function, suggestive of an interaction between the process of sleep and the immune system (Bryant et al. 2004). This is of particular importance when considering the fact that sleep duration and sleep quality is generally reducing amongst the population (Jean-Louis et al. 2000). This could also be a particular problem in groups such as the older population, who as well as having a well documented reduction in sleep duration and quality with age (Neubauer 1999), also suffer from a reduced functional capacity of their immune system as indicated by an increased susceptibility to infection and reduced vaccination responses, termed immunosenescence (see Chaps. 1 and 2).

12.4.1 Circadian Rhythmicity of the Immune System

It is becoming clear that several components of the immune system show circadian rhythmicity. It could therefore be argued that immune factors are both regulated by and are regulators of sleep. Various studies have shown a circadian rhythm in the number of total lymphocytes, natural killer (NK) cells, antibodies and cell-mediated immune responses in the peripheral blood. Levels of lymphocytes and monocytes in the blood are at their maximal during the night and are at their lowest levels after waking (Born et al. 1997; Dimitrov et al. 2009). In particular, numbers of total T cells (CD3+), CD4+ T cells, total B cells (CD20+) and activated T cells (CD25+) are higher during the night, whereas T regulatory cells, cytotoxic CD8+ T cells and $\gamma\delta$ T cells are higher around noon (Mazzoccoli et al. 2010; Dimitrov et al. 2009). NK cells reach their highest level in the afternoon, with a decrease in number and activity around midnight (Born et al. 1997; Dimitrov et al. 2009). This rhythmicity is due in large part to corresponding changes during the sleep–wake cycle in the levels of various cytokines. TNF-α and IL-1β levels peak during the night at the onset of SWS (Gudewill et al. 1992; Moldofsky et al. 1986). The production of other pro-inflammatory cytokines, IL-2 and IL-12, is also higher during night time, whereas release of the anti-inflammatory cytokine IL-10 peaks during daytime (Lissoni et al. 1998).

These data suggest that pro-inflammatory cytokines appear to be closely associated with the sleep–wake cycle. Indeed, it has been shown that TNF-α and IL-1β have a direct role in sleep regulation. Administration of TNF-α or IL-1β increases duration of SWS in animals (Dickstein et al. 1999; Opp and Imeri 2001; Shoham et al. 1987), whereas a reduction in the intra-cerebral levels of these cytokines reduces SWS duration (Opp et al. 1992, 1995; Takahashi et al. 1995a, b). Also, blocking the actions of either of these cytokines with the use of antagonistic antibodies reduces overall sleep duration (Opp and Krueger 1994; Takahashi et al. 1995a). Moreover, mice deficient in either IL1-R1 or TNF receptor have significantly reduced sleep durations when compared with control animals (Fang et al. 1997, 1998). TNF-α and IL-1β both enhance sleep by stimulating the activity of the transcriptional factor NFκB and it has been show that factors that inhibit NFκB activation, can inhibit normal and IL-1β induced sleep in rabbits (Kubota et al. 2000). Additionally, NFκB can itself promote the production of TNF-α and IL-1β, thus forming a positive-feedback loop. Furthermore, NFκB promotes the production of IL-2, which itself promotes sleep (Kubota et al. 2001).

12.4.2 The Interaction Between Sleep, Hormones and the Immune System

The effect of sleep on the immune system is likely to be related to circadian hormonal variations as evidenced by the circadian rhythmicity observed in various aspects of the immune system. For instance, the peak in total circulating lymphocytes seen during the sleep–wake cycle coincides with the nadir for cortisol levels which is reached during night time (Mazzoccoli et al. 2010). Cortisol promotes awakening and is known to have powerful immune suppressive properties, such as antagonizing the actions of IL-1β (Opp and Imeri 2001) and suppressing cytokine production by leukocytes. Moreover, cortisol also influences lymphocyte and monocyte numbers in peripheral blood by reducing the trafficking of these cells to tissues and secondary lymphoid organs (Fauci 1976). It is therefore possible that the diurnal variation in cortisol levels is a primary influence of lymphocyte numbers and levels of pro-inflammatory cytokines. Cortisol is also known to directly inhibit NK cell activity (Callewaert et al. 1991; Zhou et al. 1997), but a similar relationship between cortisol levels and NK cell activity or numbers is not seen.

The sympathetic nervous system also shows circadian variation in the circulating levels of adrenaline, which has distinct and opposing actions to that of cortisol, for example on the distribution of circulating lymphocytes (Dimitrov et al. 2009). Naïve CD4 and CD8 T cells show pronounced circadian rhythms with the daytime nadir driven by cortisol, whereas numbers of mature effector CD8 T cells were at their highest during daytime driven by adrenaline levels. Cortisol is thought to act via stimulation of redistribution of naïve T cells to the bone marrow, whereas catecholamines achieve their effects via mobilization of CD8 T cells from the marginal pool (Dimitrov et al. 2009).

GHRH levels seem to support sleep-inducing IL-1β activity, and promote the release of GH, a sleep-inducing hormone which peaks at the beginning of SWS (Obal et al. 1995, 1997; Redwine et al. 2000). Many studies demonstrate that GH has an important role in immune regulation, stimulating T and B cell proliferation and immunoglobulin synthesis as well as enhancing the maturation of myeloid progenitor cells and modulating cytokine responses (Meazza et al. 2004). GH fluctuations may therefore also contribute to the observed rhythms in lymphocyte numbers and cytokines. Prolactin is also an immune regulatory hormone, which peaks during nighttime and can directly affect NK cell function promoting their capacity to proliferate and to lyse target cells (Sun et al. 2004). However, there is no obvious correlation between prolactin levels and NK activity during a 24-h period. In addition, it has been shown that increased concentrations of prolactin and GH, as well as a decrease in cortisol (the normal hormonal changes characterizing early nocturnal sleep in which SWS predominates), synergistically induce a shift towards increased Th1 activity, (IFN-γ/IL-2 producing T helper cells), while an opposite shift towards Th2 cytokines (IL-4, IL-13) is reached during late sleep, dominated by REM stage sleep (Dimitrov et al. 2004a, b).

Melatonin has also been shown to have an immune modulatory role. In particular, it can lead to monocyte activation, increasing their production of IL-1, IL-6, TNF-α, IL-12 and reactive oxygen species (ROS) (Morrey et al. 1994). Also, melatonin can also activate Th1 lymphocytes increasing their production of IL-12 (Garcia-Maurino et al. 1999). In contrast, the LPS-mediated production of TNF-α and IL-8 by neutrophils is inhibited by melatonin (Silva et al. 2004).

12.5 The Effects of Sleep Deprivation on the Immune System

In order to better understand the relationship between sleep and the immune system, the effects of sleep deprivation on immune cell numbers and function has been studied. This may also help to understand how situations in which sleep deprivation

Table 12.1 The effect of sleep deprivation on circulating number and function of immune cells

Cell type	Number	Function
Neutrophils	↑ (Boudjeltia et al. 2008; Everson 2005; Kerkhofs et al. 2007)	↑ (Boyum et al. 1996)
	↓ (Boyum et al. 1996)	↔ (Palmblad et al. 1979)
NK cells	↑ (Born et al. 1997; Dinges et al. 1994)	↓ (Dinges et al. 1994)
	↓ (Dinges et al. 1994; Heiser et al. 2000; Ozturk et al. 1999)	↓ (Irwin et al. 1996; Moldofsky et al. 1989)
Monocytes	↑ (Born et al. 1997; Boyum et al. 1996; Dinges et al. 1994)	↓ (Palmblad et al. 1976)
T cells	↑ (Born et al. 1997)	↓ (Born et al. 1997; Irwin et al. 1996; Palmblad et al. 1979)
	↓ (Born et al. 1997; Boyum et al. 1996; Dinges et al. 1994)	↔ (Dinges et al. 1994; Moldofsky et al. 1989)
	↔ (Heiser et al. 2000; Ozturk et al. 1999)	
Regulatory T cells	↔ (Bollinger et al. 2009)	↓ (Bollinger et al. 2009)
B cells	↓ (Boyum et al. 1996)	↓ (Boyum et al. 1996)
	↔ (Dinges et al. 1994; Heiser et al. 2000)	↔ (Ozturk et al. 1999)

↓ = decreased, ↑ = increased, ↔ = no change

occurs (such as in old age) can affect immunity. However, the existing literature varies considerably in its conclusions regarding sleep deprivation and immunity, probably due to differences in the amount of sleep debt used in the different studies. Also, the vast majority of studies have focused upon sleep deprivation in young adults or rodents, with little or no data for older subjects.

12.5.1 Immune Cell Numbers

The problem of differing study design is apparent in studies which have examined immune cell number following sleep deprivation (Table 12.1). However, numerous human studies have demonstrated that sleep deprivation leads to a progressive increase in circulating granulocyte numbers. In rats, total sleep deprivation for up to 5 days lead to a progressive increase in circulating leukocytes which seems to be mainly due to an increase in immature neutrophilic granulocytes (Everson 2005). Similar findings have been made in human studies. For example, healthy young men restricted to 4 h sleep per night for three consecutive nights had a significant increase in circulating neutrophils (Boudjeltia et al. 2008; Kerkhof et al. 2007). Unlike granulocytes, studies suggest that circulating lymphocyte numbers decrease following sleep deprivation. For example, NK cell number decreased after 40 h of sleep deprivation (Dinges et al. 1994). However, this finding seems to be dependent on the duration of sleep deprivation evidenced by the fact that the same study showed an increase in circulating NK cell numbers following 64 h of sleep deprivation (Dinges et al. 1994). Similar discrepant data have been reported when examining circulating numbers of other immune cell types including monocytes, T cells and B cells (Born et al. 1997; Boyum et al. 1996; Dinges et al. 1994; Heiser et al. 2000; Ozturk et al. 1999). However, the functional consequence of such changes in the number of circulating immune cells is unclear and what is undoubtedly more relevant to immunity is to assess the effect of sleep deprivation on immune cell function.

12.5.2 Immune Cell Function

Again, due to differences in the degree of sleep deprivation examined, as well as differences in sampling times and assays used, the results for immune function are often contradictory (Table 12.1). However, several authors have shown that 48 h of sleep deprivation results in decreased mitogen-induced lymphocyte proliferation (Born et al. 1997; Irwin et al. 1996; Palmblad et al. 1979), though there are also studies that have demonstrated no change in PHA-induced lymphocyte proliferation following sleep deprivation (Dinges et al. 1994; Moldofsky et al. 1989). It has been shown in humans that partial sleep deprivation leads to a reduction in the proliferation of CD4+CD25- regulatory T cells (Bollinger et al. 2009). Studies of neutrophil and NK cell function have also yielded differing results depending on the duration of sleep deprivation. In the case of neutrophils, while one study showed that sleep deprivation for one night led to an increase in neutrophil function (Boyum et al. 1996), an alternative study found no change in neutrophil function following 48 h of sleep deprivation (Palmblad et al. 1979). It is possible however that these differences are due to the use of different end points as well as differences in the duration of sleep deprivation studied. In the case of NK cells, while short periods of sleep deprivation (36 h) resulted in reduced NK cell function (Moldofsky et al. 1989), longer periods of sleep deprivation (60 h) resulted in increased NK cell function (Dinges et al. 1994).

12.5.3 B Cell Mediated Immunity

One aspect of immunity that has received considerable attention is B cell mediated immunity, specifically production of antibodies during vaccination responses. In humans, it is reported that immunoglobulin levels are lower after 5–7 days of continuous exercise, calorie deficiency and sleep restriction (Boyum et al. 1996). In addition, humans vaccinated against the influenza A virus during a period of sleep deprivation, had antibody titres less than half those of their non-sleep deprived counterparts (Spiegel et al. 2002). In contrast, levels of IgG, IgA, IgM and some complement factors were all shown to be higher in people spending 24 h without sleep, indicating that acute sleep deprivation could even enhance constitutive humoral immunity (Hui et al. 2007).

The overall results for effects of sleep deprivation on the immune system are therefore varied, probably due to differences in study methodology and implemented sleep debt. The majority of studies employ 1–3 days of sleep deprivation, an acute period of sleep disruption. The data may thus have less relevance potentially to sleep disruption in old age which is chronic and likely to have a negative effect on immunity (Everson 1993; Landis and Whitney 1997; Rechtschaffen et al. 1983).

12.6 The Ageing Immune System

During the process of ageing there is a functional deterioration of the immune system, immunosenescence, described in detail in the first two chapters of this book and therefore covered only very briefly here. The effects of immunosenescence are apparent in both the innate and the adaptive arms of the immune system. The following section will briefly outline some of the age-associated changes in innate and adaptive immunity to aid discussion of the possible role that sleep disruption might play in immunosenescence.

12.6.1 Innate Immunity

Neutrophils are the first line of defence against rapidly dividing bacteria, as well as parasitic and fungal infections. Ageing is not accompanied by a reduction in circulating neutrophil numbers, or the ability to produce neutrophilia in response to infection (Butcher et al. 2005). However, a number of studies have shown that several aspects of neutrophil functional activity are reduced with ageing including neutrophil chemotaxis (Wenisch et al. 2000), phagocytosis of bacteria (Butcher et al. 2001), and the ability to produce an oxidative burst (Tortorella et al. 1996). This is thought to be, in part, due to a reduced response to priming agents such as granulocyte-macrophage colony stimulating factor (GM-CSF) (Fortin et al. 2007).

NK cells are innate cytotoxic lymphocytes that mediate MHC-independent cytotoxicity against certain malignancies and viral infections. Although there is a significant increase in NK cell number with age, attributed to the CD56^{dim} population of NK cells (Borrego et al. 1999; Miyaji et al. 2000; Vitale et al. 1992), circulating NK cells from aged subjects exhibit decreased cytotoxicity on a per cell basis (Facchini et al. 1987; Vitale et al. 1992). It has also been shown that NK cells from aged individuals have a reduced production of IFN-γ, MIP-1α, RANTES and IL-8 production in response to IL-2 (Krishnaraj and Bhooma 1996; Mariani et al. 2002). Monocytes are components of the innate immune system which differentiate into antigen presenting cells such as macrophages and dendritic cells (DCs) during inflammation and enter the inflamed tissue. Absolute numbers of monocytes increase with age (Della et al. 2007), but this is contrasted by a reduction in macrophage function including decreased LPS stimulated IL-6, IL-1β and TNF-α (Chelvarajan et al. 2005; Agius et al. 2009). This decrease in function appears to be due to altered toll like receptor (TLR) function (Boehmer et al. 2005). With regards to DCs, it is generally accepted that ageing results in decreased function. In humans, it has been shown that there is an age-related defect in TLR7 and TLR9-induced IFN-α production by pDCs (Jing et al. 2009). Other functions of DCs such as antigen-presenting capabilities appear to be preserved in ageing.

12.6.2 Adaptive Immunity

The thymus reaches its peak size during the first years of life and declines gradually with age thereafter, a process known as thymic involution (Steinmann 1986). The consequence of this is a reduced production of new T cells, which ultimately results in a reduced T cell receptor diversity and an increased memory:naïve cell ratio (Kohler et al. 2005; Naylor et al. 2005). Furthermore, persistent lifelong infections such as CMV, the incidence of which increases with age, can result in increased numbers of CD8+ memory T cells due to viral reactivation throughout the course of life (Khan et al. 2002). The presence of persistent viral infections such as cytomegalovirus (CMV) can result in an increase in the CD8:CD4 T cell ratio which is associated with increased mortality (Nikolich-Zugich 2008). Ageing also seems to be accompanied by an alteration in the balance of Th1 and Th2 responses, with skewing towards Th2 responses (Rink et al. 1998). In addition to these phenotypic changes, T cells from older subjects also show a reduced proliferative response to both non-specific mitogens and to antigen (Haynes and Maue 2009). This is due in large part to the shortening of telomeres (Fletcher et al. 2005; Plunkett et al. 2005) but also to the loss of co-stimulatory receptors such as CD28 (Kovaiou and Grubeck-Loebenstein 2006).

In relation to B cells, the percentage of B cells amongst peripheral blood mononuclear cells (PBMCs) is reduced with age (Breitbart et al. 2002; Chong et al. 2005; Paganelli et al. 1992) and B cells from old individuals displayed a reduced B cell receptor repertoire (BCR repertoire) (Ademokun et al. 2010). In vitro experiments in both mice (Frasca et al. 2007) and humans (Frasca et al. 2008) show that IgG secretion is reduced with age. In vivo however, there is an increase in serum IgG and IgA with age (Paganelli et al. 1992). In addition, there is also an increase in autoantibody production with age (Andersen-Ranberg et al. 2004), perhaps due to a breakdown in central tolerance. It is also well established that responses to vaccination are diminished in older recipients (Murasko et al. 2002), almost certainly due to a lack of T cell help (Murasko et al. 2002), and reduced macrophage responses leading to poor activation of DCs and reduced antigen presentation (Aydar et al. 2003).

12.7 Sleep and Immunosenescence

As discussed previously, there is increasing evidence that sleep deprivation has a negative impact on the immune system. Sleep deprivation/disruption may therefore be a particular risk in populations that are already at increased risk of infection such as the elderly. In addition, it is well established that ageing is accompanied by a functional deterioration of the immune system, termed immunosenescence (see the previous section). The fact that sleep disruption is common in older adults raises the possibility that this could contribute to immune dysfunction in these individuals. Of interest, many of the changes seen as a result of immunosenescence are also seen in studies which have examined sleep deprivation. These include reduced adaptive immune responses to vaccination, reduced T cell mediated immunity and increased innate immune activation as well as an increase in circulating levels of pro-inflammatory cytokines.

However, the fact that there appears to be a reciprocal relationship between sleep and immunity raises the question: Is sleep disruption a result of immunosenescence or does sleep disruption lead to immunosenescence? Studies of acute sleep deprivation overall show negative effects on immunity (Table 12.1) and would suggest that the latter is more likely. There is currently a paucity of data in regard to the effects of chronic sleep deprivation on immunity. As this is most relevant to the situation in old age, this gap needs to be addressed. Studies of conditions such as sleep apnoea may give some insight into the consequences of chronic sleep disruption, but again there is a lack of published research considering the effects on immunity. In the small literature on the subject, the most widely reported association is for increased levels of circulating pro-inflammatory cytokines, as a consequence of increased activation of CD8 lymphocytes and monocytes (Ryan et al. 2009). Thus, it remains a possibility that sleep disruption during old age could contribute to the phenomenon of increased basal activation of monocytes and inflammageing. However, patients with obstructive sleep apnoea showed no differences in their humoral immune responses to influenza vaccination compared to healthy volunteers (Dopp et al. 2007), suggesting that chronic sleep disruption, at least in this complex disease model, may have only minimal effect on vaccination responses. To date, there are no data concerning chronic sleep deprivation and neutrophil function, which is an important issue as infections which predominate in old age are bacterial in nature and neutrophils are the main defence against such infections.

In conclusion, ageing is accompanied by significant alterations in both the quality and quantity of sleep and epidemiological studies have shown a strong relationship between sleep duration and mortality (Kripke et al. 2002). Ageing also incurs a substantial loss of immunity resulting in increased susceptibility to infections. While studies of acute sleep deprivation, primarily in animal models, show that sleep disruption can result in immune decline, there is currently a paucity of data in humans showing a contribution of sleep disruption in old age to immunosenescence. Further research is thus required to determine such a causal relationship, which could identify a relatively simple intervention to improve health and immunity in older adults.

Acknowledgments This work is supported by funding from Research into Ageing (PH), a European Commission Marie Curie fellowship PITN-GA-2009–239665(AR).

References

Ademokun A, Wu YC, Dunn-Walters D (2010) The ageing B cell population: composition and function. Biogerontology 11:125–137

Agius, E, Lacy, KE, Vukmanovic-Stejic M, Jagger AL, Papageorgiou AP, Hall S et al (2009) Decreased TNF-alpha synthesis by macrophages restricts cutaneous immunosurveillance by memory CD4+ T cells during aging. J Exp Med 206:1929–1940

Andersen-Ranberg K, HOier-Madsen M, Wiik A, Jeune B, Hegedus L (2004) High prevalence of autoantibodies among Danish centenarians. Clin Exp Immunol 138:158–163

Arlt W, Hewison M (2004) Hormones and immune function: implications of aging. Aging Cell 3:209–216

Aserinsky E, Kleitman N (1953) Regularly occurring periods of eye motility, and concomitant phenomena, during sleep. Science 118:273–274

Aydar Y, Balogh P, Tew JG, Szakal AK (2003) Altered regulation of Fc gamma RII on aged follicular dendritic cells correlates with immunoreceptor tyrosine-based inhibition motif signaling in B cells and reduced germinal center formation. J Immunol 171:5975–5987

Beersma DG (1998) Models of human sleep regulation. Sleep Med Rev 2:31–43

Beihl DA, Liese AD, Haffner SM (2009) Sleep duration as a risk factor for incident type 2 diabetes in a multiethnic cohort. Ann Epidemiol 19:351–357

Boehmer ED, Meehan MJ, Cutro BT, Kovacs EJ (2005) Aging negatively skews macrophage. Mech Ageing Dev 126:1305–1313

Bollinger T, Bollinger A, Skrum L, Dimitrov S, Lange T, Solbach W (2009) Sleep-dependent activity of T cells and regulatory T cells. Clin Exp Immunol 155:231–238

Bonnet MH, Arand DL (1995) We are chronically sleep deprived. Sleep 18:908–911

Borbely AA (1982) A two process model of sleep regulation. Hum Neurobiol 1:195–204

Born J, Lange T, Hansen K, Molle M, Fehm HL (1997) Effects of sleep and circadian rhythm on human circulating immune cells. J Immunol 158:4454–4464

Borrego F, Alonso MC, Galiani MD, Carracedo J, Ramirez R, Ostos B et al (1999) NK phenotypic markers and IL2 response in NK cells from elderly people. Exp Gerontol 34:253–265

Boudjeltia KZ, Faraut B, Stenuit P, Esposito MJ, Dyzma M, Brohee D et al (2008) Sleep restriction increases white blood cells, mainly neutrophil count, in young healthy men: a pilot study. Vasc Health Risk Manag 4:1467–1470

Boyum A, Wiik P, Gustavsson E, Veiby OP, Reseland J, Haugen AH et al (1996) The effect of strenuous exercise, calorie deficiency and sleep deprivation on white blood cells, plasma immunoglobulins and cytokines. Scand J Immunol 43:228–235

Breitbart E, Wang X, Leka LS, Dallal GE, Meydani SN, Stollar BD (2002) Altered memory B-cell homeostasis in human aging. J Gerontol A Biol Sci Med Sci 57:B304–B311

Bryant PA, Trinder J, Curtis N (2004) Sick and tired: does sleep have a vital role in the immune system? Nat Rev Immunol 4:457–467

Butcher SK, Chahal H, Nayak L, Sinclair A, Henriquez NV, Sapey E et al (2001) Senescence in innate immune responses: reduced neutrophil phagocytic capacity and CD16 expression in elderly humans. J Leukocyte Biol 70:881–886

Butcher SK, Killampalli V, Lascelles D, Wang K, Alpar EK, Lord JM (2005) Raised cortisol: DHEAS ratios in the elderly after injury: potential impact upon neutrophil function and immunity. Aging Cell 4:319–324

Callewaert DM, Moudgil VK, Radcliff G, Waite R (1991) Hormone specific regulation of natural killer cells by cortisol. Direct inactivation of the cytotoxic function of cloned human NK cells without an effect on cellular proliferation. FEBS Lett 285:108–110

Chelvarajan RL, Collins SM, Van Willigen JM, Bondada S (2005) The unresponsiveness of aged mice to polysaccharide antigens is a result of a defect in macrophage function. J Leukocyte Biol 77:503–512

Chong Y, Ikematsu H, Yamaji K, Nishimura M, Nabeshima S, Kashiwagi S et al (2005) CD27(+) (memory) B cell decrease and apoptosis-resistant CD27(-) (naïve) B cell increase in aged humans: implications for age-related peripheral B cell developmental disturbances. Int Immunol 17:383–390

Cohen S, Doyle WJ, Alper CM, Janicki-Deverts D, Turner RB (2009) Sleep habits and susceptibility to the common cold. Arch Intern Med 169:62–67

Della BS, Bierti L, Presicce P, Arienti R, Valenti M, Saresella M et al (2007) Peripheral blood dendritic cells and monocytes are differently regulated in the elderly. Clin Immunol 122:220–228

Dickstein JB, Moldofsky H, Lue FA, Hay JB (1999) Intracerebroventricular injection of TNF-alpha promotes sleep and is recovered in cervical lymph. Am J Physiol 276:R1018–R1022

Dimitrov S, Benedict C, Heutling D, Westermann J, Born J, Lange T (2009) Cortisol and epinephrine control opposing circadian rhythms in T cell subsets. Blood 113:5134–5143

Dimitrov S, Lange T, Fehm HL, Born J (2004a) A regulatory role of prolactin, growth hormone, and corticosteroids for human T-cell production of cytokines. Brain Behav Immun 18:368–374

Dimitrov S, Lange T, Tieken S, Fehm HL, Born J (2004b) Sleep associated regulation of T helper 1/T helper 2 cytokine balance in humans. Brain Behav Immun 18:341–348

Dinges DF, Douglas SD, Zaugg L, Campbell DE, McMann JM, Whitehouse WG et al (1994) Leukocytosis and natural killer cell function parallel neurobehavioral fatigue induced by 64 hours of sleep deprivation. J Clin Invest 93:1930–1939

Dopp JM, Wiegert NA, Moran JJ, Muller D, Weber S, Hayney MS (2007) Humoral immune responses to influenza vaccination in patients with obstructive sleep apnea. Pharmacotherapy 27:1483–1489

Everson CA (1993) Sustained sleep deprivation impairs host defense. Am J Physiol 265:R1148–R1154

Everson CA (2005) Clinical assessment of blood leukocytes, serum cytokines, and serum immunoglobulins as responses to sleep deprivation in laboratory rats. Am J Physiol Regul Integr Comp Physiol 289:R1054–R1063

Everson CA, Wehr TA (1993) Nutritional and metabolic adaptations to prolonged sleep deprivation in the rat. Am J Physiol 264:R376–R387

Facchini A, Mariani E, Mariani AR, Papa S, Vitale M, Manzoli FA (1987) Increased number of circulating Leu 11 + (CD 16) large granular lymphocytes and decreased NK activity during human ageing. Clin Exp Immunol 68:340–347

Fang J, Wang Y, Krueger JM (1997) Mice lacking the TNF 55 kDa receptor fail to sleep more after TNFalpha treatment. J Neurosci 17:5949–5955

Fang J, Wang Y, Krueger JM (1998) Effects of interleukin-1 beta on sleep are mediated by the type I receptor. Am J Physiol 274:R655–R660

Fauci AS (1976) Mechanisms of corticosteroid action on lymphocyte subpopulations. II. Differential effects of in vivo hydrocortisone, prednisone and dexamethasone on in vitro expression of lymphocyte function. Clin Exp Immunol 24:54–62

Fletcher JM, Vukmanovic-Stejic M, Dunne PJ, Birch KE, Cook JE, Jackson SE et al (2005) Cytomegalovirus-specific CD4+ T cells in healthy carriers are continuously driven to replicative exhaustion. J Immunol 175:8218–8225

Foley DJ, Monjan AA, Brown SL, Simonsick EM, Wallace RB, Blazer DG. (1995) Sleep complaints among elderly persons: an epidemiologic study of three communities. Sleep 18:425–432

Fortin CF, Larbi A, Dupuis G, Lesur O, Fulop T Jr (2007) GM-CSF activates the Jak/STAT pathway to rescue polymorphonuclear neutrophils from spontaneous apoptosis in young but not elderly individuals. Biogerontology 8:173–187

Franceschi C, Capri M, Monti D, Giunta S, Olivieri F, Sevini F et al (2007) Inflammaging and anti-inflammaging: a systemic perspective on aging and longevity emerged from studies in humans. Mech Ageing Dev 128:92–105

Frasca D, Landin AM, Lechner SC, Ryan JG, Schwartz R, Riley RL et al (2008) Aging down-regulates the transcription factor E2 A, activation-induced cytidine deaminase, and Ig class switch in human B cells. J Immunol 180:5283–5290

Frasca D, Riley RL, Blomberg BB (2007) Aging murine B cells have decreased class switch induced by anti-CD40 or BAFF. Exp Gerontol 42:192–203

Garcia-Maurino S, Pozo D, Carrillo-Vico A, Calvo JR, Guerrero JM (1999) Melatonin activates Th1 lymphocytes by increasing IL-12 production. Life Sci 65:2143–2150

Golombek DA, Rosenstein RE (2010) Physiology of circadian entrainment. Physiol Rev 90:1063–1102

Gudewill S, Pollmacher T, Vedder H, Schreiber W, Fassbender K, Holsboer F (1992) Nocturnal plasma levels of cytokines in healthy men. Eur Arch Psychiatry Clin Neurosci 242:53–56

Guldner J, Schier T, Friess E, Colla M, Holsboer F, Steiger A (1997) Reduced efficacy of growth hormone-releasing hormone in modulating sleep endocrine activity in the elderly. Neurobiol Aging 18:491–495

Haynes L, Maue AC (2009) Effects of aging on T cell function. Curr Opin Immunol 21:414–417

Heiser P, Dickhaus B, Schreiber W, Clement HW, Hasse C, Hennig J et al (2000) White blood cells and cortisol after sleep deprivation and recovery sleep in humans. Eur Arch Psychiatry Clin Neurosci 250:16–23

Hofman MA, Swaab DF (2006) Living by the clock: the circadian pacemaker in older people. Ageing Res Rev 5:33–51

Hui L, Hua F, Diandong H, Hong Y (2007) Effects of sleep and sleep deprivation on immunoglobulins and complement in humans. Brain Behav Immun 21:308–310

Irwin M, McClintick J, Costlow C, Fortner M, White J, Gillin JC (1996) Partial night sleep deprivation reduces natural killer and cellular immune responses in humans. FASEB J 10:643–653

Jean-Louis G, Kripke DF, Ancoli-Israel S, Klauber MR, Sepulveda RS (2000) Sleep duration, illumination, and activity patterns in a population sample: effects of gender and ethnicity. Biol Psychiatry 47:921–927

Jing Y, Shaheen E, Drake RR, Chen N, Gravenstein S, Deng Y (2009) Aging is associated with a numerical and functional decline in plasmacytoid dendritic cells, whereas myeloid dendritic cells are relatively unaltered in human peripheral blood. Hum Immunol 70:777–784

Kerkhofs M, Boudjeltia KZ, Stenuit P, Brohee D, Cauchie P, Vanhaeverbeek M (2007) Sleep restriction increases blood neutrophils, total cholesterol and low density lipoprotein cholesterol in postmenopausal women: a preliminary study. Maturitas 56:212–215

Khan N, Shariff N, Cobbold M, Bruton R, Ainsworth JA, Sinclair AJ et al (2002) Cytomegalovirus seropositivity drives the CD8 T cell repertoire toward greater clonality in healthy elderly individuals. J Immunol 169:1984–1992

Knutson KL, Van CE, Rathouz PJ, Yan LL, Hulley SB, Liu K et al (2009) Association between sleep and blood pressure in midlife: the CARDIA sleep study. Arch Intern Med 169:1055–1061

Kohler S, Wagner U, Pierer M, Kimmig S, Oppmann B, Mowes B et al (2005) Post-thymic in vivo proliferation of naïve CD4+ T cells constrains the TCR repertoire in healthy human adults. Eur J Immunol 35:1987–1994

Kovaiou RD, Grubeck-Loebenstein B (2006) Age-associated changes within CD4+ T cells. Immunol Lett 107:8–14

Kripke DF, Garfinkel L, Wingard DL, Klauber MR, Marler MR (2002) Mortality associated with sleep duration and insomnia. Arch Gen Psychiatry 59:131–136

Krishnaraj R, Bhooma T (1996) Cytokine sensitivity of human NK cells during immunosenescence. 2. IL2-induced interferon gamma secretion. Immunol Lett 50:59–63

Krueger JM, Majde JA (2003) Humoral links between sleep and the immune system: research issues. Ann NY Acad Sci 992:9–20

Kubota T, Brown RA, Fang J, Krueger JM (2001) Interleukin-15 and interleukin-2 enhance non-REM sleep in rabbits. Am J Physiol Regul Integr Comp Physiol 281:R1004–R1012

Kubota T, Kushikata T, Fang J, Krueger JM (2000) Nuclear factor-kappaB inhibitor peptide inhibits spontaneous and interleukin-1beta-induced sleep. Am J Physiol Regul Integr Comp Physiol 279:R404–R413

Landis CA, Whitney JD (1997) Effects of 72 hours sleep deprivation on wound healing in the rat. Res Nurs Health 20:259–267

Lissoni P, Rovelli F, Brivio F, Brivio O, Fumagalli L (1998) Circadian secretions of IL-2, IL-12, IL-6 and IL-10 in relation to the light/dark rhythm of the pineal hormone melatonin in healthy humans. Nat Immunol 16:1–5
Mandai O, Guerrien A, Sockeel P, Dujardin K, Leconte P (1989) REM sleep modifications following a Morse code learning session in humans. Physiol Behav 46:639–642
Mariani E, Meneghetti A, Neri S, Ravaglia G, Forti P, Cattini L et al (2002) Chemokine production by natural killer cells from nonagenarians. Eur J Immunol 32:1524–1529
Mazzoccoli G, De CA, Greco A, Damato M, Marzulli N, Dagostino MP et al (2010) Aging related changes of circadian rhythmicity of cytotoxic lymphocyte subpopulations. J Circadian Rhythms 8:6
Meazza C, Pagani S, Travaglino P, Bozzola M (2004) Effect of growth hormone (GH) on the immune system. Pediatr Endocrinol Rev 1(Suppl 3):490–495
Miyaji C, Watanabe H, Toma H, Akisaka M, Tomiyama K, Sato Y et al (2000) Functional alteration of granulocytes, NK cells, and natural killer T cells in centenarians. Hum Immunol 61:908–916
Moldofsky H, Lue FA, Davidson JR, Gorczynski R (1989) Effects of sleep deprivation on human immune functions. FASEB J 3:1972–1977
Moldofsky H, Lue FA, Eisen J, Keystone E, Gorczynski RM (1986) The relationship of interleukin-1 and immune functions to sleep in humans. Psychosom Med 48:309–318
Morrey KM, McLachlan JA, Serkin CD, Bakouche O (1994) Activation of human monocytes by the pineal hormone melatonin. J Immunol 153:2671–2680
Murasko DM, Bernstein ED, Gardner EM, Gross P, Munk G, Dran S et al (2002) Role of humoral and cell-mediated immunity in protection from influenza disease after immunization of healthy elderly. Exp Gerontol 37:427–439
Murck H, Frieboes RM, Schier T, Steiger A (1997) Longtime administration of growth hormone-releasing hormone (GHRH) does not restore the reduced efficiency of GHRH on sleep endocrine activity in 2 old-aged subjects—a preliminary study. Pharmacopsychiatry 30:122–124
Naylor K, Li G, Vallejo AN, Lee WW, Koetz K, Bryl E et al (2005) The influence of age on T cell generation and TCR diversity. J Immunol 174:7446–7452
Neubauer DN (1999) Sleep problems in the elderly. Am Fam Physician 59:2551–2560
Nikolich-Zugich J (2008) Ageing and life-long maintenance of T-cell subsets in the face of latent persistent infections. Nat Rev Immunol 8:512–522
Obal F Jr, Bodosi B, Szilagyi A, Kacsoh B, Krueger JM (1997) Antiserum to growth hormone decreases sleep in the rat. Neuroendocrinology 66:9–16
Obal F Jr, Fang J, Payne LC, Krueger JM (1995) Growth-hormone-releasing hormone mediates the sleep-promoting activity of interleukin-1 in rats. Neuroendocrinology 61:559–565
Ohayon MM, Carskadon MA, Guilleminault C, Vitiello MV (2004) Meta-analysis of quantitative sleep parameters from childhood to old age in healthy individuals: developing normative sleep values across the human lifespan. Sleep 27:1255–1273
Opp MR, Imeri L (2001) Rat strains that differ in corticotropin-releasing hormone production exhibit different sleep-wake responses to interleukin 1. Neuroendocrinology 73:272–284
Opp MR. Krueger JM (1994) Anti-interleukin-1 beta reduces sleep and sleep rebound after sleep deprivation in rats. Am J Physiol 266:R688–R695
Opp MR, Postlethwaite AE, Seyer JM, Krueger JM (1992) Interleukin 1 receptor antagonist blocks somnogenic and pyrogenic responses to an interleukin 1 fragment. Proc Natl Acad Sci USA 89:3726–3730
Opp MR, Smith EM, Hughes TK Jr (1995) Interleukin-10 (cytokine synthesis inhibitory factor) acts in the central nervous system of rats to reduce sleep. J Neuroimmunol 60:165–168
Ozturk L, Pelin Z, Karadeniz D, Kaynak H, Cakar L, Gozukirmizi E (1999) Effects of 48 hours sleep deprivation on human immune profile. Sleep Res Online 2:107–111
Paganelli R, Quinti I, Fagiolo U, Cossarizza A, Ortolani C, Guerra E et al (1992) Changes in circulating B cells and immunoglobulin classes and subclasses in a healthy aged population. Clin Exp Immunol 90:351–354

Palmblad J, Cantell K, Strander H, Froberg J, Karlsson CG, Levi L et al (1976) Stressor exposure and immunological response in man: interferon-producing capacity and phagocytosis. J Psychosom Res 20:193–199

Palmblad J, Petrini B, Wasserman J, Akerstedt T (1979) Lymphocyte and granulocyte reactions during sleep deprivation. Psychosom Med 41:273–278

Plunkett FJ, Franzese O, Belaramani LL, Fletcher JM, Gilmour KC, Sharifi R et al (2005) The impact of telomere erosion on memory CD8+ T cells in patients with X-linked lymphoproliferative syndrome. Mech Ageing Dev 126:855–865

Rechtschaffen A, Gilliland MA, Bergmann BM, Winter JB (1983) Physiological correlates of prolonged sleep deprivation in rats. Science 221:182–184

Redwine L, Hauger RL, Gillin JC, Irwin M (2000) Effects of sleep and sleep deprivation on interleukin-6, growth hormone, cortisol, and melatonin levels in humans. J Clin Endocrinol Metab 85:3597–3603

Reppert SM, Weaver DR (2001) Molecular analysis of mammalian circadian rhythms. Annu Rev Physiol 63:647–676

Rink L, Cakman I, Kirchner H (1998) Altered cytokine production in the elderly. Mech Ageing Dev 102:199–209

Ruzsas C, Mess B (2000) Melatonin and aging. A brief survey. Neuro Endocrinol Lett 21:17–23

Ryan S, Taylor CT, McNicholas WT (2009) Systemic inflammation: a key factor in the pathogenesis of cardiovascular complications in obstructive sleep apnoea syndrome? Postgrad Med J 85:693–698

Shoham S, Davenne D, Cady AB, Dinarello CA, Krueger JM (1987) Recombinant tumor necrosis factor and interleukin 1 enhance slow-wave sleep. Am J Physiol 253:R142–R149

Silva SO, Rodrigues MR, Ximenes VF, Bueno-da-Silva AE, Amarante-Mendes GP, Campa A (2004) Neutrophils as a specific target for melatonin and kynuramines: effects on cytokine release. J Neuroimmunol 156:146–152

Somers VK, Dyken ME, Mark AL, Abboud FM (1993) Sympathetic-nerve activity during sleep in normal subjects. N Engl J Med 328:303–307

Spiegel K, Knutson K, Leproult R, Tasali E, Van CE (2005) Sleep loss: a novel risk factor for insulin resistance and Type 2 diabetes. J Appl Physiol 99:2008–2019

Spiegel K, Sheridan JF, Van CE (2002) Effect of sleep deprivation on response to immunization. J Am Med Assoc 288:1471–1472

Steiger A, Herth T, Holsboer F (1987) Sleep-electroencephalography and the secretion of cortisol and growth hormone in normal controls. Acta Endocrinol (Copenh) 116:36–42

Steinmann GG (1986) Changes in the human thymus during aging. Curr Top Pathol 75:43–88

Sun R, Li AL, Wei HM, Tian ZG (2004) Expression of prolactin receptor and response to prolactin stimulation of human NK cell lines. Cell Res 14:67–73

Takahashi S, Kapas L, Fang J, Krueger JM (1995a) An anti-tumor necrosis factor antibody suppresses sleep in rats and rabbits. Brain Res 690:241–244

Takahashi S, Tooley DD, Kapas L, Fang J, Seyer JM, Krueger JM (1995b) Inhibition of tumor necrosis factor in the brain suppresses rabbit sleep. Pflugers Arch 431:155–160

Tortorella C, Polignano A, Piazzolla G, Serrone M, Jirillo E, Antonaci S (1996) Lipopolysaccharide-, granulocyte-monocyte colony stimulating factor and pentoxifylline-mediated effects on formyl-methionyl-leucine-phenylalanine-stimulated neutrophil respiratory burst in the elderly. Microbios 85:189–198

Touitou Y (2001) Human aging and melatonin. Clinical relevance. Exp Gerontol 36:1083–1100

Van Dongen HP, Maislin G, Mullington JM, Dinges DF (2003) The cumulative cost of additional wakefulness: dose-response effects on neurobehavioral functions and sleep physiology from chronic sleep restriction and total sleep deprivation. Sleep 26:117–126

Van Someren EJ, Riemersma RF, Swaab DF (2002) Functional plasticity of the circadian timing system in old age: light exposure. Prog Brain Res 138:205–231

Van CE, Leproult R, Plat L (2000) Age-related changes in slow wave sleep and REM sleep and relationship with growth hormone and cortisol levels in healthy men. J Am Med Assoc 284:861–868

Vaz Fragoso CA, Gill TM (2007) Sleep complaints in community-living older persons: a multifactorial geriatric syndrome. J Am Geriatr Soc 55:1853–1866
Vitale M, Zamai L, Neri LM, Galanzi A, Facchini A, Rana R et al (1992) The impairment of natural killer function in the healthy aged is due to a postbinding deficient mechanism. Cell Immunol 145:1–10
Vitiello MV (1997) Sleep disorders and aging: understanding the causes. J Gerontol A Biol Sci Med Sci 52:M189–M191
Wenisch C, Patruta S, Daxbock F, Krause R, Horl W (2000) Effect of age on human neutrophil function. J Leukocyte Biol 67:40–45
Yang X, Stedra J, Cerny J (1996) Relative contribution of T and B cells to hypermutation and selection of the antibody repertoire in germinal centers of aged mice. J Exp Med 183:959–970
Zhou J, Olsen S, Moldovan J, Fu X, Sarkar FH, Moudgil VK et al (1997) Glucocorticoid regulation of natural cytotoxicity: Effects of cortisol on the phenotype and function of a cloned human natural killer cell line. Cell Immunol 178:108–116

Chapter 13
The Role of Stress and Adrenal Hormones in Immunosenescence

Moisés Evandro Bauer

13.1 Introduction

Ageing is a continuous and slow process that compromises the normal functioning of various organs and systems in both qualitative and quantitative terms. During recent years, a growing body of studies has demonstrated that ageing remodels immune functions, a process known as immunosenescence. The clinical consequences of immunosenescence may include increased susceptibility to infectious diseases, neoplasias, metabolic diseases, osteoporosis and autoimmune diseases (Castle 2000; Pawelec et al. 2010). This increased morbidity is not evenly distributed and seems to be influenced by other immune-modulating factors, including genetic background and chronic stress exposure (Bauer 2005). Indeed, psychological stress appears to be an important factor leading to earlier onset of many age-related diseases (McEwen 1998). Therefore, a better understanding of how stress is likely to promote 'biological ageing' may lead to clinical interventions or policies with a broad public health impact.

Age-related endocrine changes (endocrinosenescence) may in important ways impinge on the immunosenescence, in particular, via age-related declines in growth hormone (GH), sex hormones and dehydroepiandrosterone (DHEA). DHEA is the major secretory product of the human adrenal and is synthesized from cholesterol stores. The hormone is sulphated dehydroepiandrosterone sulfate (DHEAS) before entering the plasma, and this prohormone is converted to DHEA and its metabolites in the peripheral tissues (Canning et al. 2000). Serum DHEA levels decrease by the second decade of life in humans, with approximately 5 % of the original level in elderly people (Migeon et al. 1957), a process termed the adrenopause, and which is thought to be due to reduced cellularity in the zona reticularis of the adrenal gland. As DHEAS and DHEA have several immune-enhancing properties (further discussed in this chapter), the lack of these adrenal factors during ageing may be also

M. E. Bauer (✉)
Institute of Biomedical Research and Faculty of Biosciences, Pontifícia Universidade Católica do Rio Grande do Sul (PUCRS), Av. Ipiranga 6690, 2° andar. P.O. Box 1429, Porto Alegre, RS 90.610-000, Brazil
e-mail: mebauer@pucrs.br

J. A. Bosch et al. (eds.), *Immunosenescence*,
DOI 10.1007/978-1-4614-4776-4_13,

cogent for the immunosenescence. In contrast, there is also evidence suggesting that ageing is associated with activation of the hypothalamic–pituitary–adrenal (HPA) axis (Bauer et al. 2009; Deuschle et al. 1997; Ferrari et al. 2004), leading to increased cortisol levels in man. The HPA axis is a major stress-responsive system and higher cortisol levels are also found during chronic stress or major depression. Cortisol and DHEA secretion follows a circadian pattern with peak levels at waking and a nadir at midnight. The HPA axis is pivotal for the homeostasis of the immune system and its dysregulation has been associated with several immune-mediated diseases. For instance, HPA axis over-activation can affect susceptibility to or severity of infectious disease through the immunosuppressive effect of the glucocorticoids (GCs) (Kiecolt-Glaser et al. 1996; Vedhara et al. 1999). In contrast, blunted HPA axis responses are associated with enhanced susceptibility to autoimmune inflammatory disease (Sternberg 2002). Of note, elderly subjects show increased risk of both infectious and chronic inflammatory diseases.

This chapter summarizes current evidence suggesting that immunosenescence may be influenced by both psychological stress and stress hormones. Indeed, most cellular and molecular changes observed during immunosenescence are similarly found during chronic stress or chronic GC exposure. This chapter will focus on the role of adrenal hormones (cortisol and DHEA) in healthy ageing of the immune system as well as following chronic stress exposure. Finally, this chapter will review preliminary evidence suggesting that stress-management interventions in older adults are capable of attenuating some important features of immunosenescence.

13.2 Ageing is Associated with Increased Psychological Stress and Higher Glucocorticoid Exposure

Human ageing has been associated with psychological and behavioural changes, including difficulty in concentrating, progressive cognitive impairments and sleep disturbances (Howieson et al. 2003; Piani et al. 2004). Individually identified, these alterations are also associated with major depression. Indeed, depression is highly prevalent in age-related chronic diseases, including cardiovascular diseases, Parkinson's disease, Alzheimer's dementia, cancer and rheumatoid arthritis (Dew et al. 1998). In addition, both ageing (Gabriel et al. 2002) and major depression (Schiepers et al. 2005; Trzonkowski et al. 2004) are associated with increased levels of pro-inflammatory cytokines and could thus contribute to further immunological diseases in the frail elderly.

During the past years, our research group has reported that normal, physiological ageing is associated with significant increased psychological stress. In particular, it was found that strictly healthy (as assessed by the SENIEUR protocol) elders were significantly more stressed, anxious and depressed than young adults (Collaziol et al. 2004; Luz et al. 2003). The SENIEUR protocol defines rigorous criteria for selecting healthy individuals in immunogerontological studies (Ligthart et al. 1984). The

Table 13.1 Anti-glucocorticoid actions of DHEA, DHEAS and their metabolites. (Source: Daynes et al. 1990; Hazeldine et al. 2010; Hechter et al. 1997; Maninger et al. 2009; Padgett and Loria 1994; Solerte et al. 1999; Straub et al. 1998)

Brain	Muscle/Bone	Immune System
Better memory performance Anti-depressive Neuroprotection Increased neurogenesis Prevents neurotoxic actions of glucocorticoids	Better muscle strength Improved bone mineral density	Improved response to vaccines Anti-inflammatory (decrease TNF-α and IL-6) Higher T-cell proliferation and IL-2 production Enhanced NK cell activity Improved neutrophil function Protection to GC-induced thymic involution

health conditions are checked by clinical examination and by hematological and various biochemical parameters. Based on this protocol, it is possible to select 'strictly healthy' volunteers from populations of older adults. The literature regarding age-related psychological changes is controversial and others did not find these increased stress levels (Nolen-Hoeksema and Ahrens 2002; Stone et al. 2010). This could be due to sociodemographic or methodological issues, since clinical interviews are often required to reliably assess depression in the elderly.

In parallel to age-related stress, the SENIEUR elders had significantly higher (~45 %) salivary cortisol production throughout the day compared to young adults (Luz et al. 2003), indicative of significant activation of the HPA axis which is a normal aspect of healthy ageing (Deuschle et al. 1997; Ferrari et al. 2004). The HPA axis is pivotal for the homeostasis of the immune system and its overshooting has been associated with immune-mediated diseases, including increased susceptibility to infections (Vedhara et al. 1999) and reduced wound healing (Kiecolt-Glaser et al. 1995). Increased cortisol levels are also seen in demented patients (Magri et al. 2006), major depression (Zunszain et al. 2010) or during chronic stress (Bauer et al. 2000; Kirschbaum et al. 1995). Conversely, several age-related pathologies are also observed following excessive GC exposure and include muscle atrophy (Salehian and Kejriwal 1999), osteoporosis/hypercalcemia (Tamura et al. 2004), hyperglycemia/hyperlipidemia, atherosclerosis, type II diabetes and major depression (Juruena et al. 2003; Lee et al. 2002).

While dysregulation of the HPA axis may contribute to immunosenescence, it is unlikely to be solely responsible. The lack of certain anabolic hormones is also implicated in immunosenescence. Healthy older adults had lower DHEA levels (−54 %) throughout the day and a flatter circadian pattern compared to young adults (Luz et al. 2006). It has been suggested that DHEAS/DHEA may antagonize many physiologic changes of endogenous GCs (Hechter et al. 1997), including overcoming their immunosuppressive properties (see Table 13.1). Other antagonistic actions of DHEA supplementation include improved memory performance, bone mineral density and reduced pro-inflammatory markers (Buvat 2003; Casson et al. 1993). The lack of appropriate DHEA levels could thus be another detrimental factor underlying immunosenescence.

The antagonistic action of DHEA and cortisol suggests that the assessment of molar ratios of cortisol:DHEA (C/D) may be a more physiologically relevant measure of adrenal function (Butcher and Lord 2004; Ferrari et al. 2001; Straub et al. 2000). The measurement of isolated adrenal hormones may be thus an oversimplification and higher DHEA levels are especially protective during periods of prolonged GC hyperactivity. In contrast, we have observed higher cortisol in parallel with lower DHEA levels in healthy elders leading to markedly higher C/D ratios throughout the day. This catabolic/anabolic hormonal imbalance results in an enhanced exposure of various body systems (including brain and immune system) to the deleterious effects of GCs. Significantly, some brain regions (hippocampus) and lymphocyte subsets (especially thymocytes) are especially sensitive to cortisol because of higher expression of mineralocorticoid and GC receptors (McEwen et al. 1997). In summary, the peripheral tissues of elders may be thus more vulnerable to GC actions.

13.3 The Immunomodulatory Properties of DHEA—Impact on Ageing

The lack of appropriate DHEAS levels during ageing could promote immunosenescence. This androgen and its metabolites have reported immune-enhancing properties, in contrast to the immunosuppressive action of GCs. Previous studies have evaluated the immunomodulatory DHEA(S) effects in vitro as well as its properties during in vivo supplementation. The immunomodulatory *in vitro* effects include increased mitogen-stimulated Interleukin-2 (IL-2) production (Daynes et al. 1990), increased rodent or human lymphocyte proliferation (Padgett and Loria 1994), stimulated monocyte-mediated cytotoxicity (McLachlan et al. 1996), diminished tumour necrosis factor (TNF)-α or IL-6 production (Straub et al. 1998), increased neutrophil superoxide generation (Radford et al. 2010) and enhanced natural killer (NK) cell activity (Solerte et al. 1999). Recent data suggest that DHEA treatment in vivo can significantly ameliorate the severity of experimental autoimmune diseases by suppressing the proliferation of autoreactive T-cell, reducing pro-inflammatory cytokines and by increasing the numbers and function of CD4 + CD25 + FOXp3 + CD127− regulatory T-cells (Treg) (Auci et al. 2007; Tan et al. 2009).

DHEA(S) replacement therapy has yielded significant beneficial immunological effects for healthy elders. Indeed, it has been shown that DHEA supplementation significantly increased NK cell counts and activity and decreased IL-6 production and T-cell proliferation of the elderly (Casson et al. 1993). These results are in line with in vitro findings and highlight the potential use of DHEA(S) as anti-immunosenescence hormone. However, there is still limited evidence for the clinical significance of these findings.

It is well established that vaccine responses are significantly reduced with age, reflecting a decline in adaptive immunity (see Chap. 2 in this book). Because of its immune-enhancing properties, several studies investigated the potential of DHEA(S) as adjuvants in vaccine preparations. Initial studies reported increased adjuvant effects on the immunization of aged mice with recombinant hepatitis B surface antigen (Araneo et al. 1993) or influenza (Danenberg et al. 1995). These studies reported increased antibody titres to vaccines or even effective protection against challenge with the influenza infection (Danenberg et al. 1995). More recently, we studied the adjuvant effects of DHEAS during immunization to *Mycobaterium tuberculosis* in mice (Ribeiro et al. 2007). Only young mice coimmunized with *M. tuberculosis* heat shock protein 70 (HSP70) and DHEAS showed an early increase in specific IgG levels compared to old mice. However, splenocytes of both young and old mice that received DHEAS showed increased interferon-gamma (IFN-γ) production following priming in vitro with HSP70. These data further highlight the importance of DHEAS as hormonal adjuvant because of the role of this cytokine in the cellular response against mycobacteria. However, these animal data are in contrast to previous studies reporting DHEA(S) with minor (Degelau et al. 1997) or no adjuvant effects (Ben-Yehuda et al. 1998) during immunization to influenza in older humans. Therefore, extrapolation from studies on murine models to the human should be regarded with caution—especially because of lower circulating DHEA(S) levels in rodents.

13.4 Similarities Between Ageing and Chronic GC Exposure

In light of the evidence cited earlier, it seems reasonable to assert that increased cortisol and lower DHEA may contribute to immunological changes observed during ageing. All leucocytes and many cells in lymphoid tissues, such as the thymocytes and epithelial cells in the thymus, express receptors for the neuroendocrine products of the HPA and the sympathetic–adrenal–medullary systems.

Most primary and secondary lymphoid organs show atrophy during ageing. In particular, thymic involution is a common consequence of mammalian ageing and it precedes the malfunctioning of the immune system, resulting in a diminished capacity to generate new T-cells and to mount an adaptive immune response to new pathogens and vaccines. In addition to ageing, chronic GC exposure as observed during psychological distress (Selye 1936) or pharmacological GC treatment (Fauci 1975) also induce atrophy of the thymus, triggering apoptotic death in immature T- and B-cell precursors and mature T-cells which are exquisitely sensitive to corticosteroids (Sapolsky et al. 2000). Therefore, thymic involution is not only an exclusive phenomenon of ageing but is also associated with excessive stress hormone exposure and may be driven, in part, by age-related changes to the HPA axis.

The features of age-related thymic involution lead to a profound decline in naïve (CD45RA+) T-cell output that underlies most T-cell changes. T-cells are especially targeted during ageing and show the most consistent and largest alterations among the leukocytes. In parallel to a progressive loss of naïve T-cells, ageing is associated with increased counts in memory (CD45RO+) T-cells, an expansion of CD28− T-cells, decreased NKT numbers with impaired cytotoxicity and a robust contraction of the αβTCR repertoire (Martinez-Taboada et al. 2002; Mocchegiani et al. 2009; Naylor et al. 2005; Weng et al. 2009). A reduced TCR diversity may be explained by the specific expansion of clonotypes directed against persistent or recurring pathogens, including CMV (Hadrup et al. 2006). Interestingly, T-cells are also changed in the same way during chronic stress exposure (Biondi 2001; Bosch et al. 2009) or following GC treatment in vivo (Bauer et al. 2002; McEwen et al. 1997). Immunologists have recently characterized a new T-cell subset (CD4 + C25 + FoxP3+) with an important regulatory role in suppressing excessive or misguided immune responses that can be harmful to the host. These lymphocytes are called regulatory T-cells and are responsible for turning off immune responses against self-antigens in autoimmune disease, allergy or commensal microbes in certain inflammatory diseases (Fontenot et al. 2003; Sakaguchi 2000). Interestingly, it has been found that ageing, GCs or chronic stress can increase peripheral regulatory T-cell numbers (Hoglund et al. 2006; Navarro et al. 2006; Trzonkowski et al. 2006). While the underlying mechanisms are not completely understood, previous work has observed differences in steroid signalling associated with increased numbers of regulatory T-cells. In particular, it has been shown that murine regulatory T-cells, expressed higher GR (glucocorticoid receptor) levels, were more resistant to GC-induced apoptosis (with high Bcl-2 levels) and expressed high levels of GC-induced TNF receptor (GITR) than CD4 + CD25 - cells (Chen et al. 2004), an important receptor associated with its suppression action. In spite of the several similarities among age- and stress-related immunological alterations, only a few studies have explicitly addressed the role of stress factors on human immunosenescence.

T-cells are a pivotal part of cell-mediated immunity and immunosenescence involves impaired humoral responses and blunted T-cell proliferation to mitogens. The latter is one of the most documented age-related immune changes (Liu et al. 1997). Yet, these changes are not exclusive to ageing, and stress or GC treatment is also associated with decrements of T-cell proliferation (Biondi 2001; Sapolsky et al. 2000). Indeed, we have observed that healthy SENIEUR elders were significantly more stressed, had activated HPA axis and significantly lower (−53.6 %) T-cell proliferation compared to young adults (Luz et al. 2006), with salivary cortisol levels negatively correlated with T-cell proliferation.

The effector phases of both innate and acquired immunity are in large part mediated by cytokines. Different subpopulations of CD4+ T-cells synthesize specific cytokines and have been designated Th1 (IFN-γ, IL-2, lymphotoxin α) or Th2 (IL-4, IL-10) cells. Th1 cytokines provide help for cell-mediated responses and IgG2a antibody class switching, whereas Th2 cytokines help B-cells as well as IgA, IgE and IgG1 antibody class switching. Both human and mouse models have demonstrated that ageing is associated with a Th1-to-Th2 shift in cytokine production (Ginaldi et al. 2001). However, again, this is not an age-specific phenomenon, but is also seen during stress (Biondi 2001; Glaser et al. 2001) or GC treatment (Galon et al. 2002).

Previous work suggested possible links between endocrine senescence and immunosenescence, whereby age-related increases in (inflammatory) cytokines affect the release of hormones and, vice versa, the hormonal changes associated with age influence cytokine networks (Straub et al. 2000). Indeed, it has long been known that pro-inflammatory cytokines can readily activate the HPA axis during infection

(Besedovsky et al. 1977) and after cytokine administration (Mastorakos et al. 1993). More recent data also show that systemic production of active cortisol from inactive cortisone, via increased expression and activation of the enzyme 11βHSD1 (Cooper et al. 2001), is also influenced by inflammatory cytokines. Other studies have linked the age-related decline in DHEA production to increased serum levels of IL-6 (Daynes et al. 1993; Straub et al. 1998). The relative GC excess resulting from the increased cortisol:DHEA ratio could be associated with accelerated features of low-grade inflammation in the elderly, a process known as inflammageing. Inflammageing appears to be a common age-related phenomenon (particular in unhealthy populations), and is related to frailty, morbidity and mortality in the elderly (Franceschi et al. 2000). Indeed, chronic inflammation is considered to be involved in the pathogenesis of major age-related diseases, including Alzheimer's disease, atherosclerosis, diabetes, major depression, sarcopenia and cancer. In addition, low-grade increases in levels of circulating TNF-α, IL-6, soluble IL-2 receptor and CRP as well as low levels of albumin and cholesterol, which also act as inflammatory markers, are strong predictors of all-cause mortality risk in several longitudinal studies of elderly cohorts. However, low-grade inflammation could not be observed in very healthy elders (Luz et al. 2003) or centenarians, suggesting that inflammageing is more likely a feature of unsuccessful ageing, or of morbidities frequently observed during ageing, including repeated infections, senescent cells and reduced sex hormone levels.

One question remains to be answered: How does an anti-inflammatory hormone (GCs) promote low-grade inflammation? First, increased GC levels would lead to increased abdominal fat (as seen during ageing or GC treatment) and development of metabolic syndrome. Adipocytes and infiltrating macrophages secrete various adipokines (e.g. leptin, TNF-α, IL-6, IL-18) that reach the circulation and may thus contribute to inflammageing (Ouchi et al. 2011). Secondly, increased GC levels would render immune cells more resistant to steroids (discussed in the next section). The age-related acquired steroid resistance would render cells poorly responsive to anti-inflammatory actions of endogenous GCs.

13.5 Decreased Glucocorticoid Signalling During Ageing and Chronic Stress

The GC effects on the immune system are mediated via both intracellular and membrane GC receptors (Gold et al. 2001; McEwen et al. 1997). The presence of these receptors indicates that the immune system is prepared for HPA axis activation and the subsequent elevation in endogenous GCs. However, the functional effect of a stress hormone will depend on the sensitivity of the target tissue for that particular hormone. For instance, the number and activity of specific receptors for these signalling molecules on the target organ will ultimately direct the physiologic effect of the stressor.

We have investigated the lymphocyte sensitivity to both synthetic (dexamethasone, DEX) and naturally occurring steroids (cortisol and DHEA) to examine whether ageing was associated with alterations in neuroendocrine immunoregulation (Luz et al. 2006). It was found that healthy elders had a reduced (−19 %) in vitro lymphocyte sensitivity to DEX (a GR agonist) when compared to young adults. This phenomenon has previously been observed during chronic stress (Bauer et al. 2000; Rohleder et al. 2002), major depression (Bauer et al. 2002, 2003; Truckenmiller et al. 2005), bipolar disorder (Knijff et al. 2006) or in clinical situations where GCs are chronically administered, including treatment of autoimmune diseases, organ transplantation and allergies. These observations are in line with previous observations indicating that ageing is associated with changes in GC sensitivity of pro-inflammatory cytokine (TNF-α and IL-6) production following the Trier psychosocial stress test (Rohleder et al. 2002). In particular, monocytes of healthy older men had a higher sensitivity to DEX treatment in vitro at baseline and showed a reduced sensitivity to this steroid following acute stress exposure (speech coupled to mental arithmetic task) (Rohleder et al. 2002). These data suggest that psychological factors may be implicated in regulating peripheral GC signalling during healthy ageing. Altered steroid immunoregulation may have important therapeutic implications in clinical situations where GCs are administered, including treatment of autoimmune diseases, organ transplantation and allergies. Clinicians should consider both patient's age and psychological status in prescribing steroids as anti-inflammatory drugs.

A reduced sensitivity to GCs can also be demonstrated at the central level during ageing. Indeed, higher cortisol levels in old rather than in young subjects have been described during some pharmacological challenges, such as the DEX suppression test, stimulation by human or ovine corticotrophin-releasing hormone or by physostigmine (Raskind et al. 1994).

13.6 Chronic Stress may Lead to Premature Ageing of the Immune System

Compelling evidence indicates that chronic psychological stress may cause the premature ageing of several cellular and molecular components of the immune system (Bauer 2008; Gouin et al. 2008). Psychological stress has been widely associated with suppressive immune functions. Therefore, chronic stress may have an important impact in the immune system of older populations. One of the common models of chronic stress is caregiving for a family member with dementia. Care of the chronically ill is a demanding task that is associated with increased stress, depression and poorer immune function (Redinbaugh et al. 1995). Furthermore, providing care for a relative with dementia typically falls on the partners who are themselves older and often ill prepared for the physical and emotional demands placed upon them.

Several studies have implicated caregiving as a risk factor for health of elderly populations and the daily stress experienced by the caregivers of Alzheimer patients may lead to premature ageing of neuroendocrine and immune systems. Compared with noncaregivers, subjects who provide care to a spouse with a stroke or dementia report more infectious illness episodes (Kiecolt-Glaser et al. 1991), have poorer immune responses to influenza virus (Kiecolt-Glaser et al. 1996; Vedhara et al. 1999) and pneumococcal pneumonia vaccines (Glaser et al. 2000), present a slow wound healing (Kiecolt-Glaser et al. 1995; also see the chapter by Engeland in this volume), are at greater risk for developing mild hypertension (Shaw et al. 1999) and may be at greater risk for coronary heart disease (Vitaliano et al. 2002). In addition, a prospective longitudinal study found that the relative risk for mortality among caregivers was significantly higher (63 %) than noncaregiving controls (Schulz and Beach 1999). A previous study has indicated that IL-6 may be involved in this increased morbidity in caregiving populations (Kiecolt-Glaser et al. 2003). It remains to be investigated, however, how the extent of these changes may be related to neuroendocrine alterations observed during ageing.

We have previously demonstrated that elder caregivers of demented patients had a blunted T-cell proliferation in association with increased cortisol levels compared to nonstressed elders (Bauer et al. 2000). Furthermore, lymphocytes of older caregivers were more resistant to GC treatment in vitro compared to noncaregiver elders. When stressed healthy elders were compared to healthy non-stressed elderly and young adults, these immunological changes were found to correlate with increased cortisol levels (Bauer 2005). These data suggest that chronic stress and cortisol have similarities with human immunosenescence. Indeed, psychological stress (both perceived stress and chronicity of stress) has been associated with higher oxidative stress, lower telomerase activity and shorter telomere length, which are known determinants of cell senescence and longevity (Epel et al. 2004). Therefore, it can be speculated that chronic stress may lead to premature ageing of key allostatic systems and senescent features can be observed at various levels (Fig. 13.1). Recent data produced by our laboratory have suggested that the maintenance of health during ageing may protect elders from the effects of chronic stress exposure (Jeckel et al. 2010). We have investigated whether a stringent health status (SENIEUR) would protect caregivers from the effects of chronic stress exposure (i.e. caregiving for a spouse with dementia) and corresponding psychoneuroimmunological changes. Notwithstanding the elder SENIEUR caregivers were significantly distressed, their salivary cortisol levels were similar to those of nonstressed elders. Therefore, healthy caregivers would be protected from the deleterious effects of cortisol excess normally observed during stress. We hypothesized that lymphocytes from SENIEUR caregivers would be more responsive as compared cells obtained from non-selected stressed populations. Indeed, healthy caregivers had increased T-cell proliferation and increased cellular sensitivities to GCs when compared to nonstressed elders. Taken together, these results suggest that strictly healthy ageing may buffer or attenuate many deleterious neuroendocrine and immunological effects associated with chronic stress exposure. However, the underlying mechanisms by which the maintenance of health may buffer the impact of chronic stress on the HPA and immune systems remain to be established.

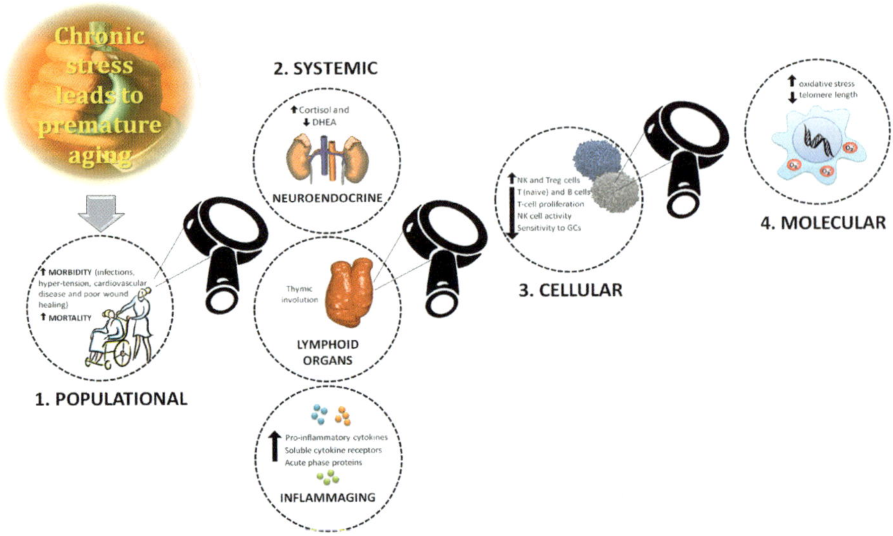

Fig. 13.1 Psychological stress may lead to premature aging at various levels. Arrows pointing up indicate increased activities, while arrows pointing down indicate blunted activities. *Treg* Regulatory T cells; *GCs* Glucocorticoids

13.7 Risk Factors for Stress-induced Premature Ageing: Individual Differences

The changes observed in allostatic systems following chronic stress exposure are not uniform, suggesting that protective or susceptibility factors, such as stress appraisal and other individual differences could be involved. A key question thus remains to be fully answered: Who is the most likely to suffer accelerated ageing, including immunosenescence, from chronic stress exposure? In animal studies, anxiety proneness is linked to greater reactivity to stress, which in turn is linked to ageing. In rats, freezing or slower maze performance predicts a premature ageing syndrome, cognitive decline, lower antioxidants and higher oxidative stress (Gilad and Gilad 1995; Zhao et al. 2009). It is known that some personality traits, coping skills, lack of social support and health-related behaviours are associated with poorer resilience to chronic stress exposure. Personality and coping styles underlie individual differences in appraisal and response to stressful situations, and both have been associated with the onset and course of chronic and progressive health problems, including cardiac morbidity and mortality (Kiecolt-Glaser et al. 2002). People with a more inhibited personality type, associated with anxiety or low self-esteem, are prone to high cortisol reactivity to acute stress or lack of a habituation to repeated stress (Pruessner et al. 2005). Elders have been shown to be more vulnerable to negative emotions due to smaller social support networks (Carstensen 1992). Caregivers of dementia patients who reported lower levels of social support and higher levels of distress on study entry showed the greatest negative changes in immune function 1 year later (Kiecolt-Glaser et al. 1991). Distressed subjects are also more likely to have health habits that put them at greater risk for diseases, including poorer sleep, a greater propensity for alcohol and drug abuse, poorer nutrition and less exercise, and these health behaviours have cardiovascular, immunological and endocrinological consequences (Kiecolt-Glaser et al. 2002).

Perhaps more instructive than knowing who is at higher risk, is information about who is more protected from the impact of chronic stress exposure. It has been shown that older adults who engage in active behaviours to overcome their physical problems experience lower levels of emotional distress (Wrosch et al. 2002) and have a lower secretion of salivary cortisol (Wrosch et al. 2007). A potent resilience factor for health outcomes may be the induction and maintenance of positive emotion through personality and coping styles. Individuals reporting high trait positive affect have lower cortisol levels, reduced inflammatory markers (e.g. IL-6, CRP) and favourable heart rate and blood pressure (Steptoe et al. 2009). At the same time, positive affect is associated with protective psychosocial factors such as greater social connectedness, perceived social support, optimism and preference for adaptive coping responses. Positive affect may be part of a broader profile of psychosocial resilience that reduces risk of adverse physical health outcomes. Finally, stressed individuals with better social support developed stronger immune responses to vaccination (Glaser et al. 1992).

13.8 Psychological/Behavioural Interventions may Attenuate Immunosenescence

Considering that chronic stress could lead to premature ageing of the immune system, in addition to its other detrimental effects, it is reasonable to investigate if stress-management interventions can attenuate or reverse some features of ageing. By reducing stress perception and promoting healthy behaviours, one may also induce improvements in cortisol.DHEA (C/D) balance, by reducing cortisol and increasing anabolic hormones (like DHEA), enhanced vagal tone and better immune responses. Interventions that include health behaviour changes are likely to be more effective in attenuating 'biological ageing' of the immune system. Increasing fitness is probably one of the most potent interventions in restoring C/D balance and immunity and can improve well-being. Long-term moderate exercise can decrease cortisol and increase DHEA, GH and IGF-1 levels (Cotman and Berchtold 2002), as well as reduce anxiety (Petruzzello et al. 1991) and depression (Barbour et al. 2007). Moderate exercise training can also increase several parameters in the elderly including greater T-cell proliferation, a reduced frequency of antigen-experienced and senescent T-cells (i.e. CD45RO+, KLRG1+, CD57+, CD28−), enhanced IL-2 production and T-cell expression of the IL-2 receptor, longer chromosome telomere lengths in blood leukocytes and in vivo immune responses to vaccines and recall antigens (Simpson and Guy 2010) (a further discussion can be found in the chapter by R.J. Simpson in this book). Previous studies have not addressed, however, if these enhanced immune changes were due to restored C/D levels. Other potential mechanisms of exercise-related immunological effects include changes in body fat distribution, lipid metabolism, cholesterol stores and peripheral circulation, as well as improved psychological well-being.

Psychosocial interventions have been effective in attenuating stress (Schulz et al. 2002) and improving adrenal hormones in the elderly. For instance, an enrichment program for elders increased significantly DHEA, testosterone, estradiol and GH levels (Arnetz et al. 1983). The enrichment or activation program provided an intervention that aimed to increase the elders' social activation, competence and independence and to counteract social isolation and passivity. A randomized controlled trial showed that older adults who practiced relaxation had reduced antibody titres to latent HSV-1, indicating that a lifestyle intervention resulted in lower levels of antigenic stimulation (Gouin et al. 2008). This study suggests that stress-buffering strategies can lead to an improvement in cellular immune-mediated control of latent viruses.

An important field of interest is the possibility to influence the quality of life through interventions with acupuncture. Acupuncture is certainly the most popular intervention of Traditional Chinese Medicine in western countries. We have recently investigated the effects of acupuncture on stress-related psychological symptoms and cellular immunity in the elderly (Pavao et al. 2010). The acupuncture treatment consisted of six sessions and the procedures included the insertion of needles at bilateral acupoints (LI4, SP6 and ST36) in healthy elders and young adults. Repeated applied acupuncture was able to significantly attenuate psychological distress (−15 to −47 %) as well as increased T-cell proliferation (~50 %), with greater intensity in the elderly group (Pavao et al. 2010). Notably, the T-cell proliferation of the elderly group reached similar levels of those found in the young adults. Acupuncture may exert its effects by influencing neurotransmitter and hormonal pathways underlying emotional states. The C/D levels were not investigated in this study and the underlying mechanisms have, as yet, to be clarified.

13.9 Conclusions and Future Perspectives

When the effects of age-related diseases are controlled for, normal physiological ageing is associated with changes in allostatic systems (endocrine and immune) that play major roles in the adaptation of the organism to outside forces that are threatening the homeostasis of the internal milieu. In particular, ageing is associated with an increase psychological stress and activation of the HPA axis. The age-related increase in the C/D ratio is also proposed as a causative factor in immunological changes observed during ageing. Over weeks, months or even years, exposure to increased secretion of stress hormones causes allostatic load ('wear and tear') and has pathophysiologic consequences (McEwen 1998). Given the findings that even short-term HPA axis activation may impair cognitive function (Lupien et al. 1994) and

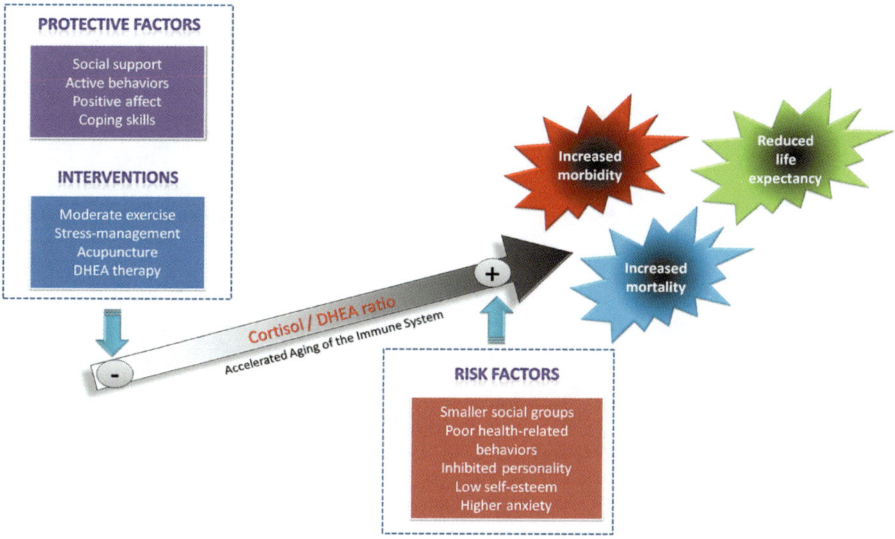

Fig. 13.2 Protective and risk factors associated with accelerated aging of the immune system. Various risk factors may lead to premature ageing of the immune system because of stress-related increase in the cortisol/dehydroepiandrosterone ratio. The accelerated immunosenescence will be ultimately implicated with increasing morbidity, mortality and reducing life expectancy. Conversely, the maintenance of active behaviours, moderate exercise, social support, personality (positive affect), coping skills and engagement in stress-management interventions may protect elders from detrimental effects of chronic stress exposure

induce sleep disturbances (Starkman et al. 1981), conditions frequently associated in the elderly, psychological or pharmacological strategies attenuating or preventing increased HPA axis activation may be of considerable benefit for the elderly.

Not only could stress management and psychosocial support promote a better quality of life for the elderly, it may also reduce hospitalization costs for health care providers. Preliminary evidence indicates that acupuncture may alleviate stress and attenuate some immunosenescence features. Finally, the maintenance of active behaviours, moderate exercise, social support, personality (positive affect) and coping skills may protect elders from detrimental effects of chronic stress exposure (Fig. 13.2). Further studies in systems biology are needed to analyze the role and relationships of health-related behaviours on immunity that might promote better coping with ageing and stress exposure. We are currently entering a new era of investigation in biology of ageing in which a systemic approach will replace reductionism in order to better explain the pace of ageing and development of age-related diseases.

Acknowledgments This study was supported by grants from Conselho Nacional de Desenvolvimento Científico e Tecnológico (CNPq) (Brazil).

References

Araneo B, Woods M, Daynes R (1993) Reversal of the immunosenescent phenotype by dehydroepiandrosterone: hormone treatment provides an adjuvant effect on the immunization of aged mice with recombinant hepatitis B surface antigen. J Infect Dis 167:830–840

Arnetz BB, Theorell T, Levi L, Kallner A, Eneroth P (1983) An experimental study of social isolation of elderly people: psychoendocrine and metabolic effects. Psychosom Med 45(5):395–406

Auci D, Kaler L, Subramanian S, Huang Y, Frincke J, Reading C et al (2007) A new orally bioavailable synthetic androstene inhibits collagen-induced arthritis in the mouse: androstene hormones as regulators of regulatory T cells. Ann NY Acad Sci 1110:630–640

Barbour KA, Edenfield TM, Blumenthal JA (2007) Exercise as a treatment for depression and other psychiatric disorders: a review. J Cardiopulm Rehabil Prev 27(6):359–367

Bauer ME (2005) Stress, glucocorticoids and ageing of the immune system. Stress 8(1):69–83

Bauer, ME (2008) Chronic stress and immunosenescence: a review. Neuroimmunomodulation 15(4–6):241–250

Bauer ME, Jeckel CM, Luz C (2009) The role of stress factors during aging of the immune system. Ann NY Acad Sci 1153:139–152

Bauer ME, Papadopoulos A, Poon L, Perks P, Lightman S, Checkley S et al (2002) Dexamethasone-induced effects on lymphocyte distribution and expression of adhesion molecules in treatment resistant major depression. Psychiat Res 113:1–15

Bauer ME, Papadopoulos A, Poon L, Perks P, Lightman S, Checkley S et al (2003) Altered glucocorticoid immunoregulation in treatment resistant depression. Psychoneuroendocrinol 28:49–65

Bauer ME, Vedhara K, Perks P, Wilcock G, Lightman S, Shanks N (2000) Chronic stress in caregivers of dementia patients is associated with reduced lymphocyte sensitivity to glucocorticoids. J Neuroimmunol 103:84–92

Ben-Yehuda A, Danenberg H, Zakay-Rones Z, Gross D, Friedman G (1998) The influence of sequential annual vaccination and DHEA administration on the efficacy of the immune response to influenza vaccine in the elderly. Mech Ageing Dev 102:299–306

Besedovsky H, Sorkin E, Felix D, Haas H (1977) Hypothalamic changes during the immune response. Eur J Immunol 7:323–325

Biondi M (2001) Effects of stress on immune functions: an overview. In: Ader R, Felten D, Cohen N (eds) Psychoneuroimmunonology, vol 2. Academic Press, San Diego, pp 189–226

Bosch JA, Fischer JE, Fischer JC (2009) Psychologically adverse work conditions are associated with CD8 + T cell differentiation indicative of immunosenescence. Brain Behav Immun 23(4):527–534

Butcher SK, Lord JM (2004) Stress responses and innate immunity: aging as a contributory factor. Aging Cell 3(4):151–160

Buvat J (2003) Androgen therapy with dehydroepiandrosterone. World J Urol 21(5):346–355

Canning MO, Grotenhuis K, de Wit HJ, Drexhage HA (2000) Opposing effects of dehydroepiandrosterone and dexamethasone on the generation of monocyte-derived dendritic cells. Eur J Endocrinol 143(5):687–695

Carstensen LL (1992) Social and emotional patterns in adulthood: support for socioemotional selectivity theory. Psychol Aging 7(3):331–338

Casson PR, Andersen RN, Herrod HG, Stentz FB, Straughn AB, Abraham GE et al (1993) Oral dehydroepiandrosterone in physiologic doses modulates immune function in postmenopausal women. Am J Obstet Gynecol 169(6):1536–1539

Castle SC (2000) Clinical relevance of age-related immune dysfunction. Clin Infect Dis 31(2):578–585

Chen X, Murakami T, Oppenheim JJ, Howard OM (2004) Differential response of murine CD4 + CD25 + and CD4 + CD25- T cells to dexamethasone-induced cell death. Eur J Immunol 34(3):859–869

Collaziol D, Luz C, Dornelles F, Cruz I, Bauer ME (2004) Psychoneuroendocrine correlates of lymphocyte subsets during healthy ageing. Mech Ageing Dev 125:219–227

Cooper MS, Bujalska I, Rabbitt E, Walker EA, Bland R, Sheppard MC et al (2001) Modulation of 11beta-hydroxysteroid dehydrogenase isozymes by proinflammatory cytokines in osteoblasts: an autocrine switch from glucocorticoid inactivation to activation. J Bone Miner Res 16(6):1037–1044

Cotman CW, Berchtold NC (2002) Exercise: a behavioral intervention to enhance brain health and plasticity. Trends Neurosci 25(6):295–301

Danenberg H, Ben-Yehuda A, Zakay-Rones Z, Friedman G (1995) Dehydroepiandrosterone (DHEA) treatment reverses the impaired immune response of old mice to influenza vaccination and protects from influenza infection. Vaccine 13:1445–1448

Daynes R, Araneo B, Ershler W, Maloney C, Li G-Z, Ryu S-Y (1993) Altered regulation of IL-6 production with normal aging—possible linkage to the age-associated decline in dehydroepiandrosterone and its sulfated derivative. J Immunol 150:5219–5230

Daynes R, Dudley D, Araneo B (1990) Regulation of murine lymphokine production in vivo. II. Dehydroepiandrosterone is a natural enhancer of interleukin-2 synthesis by helper T cells. Eur J Immunol 20:793–802

Degelau J, Guay D, Hallgren H (1997) The effect of DHEAS on influenza vaccination in aging adults. J Amer Geriatr Soc 45:747–751

Deuschle M, Gotthardt U, Schweiger U, Weber B, Korner A, Schmider J et al (1997) With aging in humans the activity of the hypothalamus-pituitary-adrenal system increases and its amplitude flattens. Life Sci 61:2239–2246

Dew MA, Reynolds CF, Frank E, Begley AE, Miller MD, Cornes C et al (1998) Effects of age at onset of first lifetime episode of recurrent major depression on treatment response and illness course in elderly patients. Am J Psychiatry 155(6):759–799

Epel ES, Blackburn EH, Lin J, Dhabhar FS, Adler NE, Morrow JD et al (2004) Accelerated telomere shortening in response to life stress. Proc Natl Acad Sci USA 101(49):17312–17315

Fauci A (1975) Mechanisms of corticosteroid action on lymphocyte subpopulations. Immunol 28:669–679

Ferrari E, Cravello L, Muzzoni B, Casaritti D, Paltro M, Sorlete SB et al (2001) Age-related changes of the hypothalamic-pytuitary-adrenal axis: phathophysiological correlates. Eur J Endocrinol 144:319–329

Ferrari E, Mirani M, Barili L, Falvo F, Solerte SB, Cravello L et al (2004) Cognitive and affective disorders in the elderly: a neuroendocrine study. Arch Gerontol Geriat Suppl(9):171–182

Fontenot JD, Gavin MA, Rudensky AY (2003) Foxp3 programs the development and function of CD4 + CD25 + regulatory T cells. Nat Immunol 4(4):330–336

Franceschi C, Bonafe M, Valensin S, Olivieri F, De Luca M, Ottaviani E et al (2000) Inflamm-aging. An evolutionary perspective on immunosenescence. Ann NY Acad Sci 908:244–254

Gabriel P, Cakman I, Rink L (2002) Overproduction of monokines by leukocytes after stimulation with lipopolysaccharide in the elderly. Exp Gerontol 37:235–247

Galon J, Franchimont D, Hiroi N, Frey G, Boetner A, Ehrart-Bornstein M et al (2002) Gene profiling reveals unknown enhancing and suppressive actions of glucocorticoids on immune cells. FASEB J 16:61–71

Gilad GM, Gilad VH (1995) Strain, stress, neurodegeneration and longevity. Mech Ageing Dev 78(2):75–83

Ginaldi L, Loreto MF, Corsi MP, Modesti M, De Martinis M (2001) Immunosenescence and infectious diseases. Microbes Infect 3:851–857

Glaser R, Kiecolt-Glaser JK, Bonneau RH, Malarkey W, Kennedy S, Hughes J (1992) Stress-induced modulation of the immune response to recombinant hepatitis B vaccine. Psychosom Med 54(1):22–29

Glaser R, McCallum R, Laskowski BF, Malarkey WB, Sheridan J, Kiecolt-Glaser J (2001) Evidence for a shift in the Th-1 to Th-2 cytokine response associated with chronic stress and aging. J Gerontol Med Sci 56A:M477–M482

Glaser R, Sheridan J, Malarkey WB, MacCallum RC, Kiecolt-Glaser JK (2000) Chronic stress modulates the immune response to a pneumococcal pneumonia vaccine. Psychosom Med 62(6):804–807

Gold R, Buttgereit F, Toyka KV (2001) Mechanism of action of glucocorticosteroid hormones: possible implications for therapy of neuroimmunological disorders. J Neuroimmunol 117(1–2):1–8
Gouin JP, Hantsoo L, Kiecolt-Glaser JK (2008) Immune dysregulation and chronic stress among older adults: a review. Neuroimmunomodulation 15(4–6):251–259
Hadrup SR, Strindhall J, Kollgaard T, Seremet T, Johansson B, Pawelec G et al (2006) Longitudinal studies of clonally expanded CD8 T cells reveal a repertoire shrinkage predicting mortality and an increased number of dysfunctional cytomegalovirus-specific T cells in the very elderly. J Immunol 176(4):2645–2653
Hazeldine J, Arlt W, Lord JM (2010) Dehydroepiandrosterone as a regulator of immune cell function. J Steroid Biochem Mol Biol 120(2–3):127–136
Hechter O, Grossman A, Chatterton RT (1997) Relationship of dehydroepiandrosterone and cortisol in disease. Med Hypotheses 49:85–91
Hoglund CO, Axen J, Kemi C, Jernelov S, Grunewald J, Muller-Suur C et al (2006) Changes in immune regulation in response to examination stress in atopic and healthy individuals. Clin Exp Allergy 36(8):982–992
Howieson DB, Camicioli R, Quinn J, Silbert LC, Care B, Moore MM et al (2003) Natural history of cognitive decline in the old. Neurology 60(9):1489–1494
Jeckel CM, Lopes RP, Berleze MC, Luz C, Feix L, Argimon II et al (2010) Neuroendocrine and immunological correlates of chronic stress in 'strictly healthy' populations. Neuroimmunomodulation 17(1):9–18
Juruena MF, Cleare AJ, Bauer ME, Pariante CM (2003) Molecular mechanisms of glucocorticoid receptor sensitivity and relevance to affective disorders. Acta Neuropsychiatr 15:354–367
Kiecolt-Glaser JK, Dura JR, Speicher CE, Trask OJ, Glaser R (1991) Spousal caregivers of dementia victims: longitudinal changes in immunity and health. Psychosom Med 53(4):345–362
Kiecolt-Glaser J, Glaser R, Gravenstein S, Malarkey W, Sheridan J (1996) Chronic stress alters the immune response to influenza virus vaccine in older adults. Proc Natl Acad Sci USA 93(7):3043–3047
Kiecolt-Glaser J, Marucha P, Malarkey W, Mercado A, Glaser R (1995) Slowing of wound healing by psychological stress. Lancet 346(8984):1194–1196
Kiecolt-Glaser JK, McGuire L, Robles TF, Glaser R (2002) Emotions, morbidity, and mortality: new perspectives from psychoneuroimmunology. Annu Rev Psychol 53:83–107
Kiecolt-Glaser JK, Preacher KJ, MacCallum RC, Atkinson C, Malarkey WB, Glaser R (2003) Chronic stress and age-related increases in the proinflammatory cytokine IL-6. Proc Natl Acad Sci USA 100(15):9090–9095
Kirschbaum C, Prussner J, Stone A, Federenko I, Gaab J, Lintz D et al (1995) Persistent high cortisol responses to repeated psychological stress in a subpopulation of healthy men. Psychol Med 57:468–474
Knijff EM, Breunis MN, van Geest MC, Kupka RW, Ruwhof C, de Wit HJ et al (2006) A relative resistance of T cells to dexamethasone in bipolar disorder. Bipolar Disord 8(6):740–750
Lee AL, Ogle WO, Sapolsky RM (2002) Stress and depression: possible links to neuron death in the hippocampus. Bipolar Disord 4(2):117–128
Ligthart G, Corberand J, Fournier C, Galanaud P, Humans W, Kennes B et al (1984) Admission criteria for immunogerontological studies in man: the SENIEUR protocol. Mech Ageing Dev 28:47–55
Liu J, Wang S, Liu H, Yang L, Nan G (1997) The monitoring biomarker for immune function os lymphocytes in the elderly. Mech Ageing Dev 94:177–182
Lupien S, Lecours A, Lussier I, Schwartz G, Nair N, Meaney M (1994) Basal cortisol levels and cognitive deficits in human aging. J Neurosci 14:2893–2903
Luz C, Collaziol D, Preissler T, da Cruz IM, Glock L, Bauer ME (2006) Healthy aging is associated with unaltered production of immunoreactive growth hormone but impaired neuroimmunomodulation. Neuroimmunomodulation 13(3):160–169
Luz C, Dornelles F, Preissler T, Collaziol D, Cruz I, Bauer ME (2003) Impact of psychological and endocrine factors on cytokine production of healthy elderly people. Mech Ageing Dev 124:887–895

Magri F, Cravello L, Barili L, Sarra S, Cinchetti W, Salmoiraghi F et al (2006) Stress and dementia: the role of the hypothalamicpituitary-adrenal axis. Aging Clin Exp Res 18(2):167–170

Maninger N, Wolkowitz OM, Reus VI, Epel ES, Mellon, SH (2009) Neurobiological and neuropsychiatric effects of dehydroepiandrosterone (DHEA) and DHEA sulfate (DHEAS) Front Neuroendocrinol 30(1):65–91

Martinez-Taboada V, Bartolome MJ, Amado JA, Blanco R, Garcia-Unzueta MT, Rodriguez-Valverde V et al (2002) Changes in peripheral blood lymphocyte subsets in elderly subjects are associated with an impaired function of the hypothalamic-pituitary-adrenal axis. Mech Ageing Dev 123(11):1477–1486

Mastorakos G, Chrousos GP, Weber J (1993) Recombinant IL-6 activates the hypothalamic-pituitary-adrenal axis in humans. J Clin Endocrinol Metab 77:1690–1694

McEwen B (1998) Protective and damaging effects of stress mediators. New Engl J Med 338:171–179

McEwen B, Biron C, Brunson K, Bulloch K, Chambers W, Dhabhar F et al (1997) The role of adrenocorticosteroids as modulators of immune function in health and disease: neural, endocrine and immune interactions. Brain Res Rev 23:79–133

McLachlan JA, Serkin CD, Bakouche O (1996) Dehydroepiandrosterone modulation of lipopolysaccharide-stimulated monocyte cytotoxicity. J Immunol 156(1):328–335

Migeon C, Keller A, Lawrence B, Shepard T (1957) Dehydroepiandrosterone and androsterone levels in human plasma: effects of age, sex, day to day diurnal variations. J Clin Endocrinol Metab 17:1051

Mocchegiani E, Giacconi R, Cipriano C, Malavolta M (2009) NK and NKT cells in aging and longevity: role of zinc and metallothioneins. J Clin Immunol 29(4):416–425

Navarro J, Aristimuno C, Sanchez-Ramon S, Vigil D, Martinez-Gines ML, Fernandez-Cruz E et al (2006) Circulating dendritic cells subsets and regulatory T-cells at multiple sclerosis relapse: differential short-term changes on corticosteroids therapy. J Neuroimmunol 176(1–2):153–161

Naylor K, Li G, Vallejo AN, Lee WW, Koetz K, Bryl E et al (2005) The influence of age on T cell generation and TCR diversity. J Immunol 174(11):7446–7452

Nolen-Hoeksema S, Ahrens C (2002) Age differences and similarities in the correlates of depressive symptoms. Psychol Aging 17(1):116–124

Ouchi N, Parker JL, Lugus JJ, Walsh K (2011) Adipokines in inflammation and metabolic disease. Nat Rev Immunol 11(2):85–97

Padgett DA, Loria R (1994) In vitro potentiation of lymphocyte activation by dehydroepiandrosterone, androstenediol, and androstenetriol. J Immunol 153:1544–1552

Pavao TS, Vianna P, Pillat MM, Machado AB, Bauer ME (2010) Acupuncture is effective to attenuate stress and stimulate lymphocyte proliferation in the elderly. Neurosci Lett 484(1):47–50

Pawelec G, Derhovanessian E, Larbi A (2010) Immunosenescence and cancer. Crit Rev Oncol Hematol 75(2):165–172

Petruzzello SJ, Landers DM, Hatfield BD, Kubitz KA, Salazar W (1991) A meta-analysis on the anxiety-reducing effects of acute and chronic exercise. Outcomes and mechanisms. Sports Med 11(3):143–182

Piani A, Brotini S, Dolso P, Budai R, Gigli GL (2004) Sleep disturbances in elderly: a subjective evaluation over 65. Arch Gerontol Geriat Suppl(9):325–331

Pruessner JC, Baldwin MW, Dedovic K, Renwick R, Mahani NK, Lord C et al (2005) Self-esteem, locus of control, hippocampal volume, and cortisol regulation in young and old adulthood. Neuroimage 28(4):815–826

Radford DJ, Wang K, McNelis JC, Taylor AE, Hechenberger G, Hofmann J et al (2010) Dehdyroepiandrosterone sulfate directly activates protein kinase C-beta to increase human neutrophil superoxide generation. Mol Endocrinol 24(4):813–821

Raskind M, Peskind E, Wilkinson C (1994) Hypothalamic-pituitary-adrenal axis regulation and human aging. Ann NY Acad Sci 746:327–335

Redinbaugh E, McCallum R, Kiecolt-Glaser J (1995) Recurrent syndromal depression in caregivers. Psychol Aging:358–368

Ribeiro F, Lopes RP, Nunes CP, Maito F, Bonorino C, Bauer ME (2007) Dehydroepiandrosterone sulphate enhances IgG and Interferon-gamma production during immunization to tuberculosis in young but not aged mice. Biogerontology 8(2):209–220

Rohleder N, Kudielka BM, Hellhammer DH, Wolf JM, Kirschbaum C (2002) Age and sex steroid-related changes in glucocorticoid sensitivity of pro-inflammatory cytokine production after psychosocial stress. J Neuroimmunol 126(1–2):69–77

Sakaguchi S (2000) Regulatory T cells: key controllers of immunologic self-tolerance. Cell 101(5):455–458

Salehian MB, Kejriwal MK (1999) Glucocorticoid-induced muscle atrophy: mechanisms and therapeutic strategies. Endocr Pract 5(5):277–281

Sapolsky RM, Romero LM, Munck AU (2000) How do glucocorticoids influence stress responses? Integrating permissive, suppressive, stimulatory, and preparative actions. Endocr Rev 21(1):55–89

Schiepers OJ, Wichers MC, Maes M (2005) Cytokines and major depression. Prog Neuropsychopharmacol Biol Psychiatry 29(2):201–217

Schulz R, Beach SR (1999) Caregiving as a risk factor for mortality: the caregiver health effects study. Jama 282(23):2215–2219

Schulz R, O'Brien A, Czaja S, Ory M, Norris R, Martire LM et al (2002) Dementia caregiver intervention research: in search of clinical significance. Gerontologist 42(5):589–602

Selye H (1936) Syndrome produced by diverse nocuous agents. Nature 138:32

Shaw WS, Patterson TL, Ziegler MG, Dimsdale JE, Semple SJ, Grant I (1999) Accelerated risk of hypertensive blood pressure recordings among Alzheimer caregivers. J Psychosom Res 46(3):215–227

Simpson RJ, Guy K (2010) Coupling aging immunity with a sedentary lifestyle: has the damage already been done?—a mini-review. Gerontology 56(5):449–458

Solerte SB, Fioravanti M, Vignati G, Giustina A, Cravello L, Ferrari E (1999) Dehydroepiandrosterone sulfate enhances natural killer cell cytotoxicity in humans via locally generated immunoreactive insulin-like growth factor I. J Clin Endocrinol Metab 84(9):3260–3267

Starkman M, Schteingart D, Schork M (1981) Depressed mood and other psychiatric manifestations of Cushing's syndrome: relationship to hormone levels. Psychosom Med 43:3–18

Steptoe A, Dockray S, Wardle J (2009) Positive affect and psychobiological processes relevant to health. J Pers 77(6):1747–1776

Sternberg E (2002) Neuroendocrine regulation of autoimmune/inflammatory disease. J Endocrinol 169:429–435

Stone AA, Schwartz JE, Broderick JE, Deaton A (2010) A snapshot of the age distribution of psychological well-being in the United States. Proc Natl Acad Sci USA 107(22):9985–9990

Straub RH, Konecna L, Hrach S, Rothe G, Kreutz M, Scholmerich J et al (1998) Serum dehydroepiandrosterone (DHEA) and DHEA sulfate are negatively correlated with serum interleukin-6 (IL-6), and DHEA inhibits IL-6 secretion from mononuclear cells in man in vitro: possible link between endocrinosenescence and immunosenescence. J Clin Endocrinol Metab 83(6):2012–2017

Straub RH, Miller L, Scholmerich J, Zietz B (2000) Cytokines and hormones as possible links between endocrinosenescence and immunosenescence. J Neuroimmunol 109:10–15

Tamura Y, Okinaga H, Takami H (2004) Glucocorticoid-induced osteoporosis. Biomed Pharmacother 58(9):500–504

Tan XD, Dou YC, Shi CW, Duan RS, Sun RP (2009) Administration of dehydroepiandrosterone ameliorates experimental autoimmune neuritis in Lewis rats. J Neuroimmunol 207(1–2):39–44

Truckenmiller ME, Princiotta MF, Norbury CC, Bonneau RH (2005) Corticosterone impairs MHC class I antigen presentation by dendritic cells via reduction of peptide generation. J Neuroimmunol 160(1–2):48–60

Trzonkowski P, Mysliwska J, Godlewska B, Szmit E, Lukaszuk K, Wieckiewicz J et al (2004) Immune consequences of the spontaneous pro-inflammatory status in depressed elderly patients. Brain Behav Immun 18(2):135–148

Trzonkowski P, Szmit E, Mysliwska J, Mysliwski A (2006) CD4 + CD25 + T regulatory cells inhibit cytotoxic activity of CTL and NK cells in humans-impact of immunosenescence. Clin Immunol 119(3):307–316

Vedhara K, Cox N, Wilcock G, Perks P, Lightman S, Shanks N (1999) Chronic stress in elderly carers of dementia patients and antibody response to influenza vaccination. Lancet 353:627–631

Vitaliano PP, Scanlan JM, Zhang J, Savage MV, Hirsch IB, Siegler IC (2002) A path model of chronic stress, the metabolic syndrome, and coronary heart disease. Psychosom Med 64(3):418–435

Weng NP, Akbar AN, Goronzy J (2009) CD28(−) T cells: their role in the age-associated decline of immune function. Trends Immunol 30(7):306–312

Wrosch C, Schulz R, Heckhausen J (2002) Health stresses and depressive symptomatology in the elderly: the importance of health engagement control strategies. Health Psychol 21(4):340–348

Wrosch C, Schulz R, Miller GE, Lupien S, Dunne E (2007) Physical health problems, depressive mood, and cortisol secretion in old age: buffer effects of health engagement control strategies. Health Psychol 26(3):341–349

Zhao Q, Rank G, Tan YT, Li H, Moritz RL, Simpson RJ et al (2009) PRMT5-mediated methylation of histone H4R3 recruits DNMT3 A, coupling histone and DNA methylation in gene silencing. Nat Struct Mol Biol 16(3):304–311

Zunszain PA, Anacker C, Cattaneo A, Carvalho LA, Pariante CM (2010) Glucocorticoids, cytokines and brain abnormalities in depression. Prog Neuropsychopharmacol Biol Psychiatry

Index

J. A. Bosch et al. (eds.), *Immunosenescence*,
DOI 10.1007/978-1-4614-4776-4, © Springer Science+Business Media New York 2013

Printed by Printforce, the Netherlands